Hyperbolic Triangle Centers

Fundamental Theories of Physics

An International Book Series on The Fundamental Theories of Physics: Their Clarification, Development and Application

Volume 166

For other titles published in this series, go to http://www.springer.com/series/6001

A.A. Ungar

Hyperbolic Triangle Centers

The Special Relativistic Approach

Springer

Prof. A.A. Ungar
Dept. Mathematics 2750
North Dakota State University
58108-6050 Fargo, ND
USA
Abraham.Ungar@ndsu.edu

ISBN 978-94-007-3265-0 ISBN 978-90-481-8637-2 (eBook)
DOI 10.1007/978-90-481-8637-2
Springer Dordrecht Heidelberg London New York

Softcover reprint of the hardcover 1st edition 2010

Cover design: eStudio Calamar S.L.

Printed on acid-free paper

Springer is part of Springer Science+Business Media (www.springer.com)

To my Daughters and Sons
Tamar, Ziva, Ilan and Ofer
and to my Grandchildren

Preface

The seeds of this book were planted in 1988 [55], when the author discovered that the seemingly structureless Einstein addition of relativistically admissible velocities possesses rich nonassociative algebraic structures that became known as a *gyrocommutative gyrogroup* and a *gyrovector space*. Einstein gyrovector spaces turn out to form the algebraic setting for the Cartesian–Beltrami–Klein ball model of the hyperbolic geometry of János Bolyai and Nikolai Ivanovich Lobachevsky, just as vector spaces form the algebraic setting for the standard Cartesian model of Euclidean geometry.

This book presents the novel approach to triangle centers in hyperbolic geometry that Einstein's special theory of relativity offers. Writing the book became possible following the adaption of Cartesian coordinates and vector algebra for use in hyperbolic geometry in its forerunners, the author's four earlier books [58, 60, 63, 64]. This adaption enables in this book the Möbius barycentric coordinates in Euclidean geometry to be embedded into hyperbolic geometry. The resulting hyperbolic barycentric coordinates form a tool for determining various hyperbolic triangle centers and relations between them, just as Euclidean barycentric coordinates form a tool for determining various Euclidean triangle centers and relations between them.

Following Möbius idea, let $\mathbb{R}^3$ be the Euclidean 3-space whose points represent Newtonian velocities in classical mechanics. The barycentric coordinates of a point $P \in \mathbb{R}^3$ with respect to a triangle in $\mathbb{R}^3$ are viewed as masses suspended at the triangle vertices in such a way that P is their center of momentum. Clearly, the masses are determined by P up to a multiplicative common factor, so that barycentric coordinates of a point with respect to a triangle are homogeneous.

Incorporating Einstein's ideas, let $\mathbb{R}^3_c$ be the c-ball of $\mathbb{R}^3$, that is, the ball of $\mathbb{R}^3$ that contains all vectors of $\mathbb{R}^3$ with magnitude smaller than the vacuum speed of light, c. Viewing the points of the ball as Einsteinian velocities in relativistic mechanics, we obtain hyperbolic barycentric coordinates that are fully analogous to their Euclidean counterpart. Thus, the hyperbolic barycentric coordinates of a point $P \in \mathbb{R}^3_c$ with respect to a hyperbolic triangle in $\mathbb{R}^3_c$ are viewed as Einsteinian, relativistic masses suspended at the triangle vertices in such a way that P is their center of momentum. Clearly, the relativistic masses are determined by P up to a multiplicative common factor, so that hyperbolic barycentric coordinates of a point with respect to a hyperbolic triangle are homogeneous.

Seemingly "unfortunately", (i) the Einstein relativistic mass is velocity dependent so that, as a result, (ii) it seemingly does not sit under the umbrella of the Minkowskian four-vector formalism of Einstein's special theory of relativity. However, it is demonstrated in this book that (i) the velocity dependence of the Einstein relativistic mass is precisely what is needed to successfully determine various triangle centers in hyperbolic geometry and that, contrary to general belief, (ii) the Einstein relativistic mass meshes up extraordinarily well with the Minkowskian four-vector formalism of Einstein's special relativity.

Accordingly, this book takes an in-depth look at one of the aspects of gyrocommutative gyrogroups and gyrovector spaces where Einstein's special relativity and the hyperbolic geometry of Bolyai and Lobachevsky meet.

A most convincing way to describe the success of the author's adaption of Cartesian coordinates and vector algebra for use in hyperbolic geometry is presented by the renowned historian of relativity physics Scott Walter in his review of the author's 2001 book [58], which is the first forerunner of this book. Therefore, part of Scott Walter's review is quoted below.

> Over the years, there have been a handful of attempts to promote the non-Euclidean style for use in problem solving in relativity and electrodynamics, the failure of which to attract any substantial following, compounded by the absence of any positive results must give pause to anyone considering a similar undertaking. Until recently, no one was in a position to offer an improvement on the tools available since 1912. In his [2001] book, Ungar furnishes the crucial missing element from the panoply of the non-Euclidean style: an elegant nonassociative algebraic formalism that fully exploits the structure of Einstein's law of velocity composition. The formalism relies on what the author calls the "missing link" between Einstein's velocity addition formula and ordinary vector addition: Thomas precession ...
>
> Scott Walter, 2002 [71]

Indeed, the special relativistic effect known as *Thomas precession* is mathematically abstracted into a version called *Thomas gyration*. The latter, in turn, justifies the prefix "gyro" that is extensively used in this book. Thus, as a matter of a few examples, Einstein's velocity addition is neither commutative nor associative, but it turns out to be both *gyrocommutative* and *gyroassociative*, thus giving rise to the algebraic structures known as gyrogroups and gyrovector spaces. Remarkably, the mere introduction of *gyrations* turns Euclidean geometry, the geometry of classical mechanics, into hyperbolic geometry, the geometry of relativistic mechanics. This remarkable result gives rise to the gyro-language of this book, in which one prefixes a gyro to a classical term to mean the analogous modern term.

As a mathematical prerequisite for a fruitful reading of this book, it is assumed familiarity with Euclidean geometry from the point of view of vectors and with basic elements of linear algebra and classical mechanics. In particular, there is no demand upon readers of this book as to a prior acquaintance with either special relativity, nonassociative algebra or hyperbolic geometry.

The immediate purpose of this book is to build a momentum toward the hunt of more hyperbolic triangle centers and the study of relationships between them. It is hoped that this book will be in the forefront of shaping and popularizing the future of the study under one umbrella of both (i) the special theory of relativity of Einstein and (ii) the hyperbolic geometry of Bolyai and Lobachevsky.

Fargo, ND, USA *A.A. Ungar*

Contents

Part III Hyperbolic Triangle Centers

Acronyms

$\oplus$	A gyrogroup operation, Definition 1.5, p. 12. Particularly, Einstein addition, (1.2). p. 4
$\boxplus$	A gyrogroup cooperation, Definition 1.7, p. 13. Particularly, Einstein coaddition (1.37), p. 13
$\otimes$	Scalar multiplication in a gyrovector space. Particularly, Einstein scalar multiplication in an Einstein gyrovector space, (2.2), p. 46
$A := B$	(or, equivalently, $B =: A$) A is defined by B
$A, B, C, \ldots$	Points of $\mathbb{R}^n$ or $\mathbb{R}^n_s$
AB	Segment (gyrosegment) joining A and B in $\mathbb{R}^n$ (in $\mathbb{R}^n_s$)
ABC	A triangle (gyrotriangle) with vertices A, B and C in $\mathbb{R}^n$ (in $\mathbb{R}^n_s$)
$\mathbf{a}, \mathbf{b}, \mathbf{c}, \ldots$	Vectors of $\mathbb{R}^n$ or gyrovectors of $\mathbb{R}^n_s$
$\mathbf{a}_{jk}$	$\mathbf{a}_{jk} = \ominus A_j \oplus A_k$
a_{jk}	$a_{jk} = \Vert\mathbf{a}_{jk}\Vert$
c	Vacuum speed of light (unless indicated otherwise)
$\mathrm{Aut}(G, \oplus)$	Automorphism group of the groupoid $(G, \oplus)$
$\mathrm{Aut}_o(G, \oplus)$	A subgroup of the automorphism group of a gyrogroup $(G, \oplus)$ that contains all the gyrations of the gyrogroup
F_k	Tangency points on a gyrotriangle ingyrocircle, Fig. 7.14, p. 192
G_e	Gergonne gyropoint, Fig. 7.14, p. 192
$\mathrm{gyr}[a, b]$	The gyroautomorphism (or, gyration) generated by two elements, a and b, of a gyrogroup
$\mathrm{gyr}^{-1}[a, b]$	The inverse gyroautomorphism of gyroautomorphism $\mathrm{gyr}[a, b]$
H	Orthogyrocenter, Fig. 7.17, p. 204; Orthocenter
I	Ingyrocenter, Fig. 7.11, p. 185; Incenter
N_a	Nagel gyropoint, Fig. 8.4, p. 237
O	Circumgyrocenter, Fig. 7.1, p. 154; Circumcenter, Fig. 7.2, p. 159; Origin of a Cartesian coordinate system
Ω	A Brocard Gyropoint, Fig. 9.8, p. 293
$\mathbb{R}^n$	The Euclidean n-space. Equivalently, space of n-tuples of real numbers
$\mathbb{R}^n_s$	The open ball of the Euclidean n-space with center $\mathbf{0}$ and radius $s > 0$

$\mathbb{R}_c^3$	The set of all relativistically admissible velocities
T_k, T_{ij}	Tangency points on exgyrocircles, Fig. 8.4, p. 237, and Fig. 8.6, p. 251; on excircles, Fig. 8.5, p. 249
R	Circumgyroradius; Circumradius; Rotation of $\mathbb{R}_s^n$ about its center
r	Ingyroradius, Sect. 7.11; Inradius
r_k	Exgyroradii; Exradii

Part I
The Special Relativistic Approach to Hyperbolic Geometry

In Part I of the book we set the stage for Part II of the book by presenting the special relativistic approach to the Cartesian–Beltrami–Klein ball model of the hyperbolic geometry of Bolyai and Lobachevsky.

Specifically, we show in this part of the book that

1. Einstein's velocity addition law gives rise to the Einstein addition that possesses a grouplike structure called a *gyrocommutative gyrogroup*; that
2. Einstein gyrogroups admit scalar multiplication that turns them into gyrovector spaces; and that
3. The Einstein relativistic, velocity dependent mass conforms with the Minkowskian formalism of special relativity theory, contrary to general belief

The importance of Einstein gyrovector spaces for the mission of this book rests on the result that Einstein gyrovector spaces form the algebraic setting for the Cartesian–Beltrami–Klein ball model of hyperbolic geometry just as vector spaces form the algebraic setting for the standard Cartesian model of Euclidean geometry.

The result that Einstein's relativistic mass conforms with the Minkowskian formalism of special relativity legitimizes its use in the special relativistic approach to the determination of hyperbolic triangle centers. Conversely, the success of Einstein's relativistic mass in the determination of hyperbolic triangle centers legitimizes Einstein's relativistic mass in the contest of special relativity theory.

Chapter 1
Einstein Gyrogroups

Abstract Einstein's addition law of three-dimensional relativistically admissible velocities is the corner stone of Einstein's three-vector formalism of the special theory of relativity that he founded in 1905. In this chapter, we present Einstein addition along with its role in nonassociative algebra and hyperbolic geometry. We make no demands upon readers of this book as to a prior acquaintance with either special relativity, nonassociative algebra or hyperbolic geometry.

1.1 Introduction

Einstein's addition law of three-dimensional relativistically admissible velocities appears in [58, 60, 63, 64] as the corner stone of Einstein's three-vector formalism of the special theory of relativity that he founded in 1905 [12, 34]. Einstein addition is presented in this chapter, along with the nonassociative algebraic structures that it encodes. These are the gyrocommutative gyrogroup structure and the gyrovector space structure that form a natural generalization of the common commutative group structure and vector space structure from associative algebra into nonassociative algebra.

It will turn out that the resulting Einstein gyrovector spaces, studied in Chap. 2, form the setting for the Cartesian–Beltrami–Klein ball model of hyperbolic geometry just as vector spaces form the setting for the standard Cartesian model of Euclidean geometry.

We make no demands upon readers of this book as to a prior acquaintance with either the hyperbolic geometry of Bolyai and Lobachevsky or the special theory of relativity of Einstein. Rather, we will present the modern and unknown in terms of analogies that they share with the classical and familiar. Accordingly, as a mathematical prerequisite for a fruitful reading of this book, familiarity with Euclidean geometry from the point of view of vectors and with basic elements of linear algebra and classical mechanics is assumed.

A.A. Ungar, *Hyperbolic Triangle Centers*,
Fundamental Theories of Physics 166,
DOI 10.1007/978-90-481-8637-2_1,

1.2 Einstein Velocity Addition

Let c be any positive constant and let $(\mathbb{R}^n, +, \cdot)$ be the Euclidean n-space, $n = 1, 2, 3, \ldots$, equipped with the common vector addition, $+$, and inner product, $\cdot$. Furthermore, let

$$\mathbb{R}^n_c = \{\mathbf{v} \in \mathbb{R}^n : \|\mathbf{v}\| < c\} \tag{1.1}$$

be the c-ball of all relativistically admissible velocities of material particles. It is the open ball of radius c, centered at the origin of $\mathbb{R}^n$, consisting of all vectors $\mathbf{v}$ in $\mathbb{R}^n$ with magnitude $\|\mathbf{v}\|$ smaller than c.

Einstein velocity addition is a binary operation, $\oplus$, in the c-ball $\mathbb{R}^n_c$ of all relativistically admissible velocities, given by the equation [58], [49, (2.9.2)], [40, p. 55], [18],

$$\mathbf{u}\oplus\mathbf{v} = \frac{1}{1+\frac{\mathbf{u}\cdot\mathbf{v}}{c^2}}\left\{\mathbf{u} + \frac{1}{\gamma_{\mathbf{u}}}\mathbf{v} + \frac{1}{c^2}\frac{\gamma_{\mathbf{u}}}{1+\gamma_{\mathbf{u}}}(\mathbf{u}\cdot\mathbf{v})\mathbf{u}\right\} \tag{1.2}$$

for all $\mathbf{u}, \mathbf{v} \in \mathbb{R}^n_c$, where $\gamma_{\mathbf{u}}$ is the gamma factor given by the equation

$$\gamma_{\mathbf{v}} = \frac{1}{\sqrt{1-\frac{\|\mathbf{v}\|^2}{c^2}}} = \frac{1}{\sqrt{1-\frac{\mathbf{v}^2}{c^2}}} \tag{1.3}$$

Here $\mathbf{u}\cdot\mathbf{v}$ and $\|\mathbf{v}\|$ are the inner product and the norm in the ball, which the ball $\mathbb{R}^n_c$ inherits from its space $\mathbb{R}^n$, $\|\mathbf{v}\|^2 = \mathbf{v}\cdot\mathbf{v} = \mathbf{v}^2$. A nonempty set with a binary operation is called a *groupoid* so that, accordingly, the pair $(\mathbb{R}^n_c, \oplus)$ is an *Einstein groupoid*.

In the Newtonian limit of large c, $c \to \infty$, the ball $\mathbb{R}^n_c$ expands to the whole of its space $\mathbb{R}^n$, as we see from (1.1), and Einstein addition $\oplus$ in $\mathbb{R}^n_c$ reduces to the ordinary vector addition $+$ in $\mathbb{R}^n$, as we see from (1.2) and (1.3).

In physical applications, $\mathbb{R}^n = \mathbb{R}^3$ is the Euclidean 3-space, which is the space of all classical, Newtonian velocities, and $\mathbb{R}^n_c = \mathbb{R}^3_c \subset \mathbb{R}^3$ is the c-ball of $\mathbb{R}^3$ of all relativistically admissible, Einsteinian velocities. Furthermore, the constant c represents in physical applications the vacuum speed of light. Since we are interested in applications to geometry, we allow n to be any positive integer.

Einstein addition (1.2) of relativistically admissible velocities, with $n = 3$, was introduced by Einstein in his 1905 paper [12], [13, p. 141] that founded the special theory of relativity, where the magnitudes of the two sides of Einstein addition (1.2) are presented. One has to remember here that the Euclidean 3-vector algebra was not so widely known in 1905 and, consequently, was not used by Einstein. Einstein calculated in [12] the behavior of the velocity components parallel and orthogonal to the relative velocity between inertial systems, which is as close as one can get without vectors to the vectorial version (1.2) of Einstein addition.

We naturally use the abbreviation $\mathbf{u}\ominus\mathbf{v} = \mathbf{u}\oplus(-\mathbf{v})$ for Einstein subtraction, so that, for instance, $\mathbf{v}\ominus\mathbf{v} = \mathbf{0}$, $\ominus\mathbf{v} = \mathbf{0}\ominus\mathbf{v} = -\mathbf{v}$ and, in particular,

$$\ominus(\mathbf{u}\oplus\mathbf{v}) = \ominus\mathbf{u}\ominus\mathbf{v} \tag{1.4}$$

and

$$\ominus\mathbf{u}\oplus(\mathbf{u}\oplus\mathbf{v}) = \mathbf{v} \tag{1.5}$$

for all $\mathbf{u}, \mathbf{v}$ in the ball $\mathbb{R}^n_c$, in full analogy with vector addition and subtraction in $\mathbb{R}^n$. Identity (1.4) is known as the *automorphic inverse property*, and Identity (1.5) is known as the *left cancellation law* of Einstein addition [63]. We may note that Einstein addition does not obey the naive right counterpart of the left cancellation law (1.5) since, in general,

$$(\mathbf{u}\oplus\mathbf{v})\ominus\mathbf{v} \neq \mathbf{u} \tag{1.6}$$

However, this seemingly lack of a *right cancellation law* of Einstein addition is repaired in Sect. 1.9, p. 21.

Einstein addition and the gamma factor are related by the *gamma identity*,

$$\gamma_{\mathbf{u}\oplus\mathbf{v}} = \gamma_{\mathbf{u}}\gamma_{\mathbf{v}}\left(1 + \frac{\mathbf{u}\cdot\mathbf{v}}{c^2}\right) \tag{1.7}$$

which can be equivalently written as

$$\gamma_{\ominus\mathbf{u}\oplus\mathbf{v}} = \gamma_{\mathbf{u}}\gamma_{\mathbf{v}}\left(1 - \frac{\mathbf{u}\cdot\mathbf{v}}{c^2}\right) \tag{1.8}$$

for all $\mathbf{u}, \mathbf{v} \in \mathbb{R}^n_c$. Here, (1.8) is obtained from (1.7) by replacing $\mathbf{u}$ by $\ominus\mathbf{u} = -\mathbf{u}$ in (1.7).

A frequently used identity that follows immediately from (1.3) is

$$\frac{\mathbf{v}^2}{c^2} = \frac{\|\mathbf{v}\|^2}{c^2} = \frac{\gamma_{\mathbf{v}}^2 - 1}{\gamma_{\mathbf{v}}^2} \tag{1.9}$$

and, similarly, a useful identity that follows immediately from (1.8) is

$$\frac{\mathbf{u}\cdot\mathbf{v}}{c^2} = 1 - \frac{\gamma_{\ominus\mathbf{u}\oplus\mathbf{v}}}{\gamma_{\mathbf{u}}\gamma_{\mathbf{v}}} \tag{1.10}$$

It is the gamma identity (1.7) that signaled the emergence of hyperbolic geometry in special relativity when it was first studied by Sommerfeld [51] and Varičak [66, 67] in terms of *rapidities*, a term coined by Robb [47]. In fact, the gamma identity plays a role in hyperbolic geometry, analogous to the law of cosines in Euclidean geometry, as we will see in Sect. 6.3, p. 132. Historically, it formed the first link between special relativity and the hyperbolic geometry of Bolyai and Lobachevsky, recently leading to the novel trigonometry in hyperbolic geometry that became known as *gyrotrigonometry*, developed in [63, Chap. 12], [64, Chap. 4], [57, 62] and in Part II of this book.

Einstein addition is noncommutative. Indeed, while Einstein addition is commutative under the norm,

$$\|\mathbf{u}\oplus\mathbf{v}\| = \|\mathbf{v}\oplus\mathbf{u}\| \tag{1.11}$$

we have, in general,

$$\mathbf{u}\oplus\mathbf{v} \neq \mathbf{v}\oplus\mathbf{u} \tag{1.12}$$

for $\mathbf{u}, \mathbf{v} \in \mathbb{R}^n_c$. Moreover, Einstein addition is also nonassociative since, in general,

$$(\mathbf{u}\oplus\mathbf{v})\oplus\mathbf{w} \neq \mathbf{u}\oplus(\mathbf{v}\oplus\mathbf{w}) \tag{1.13}$$

for $\mathbf{u}, \mathbf{v}, \mathbf{w} \in \mathbb{R}^n_c$.

It seems that following the breakdown of commutativity and associativity in Einstein addition some mathematical regularity has been lost in the transition from Newton's velocity vector addition in $\mathbb{R}^n$ to Einstein's velocity addition (1.2) in $\mathbb{R}^n_c$. This is, however, not the case since Thomas gyration comes to the rescue, as we will see in Sect. 1.4. Owing to the presence of Thomas gyration, the Einstein groupoid $(\mathbb{R}^n_c, \oplus)$ has a grouplike structure [56] that we naturally call the *Einstein gyrogroup* [58]. The formal definition of the resulting abstract gyrogroup will be presented in Definition 1.5, p. 12.

1.3 Einstein Addition With Respect to Cartesian Coordinates

Like any physical law, Einstein velocity addition law (1.2) is coordinate independent. Indeed, it is presented in (1.2) in terms of vectors, noting that one of the great advantages of vectors is their ability to express results independent of any coordinate system.

However, in order to generate numerical and graphical demonstrations of physical laws, we need coordinates. Accordingly, we introduce Cartesian coordinates into the Euclidean n-space $\mathbb{R}^n$ and its ball $\mathbb{R}^n_c$, with respect to which we generate the graphs of this book. Introducing the Cartesian coordinate system Σ into $\mathbb{R}^n$ and $\mathbb{R}^n_c$, each point $P \in \mathbb{R}^n$ is given by an n-tuple

$$P = (x_1, x_2, \ldots, x_n), \qquad x_1^2 + x_2^2 + \cdots + x_n^2 < \infty \tag{1.14}$$

of real numbers, which are the coordinates, or components, of P with respect to Σ. Similarly, each point $P \in \mathbb{R}^n_c$ is given by an n-tuple

$$P = (x_1, x_2, \ldots, x_n), \qquad x_1^2 + x_2^2 + \cdots + x_n^2 < c^2 \tag{1.15}$$

of real numbers, which are the coordinates, or components of P with respect to Σ.

Equipped with a Cartesian coordinate system Σ and its standard vector addition given by component addition, along with its resulting scalar multiplication, $\mathbb{R}^n$ forms the standard Cartesian model of n-dimensional Euclidean geometry. In full analogy, equipped with a Cartesian coordinate system Σ and its Einstein addition, along with its resulting scalar multiplication (to be studied in Sect. 2.1), the ball $\mathbb{R}^n_c$ forms in this book the Cartesian–Beltrami–Klein ball model of n-dimensional hyperbolic geometry.

As an illustrative example, we present below the Einstein velocity addition law (1.2) in $\mathbb{R}_c^3$ with respect to a Cartesian coordinate system.

Let $\mathbb{R}_c^3$ be the c-ball of the Euclidean 3-space, equipped with a Cartesian coordinate system Σ. Accordingly, each point of the ball is represented by its coordinates $(x_1, x_2, x_3)^t$ (exponent t denotes transposition) with respect to Σ, satisfying the condition $x_1^2 + x_2^2 + x_3^2 < c^2$.

Furthermore, let $\mathbf{u}, \mathbf{v}, \mathbf{w} \in \mathbb{R}_c^3$ be three points in $\mathbb{R}_c^3 \subset \mathbb{R}^3$ given by their coordinates with respect to Σ,

$$\mathbf{u} = \begin{pmatrix} u_1 \\ u_2 \\ u_3 \end{pmatrix}, \qquad \mathbf{v} = \begin{pmatrix} v_1 \\ v_2 \\ v_3 \end{pmatrix}, \qquad \mathbf{w} = \begin{pmatrix} w_1 \\ w_2 \\ w_3 \end{pmatrix} \tag{1.16}$$

where

$$\mathbf{w} = \mathbf{u} \oplus \mathbf{v} \tag{1.17}$$

The dot product of $\mathbf{u}$ and $\mathbf{v}$ is given in Σ by the equation

$$\mathbf{u} \cdot \mathbf{v} = u_1 v_1 + u_2 v_2 + u_3 v_3 \tag{1.18}$$

and the squared norm $\|\mathbf{v}\|^2 = \mathbf{v} \cdot \mathbf{v}$ of $\mathbf{v}$ is given by the equation

$$\|\mathbf{v}\|^2 = v_1^2 + v_2^2 + v_3^2 \tag{1.19}$$

Hence, it follows from the coordinate independent vector representation (1.2) of Einstein addition that the coordinate dependent Einstein addition (1.17) with respect to the Cartesian coordinate system Σ takes the form

$$\begin{aligned} \begin{pmatrix} w_1 \\ w_2 \\ w_3 \end{pmatrix} &= \frac{1}{1 + \frac{u_1 v_1 + u_2 v_2 + u_3 v_3}{c^2}} \\ &\quad \times \left\{ \left[1 + \frac{1}{c^2} \frac{\gamma_{\mathbf{u}}}{1 + \gamma_{\mathbf{u}}} (u_1 v_1 + u_2 v_2 + u_3 v_3) \right] \begin{pmatrix} u_1 \\ u_2 \\ u_3 \end{pmatrix} + \frac{1}{\gamma_{\mathbf{u}}} \begin{pmatrix} v_1 \\ v_2 \\ v_3 \end{pmatrix} \right\} \end{aligned} \tag{1.20}$$

where

$$\gamma_{\mathbf{u}} = \frac{1}{\sqrt{1 - \frac{u_1^2 + u_2^2 + u_3^2}{c^2}}} \tag{1.21}$$

The three components of Einstein addition (1.17) are w_1, w_2 and w_3 in (1.20). For a two-dimensional illustration of Einstein addition (1.20) one may impose the condition $u_3 = v_3 = 0$, implying $w_3 = 0$.

In the Newtonian–Euclidean limit, $c \to \infty$, the ball $\mathbb{R}_c^3$ expands to the Euclidean 3-space $\mathbb{R}^3$, and Einstein addition (1.20) reduces to the common vector addition in $\mathbb{R}^3$,

$$\begin{pmatrix} w_1 \\ w_2 \\ w_3 \end{pmatrix} = \begin{pmatrix} u_1 \\ u_2 \\ u_3 \end{pmatrix} + \begin{pmatrix} v_1 \\ v_2 \\ v_3 \end{pmatrix} \tag{1.22}$$

1.4 Einstein Addition vs. Vector Addition

Vector addition, $+$, in $\mathbb{R}^n$ is both commutative and associative, satisfying

$$\begin{aligned} \mathbf{u}+\mathbf{v} &= \mathbf{v}+\mathbf{u}, && \text{(Commutative Law)} \\ \mathbf{u}+(\mathbf{v}+\mathbf{w}) &= (\mathbf{u}+\mathbf{v})+\mathbf{w} && \text{(Associative Law)} \end{aligned} \tag{1.23}$$

for all $\mathbf{u}, \mathbf{v}, \mathbf{w} \in \mathbb{R}^n$. In contrast, Einstein addition, $\oplus$, in $\mathbb{R}_c^n$ is neither commutative nor associative.

In order to measure the extent to which Einstein addition deviates from associativity we introduce *gyrations*, which are maps that are *trivial* in the special cases when the application of $\oplus$ is associative. For any $\mathbf{u}, \mathbf{v} \in \mathbb{R}_c^n$, the gyration $\mathrm{gyr}[\mathbf{u}, \mathbf{v}]$ is a map of the Einstein groupoid $(\mathbb{R}_c^n, \oplus)$ onto itself. Gyrations $\mathrm{gyr}[\mathbf{u}, \mathbf{v}] \in \mathrm{Aut}(\mathbb{R}_c^3, \oplus)$, $\mathbf{u}, \mathbf{v} \in \mathbb{R}_c^3$, are defined in terms of Einstein addition by the equation

$$\mathrm{gyr}[\mathbf{u}, \mathbf{v}]\mathbf{w} = \ominus(\mathbf{u}\oplus\mathbf{v})\oplus\{\mathbf{u}\oplus(\mathbf{v}\oplus\mathbf{w})\} \tag{1.24}$$

for all $\mathbf{u}, \mathbf{v}, \mathbf{w} \in \mathbb{R}_c^3$, and they turn out to be automorphisms of the Einstein groupoid $(\mathbb{R}_c^3, \oplus)$.

We recall that an automorphism of a groupoid $(S, \oplus)$ is a one-to-one map f of S onto itself that respects the binary operation, that is, $f(a\oplus b) = f(a)\oplus f(b)$ for all $a, b \in S$. The set of all automorphisms of a groupoid $(S, \oplus)$ forms a group, denoted $\mathrm{Aut}(S, \oplus)$. To emphasize that the gyrations of an Einstein gyrogroup $(\mathbb{R}_c^3, \oplus)$ are automorphisms of the gyrogroup, gyrations are also called *gyroautomorphisms*.

A gyration $\mathrm{gyr}[\mathbf{u}, \mathbf{v}]$, $\mathbf{u}, \mathbf{v} \in \mathbb{R}_c^3$, is *trivial* if $\mathrm{gyr}[\mathbf{u}, \mathbf{v}]\mathbf{w} = \mathbf{w}$ for all $\mathbf{w} \in \mathbb{R}_c^3$. Thus, for instance, the gyrations $\mathrm{gyr}[\mathbf{0}, \mathbf{v}]$, $\mathrm{gyr}[\mathbf{v}, \mathbf{v}]$ and $\mathrm{gyr}[\mathbf{v}, \ominus\mathbf{v}]$ are trivial for all $\mathbf{v} \in \mathbb{R}_c^3$, as we see from (1.24).

Einstein gyrations, which possess their own rich structure, measure the extent to which Einstein addition deviates from commutativity and associativity as we see from the gyrocommutative and the gyroassociative laws of Einstein addition in the following identities [58, 60, 63]:

$$
\begin{aligned}
\mathbf{u}\oplus\mathbf{v} &= \mathrm{gyr}[\mathbf{u},\mathbf{v}](\mathbf{v}\oplus\mathbf{u}), && \text{(Gyrocommutative Law)}\\
\mathbf{u}\oplus(\mathbf{v}\oplus\mathbf{w}) &= (\mathbf{u}\oplus\mathbf{v})\oplus \mathrm{gyr}[\mathbf{u},\mathbf{v}]\mathbf{w}, && \text{(Left Gyroassociative Law)}\\
(\mathbf{u}\oplus\mathbf{v})\oplus\mathbf{w} &= \mathbf{u}\oplus\big(\mathbf{v}\oplus \mathrm{gyr}[\mathbf{v},\mathbf{u}]\mathbf{w}\big), && \text{(Right Gyroassociative Law)}\\
\mathrm{gyr}[\mathbf{u}\oplus\mathbf{v},\mathbf{v}] &= \mathrm{gyr}[\mathbf{u},\mathbf{v}], && \text{(Gyration Left Loop Property)}\\
\mathrm{gyr}[\mathbf{u},\mathbf{v}\oplus\mathbf{u}] &= \mathrm{gyr}[\mathbf{u},\mathbf{v}], && \text{(Gyration Right Loop Property)}\\
\mathrm{gyr}[\ominus\mathbf{u},\ominus\mathbf{v}] &= \mathrm{gyr}[\mathbf{u},\mathbf{v}], && \text{(Gyration Even Property)}\\
\big(\mathrm{gyr}[\mathbf{u},\mathbf{v}]\big)^{-1} &= \mathrm{gyr}[\mathbf{v},\mathbf{u}], && \text{(Gyration Inversion Law)}
\end{aligned}
\tag{1.25}
$$

for all $\mathbf{u}, \mathbf{v}, \mathbf{w} \in \mathbb{R}^n_c$.

Einstein addition is thus regulated by gyrations to which it gives rise owing to its nonassociativity, so that Einstein addition and its gyrations are inextricably linked. The resulting gyrocommutative gyrogroup structure of Einstein addition was discovered in 1988 [55]. Interestingly, (Thomas) gyrations are the mathematical abstraction of the relativistic effect known as *Thomas precession* [63, Sect. 10.3].

The loop properties in (1.25) present important gyration identities. These two gyration identities are, however, just the tip of a giant iceberg. Many other useful gyration identities are studied in [58, 60, 63] and will be studied in the sequel.

1.5 Gyrations

Owing to its nonassociativity, Einstein addition gives rise in (1.24) to gyrations

$$
\mathrm{gyr}[\mathbf{u},\mathbf{v}] : \mathbb{R}^n_c \to \mathbb{R}^n_c \tag{1.26}
$$

for any $\mathbf{u}, \mathbf{v} \in \mathbb{R}^n_c$ in an Einstein groupoid $(\mathbb{R}^n_c, \oplus)$. Gyrations, in turn, regulate Einstein addition, endowing it with the rich structure of a gyrocommutative gyrogroup, as we will see in Sect. 1.6, and a gyrovector space, as we will see in Sect. 2.1. Clearly, gyrations measure the extent to which Einstein addition is nonassociative, where associativity corresponds to trivial gyrations.

An explicit presentation of the gyrations of Einstein groupoids $(\mathbb{R}^n_c, \oplus)$ is, therefore, desirable. Indeed, the gyration equation (1.24) can be manipulated into the equation

$$
\mathrm{gyr}[\mathbf{u},\mathbf{v}]\mathbf{w} = \mathbf{w} + \frac{A\mathbf{u} + B\mathbf{v}}{D} \tag{1.27}
$$

where

$$\begin{aligned}
A &= -\frac{1}{c^2}\frac{\gamma_{\mathbf{u}}^2}{(\gamma_{\mathbf{u}}+1)}(\gamma_{\mathbf{v}}-1)(\mathbf{u}\cdot\mathbf{w}) + \frac{1}{c^2}\gamma_{\mathbf{u}}\gamma_{\mathbf{v}}(\mathbf{v}\cdot\mathbf{w}) \\
&\quad + \frac{2}{c^4}\frac{\gamma_{\mathbf{u}}^2\gamma_{\mathbf{v}}^2}{(\gamma_{\mathbf{u}}+1)(\gamma_{\mathbf{v}}+1)}(\mathbf{u}\cdot\mathbf{v})(\mathbf{v}\cdot\mathbf{w}) \\
B &= -\frac{1}{c^2}\frac{\gamma_{\mathbf{v}}}{\gamma_{\mathbf{v}}+1}\{\gamma_{\mathbf{u}}(\gamma_{\mathbf{v}}+1)(\mathbf{u}\cdot\mathbf{w}) + (\gamma_{\mathbf{u}}-1)\gamma_{\mathbf{v}}(\mathbf{v}\cdot\mathbf{w})\} \\
D &= \gamma_{\mathbf{u}}\gamma_{\mathbf{v}}\left(1+\frac{\mathbf{u}\cdot\mathbf{v}}{c^2}\right) + 1 = \gamma_{\mathbf{u}\oplus\mathbf{v}} + 1 > 1
\end{aligned} \tag{1.28}$$

for all $\mathbf{u}, \mathbf{v}, \mathbf{w} \in \mathbb{R}_c^n$. Allowing $\mathbf{w} \in \mathbb{R}^n \supset \mathbb{R}_c^n$ in (1.27)–(1.28), gyrations $\mathrm{gyr}[\mathbf{u}, \mathbf{v}]$ are expendable from maps of $\mathbb{R}_c^n$ to linear maps of $\mathbb{R}^n$ for all $\mathbf{u}, \mathbf{v} \in \mathbb{R}_c^n$.

In each of the three special cases when (i) $\mathbf{u} = \mathbf{0}$, or (ii) $\mathbf{v} = \mathbf{0}$, or (iii) $\mathbf{u}$ and $\mathbf{v}$ are parallel in $\mathbb{R}^n$, $\mathbf{u}\|\mathbf{v}$, we have $A\mathbf{u} + B\mathbf{v} = \mathbf{0}$ so that $\mathrm{gyr}[\mathbf{u}, \mathbf{v}]$ is trivial. Thus, we have

$$\begin{aligned}
\mathrm{gyr}[\mathbf{0}, \mathbf{v}]\mathbf{w} &= \mathbf{w} \\
\mathrm{gyr}[\mathbf{u}, \mathbf{0}]\mathbf{w} &= \mathbf{w} \\
\mathrm{gyr}[\mathbf{u}, \mathbf{v}]\mathbf{w} &= \mathbf{w}, \qquad \mathbf{u}\|\mathbf{v}
\end{aligned} \tag{1.29}$$

for all $\mathbf{u}, \mathbf{v} \in \mathbb{R}_c^n$, and all $\mathbf{w} \in \mathbb{R}^n$.

It follows from (1.27) that

$$\mathrm{gyr}[\mathbf{v}, \mathbf{u}]\big(\mathrm{gyr}[\mathbf{u}, \mathbf{v}]\mathbf{w}\big) = \mathbf{w} \tag{1.30}$$

for all $\mathbf{u}, \mathbf{v} \in \mathbb{R}_c^n$, $\mathbf{w} \in \mathbb{R}^n$, so that gyrations are invertible linear maps of $\mathbb{R}^n$, the inverse, $\mathrm{gyr}^{-1}[\mathbf{u}, \mathbf{v}]$, of $\mathrm{gyr}[\mathbf{u}, \mathbf{v}]$ being $\mathrm{gyr}[\mathbf{v}, \mathbf{u}]$. We thus have the gyration inversion property

$$\mathrm{gyr}^{-1}[\mathbf{u}, \mathbf{v}] = \mathrm{gyr}[\mathbf{v}, \mathbf{u}] \tag{1.31}$$

for all $\mathbf{u}, \mathbf{v} \in \mathbb{R}_c^n$.

Gyrations keep the inner product of elements of the ball $\mathbb{R}_c^n$ invariant, that is,

$$\mathrm{gyr}[\mathbf{u}, \mathbf{v}]\mathbf{a}\cdot \mathrm{gyr}[\mathbf{u}, \mathbf{v}]\mathbf{b} = \mathbf{a}\cdot\mathbf{b} \tag{1.32}$$

for all $\mathbf{a}, \mathbf{b}, \mathbf{u}, \mathbf{v} \in \mathbb{R}_c^n$. Hence, $\mathrm{gyr}[\mathbf{u}, \mathbf{v}]$ is an *isometry* of $\mathbb{R}_c^n$, keeping the norm of elements of the ball $\mathbb{R}_c^n$ invariant,

$$\|\mathrm{gyr}[\mathbf{u}, \mathbf{v}]\mathbf{w}\| = \|\mathbf{w}\| \tag{1.33}$$

Accordingly, $\mathrm{gyr}[\mathbf{u}, \mathbf{v}]$ represents a rotation of the ball $\mathbb{R}_c^n$ about its origin for any $\mathbf{u}, \mathbf{v} \in \mathbb{R}_c^n$.

The invertible self-map $\mathrm{gyr}[\mathbf{u}, \mathbf{v}]$ of $\mathbb{R}_c^n$ respects Einstein addition in $\mathbb{R}_c^n$,

$$\mathrm{gyr}[\mathbf{u}, \mathbf{v}](\mathbf{a}\oplus\mathbf{b}) = \mathrm{gyr}[\mathbf{u}, \mathbf{v}]\mathbf{a}\oplus \mathrm{gyr}[\mathbf{u}, \mathbf{v}]\mathbf{b} \tag{1.34}$$

for all $\mathbf{a}, \mathbf{b}, \mathbf{u}, \mathbf{v} \in \mathbb{R}_c^n$, so that gyr$[\mathbf{u}, \mathbf{v}]$ is an automorphism of the Einstein groupoid $(\mathbb{R}_c^n, \oplus)$.

1.6 From Einstein Velocity Addition to Gyrogroups

Taking the key features of the Einstein groupoid $(\mathbb{R}_c^n, \oplus)$ as axioms, and guided by analogies with groups, we are led to the formal gyrogroup definition in which gyrogroups turn out to form a most natural generalization of groups. Definitions related to groups and gyrogroups thus follow.

Definition 1.1 (Groups) A groupoid $(G, +)$ is a group if its binary operation satisfies the following axioms. In G there is at least one element, 0, called a left identity, satisfying

(G1)

$$0 + a = a$$

for all $a \in G$. There is an element $0 \in G$ satisfying Axiom (G1) such that for each $a \in G$ there is an element $-a \in G$, called a left inverse of a, satisfying

(G2)

$$-a + a = 0$$

Moreover, the binary operation obeys the associative law

(G3)

$$(a + b) + c = a + (b + c)$$

for all $a, b, c \in G$.

Groups are classified into commutative and noncommutative groups.

Definition 1.2 (Commutative Groups) A group $(G, +)$ is commutative if its binary operation obeys the commutative law

(G6)

$$a + b = b + a$$

for all $a, b \in G$.

Definition 1.3 (Subgroups) A subset H of a subgroup $(G, +)$ is a subgroup of G if it is nonempty, and H is closed under group compositions and inverses in G, that is, $x, y \in H$ implies $x + y \in H$ and $-x \in H$.

Theorem 1.4 (The Subgroup Criterion) *A subset H of a group G is a subgroup if and only if* (i) *H is nonempty, and* (ii) *$x, y \in H$ implies $x - y \in H$.*

For a proof of the Subgroup Criterion see any book on group theory.

Definition 1.5 (Gyrogroups) A groupoid $(G, \oplus)$ is a gyrogroup if its binary operation satisfies the following axioms. In G there is at least one element, 0, called a left identity, satisfying

(G1)

$$0\oplus a = a$$

for all $a \in G$. There is an element $0 \in G$ satisfying Axiom (G1) such that for each $a \in G$ there is an element $\ominus a \in G$, called a left inverse of a, satisfying

(G2)

$$\ominus a\oplus a = 0$$

Moreover, for any $a, b, c \in G$ there exists a unique element $\mathrm{gyr}[a, b]c \in G$ such that the binary operation obeys the left gyroassociative law

(G3)

$$a\oplus(b\oplus c) = (a\oplus b)\oplus \mathrm{gyr}[a, b]c$$

The map $\mathrm{gyr}[a, b] : G \to G$ given by $c \mapsto \mathrm{gyr}[a, b]c$ is an automorphism of the groupoid $(G, \oplus)$, that is,

(G4)

$$\mathrm{gyr}[a, b] \in \mathrm{Aut}(G, \oplus)$$

and the automorphism $\mathrm{gyr}[a, b]$ of G is called the gyroautomorphism, or the gyration, of G generated by $a, b \in G$. The operator $\mathrm{gyr} : G \times G \to \mathrm{Aut}(G, \oplus)$ is called the gyrator of G. Finally, the gyroautomorphism $\mathrm{gyr}[a, b]$ generated by any $a, b \in G$ possesses the left loop property

(G5)

$$\mathrm{gyr}[a, b] = \mathrm{gyr}[a\oplus b, b]$$

The gyrogroup axioms (G1)–(G5) in Definition 1.5 are classified into three classes:

1. The first pair of axioms, (G1) and (G2), is a reminiscent of the group axioms.
2. The last pair of axioms, (G4) and (G5), presents the gyrator axioms.
3. The middle axiom, (G3), is a hybrid axiom linking the two pairs of axioms in (1) and (2).

As in group theory, we use the notation $a\ominus b = a\oplus(\ominus b)$ in gyrogroup theory as well.

In full analogy with groups, gyrogroups are classified into gyrocommutative and non-gyrocommutative gyrogroups.

Definition 1.6 (Gyrocommutative Gyrogroups) A gyrogroup $(G, \oplus)$ is gyrocommutative if its binary operation obeys the gyrocommutative law

(G6)

$$a \oplus b = \mathrm{gyr}[a, b](b \oplus a)$$

for all $a, b \in G$.

In order to capture analogies with groups, we introduce into the abstract gyrogroup $(G, \oplus)$ a second binary operation, $\boxplus$, called the gyrogroup *cooperation*, or coaddition.

Definition 1.7 (The Gyrogroup Cooperation (Coaddition)) Let $(G, \oplus)$ be a gyrogroup. The gyrogroup cooperation (or, coaddition), $\boxplus$, is a second binary operation in G related to the gyrogroup operation (or, addition), $\oplus$, by the equation

$$a \boxplus b = a \oplus \mathrm{gyr}[a, \ominus b]b \tag{1.35}$$

for all $a, b \in G$.

Naturally, we use the notation $a \boxminus b = a \boxplus (\ominus b)$ where $\ominus b = -b$, so that

$$a \boxminus b = a \ominus \mathrm{gyr}[a, b]b \tag{1.36}$$

The gyrogroup cooperation is commutative if and only if the gyrogroup operation is gyrocommutative, as we will see in Theorem 1.33, p. 35.

Hence, in particular, Einstein coaddition $\boxplus = \boxplus_E$ is commutative since Einstein addition $\oplus = \oplus_E$ is gyrocommutative. Indeed, Einstein coaddition $\boxplus = \boxplus_E$ is commutative, given explicitly by the equation [63, pp. 92–93]

$$\mathbf{u} \boxplus \mathbf{v} = \frac{\gamma_{\mathbf{u}} + \gamma_{\mathbf{v}}}{\gamma_{\mathbf{u}}^2 + \gamma_{\mathbf{v}}^2 + \gamma_{\mathbf{u}}\gamma_{\mathbf{v}}(1 + \frac{\mathbf{u}\cdot\mathbf{v}}{c^2}) - 1}(\gamma_{\mathbf{u}}\mathbf{u} + \gamma_{\mathbf{v}}\mathbf{v}) = 2\otimes\frac{\gamma_{\mathbf{u}}\mathbf{u} + \gamma_{\mathbf{v}}\mathbf{v}}{\gamma_{\mathbf{u}} + \gamma_{\mathbf{v}}} \tag{1.37}$$

for all $\mathbf{u}, \mathbf{v} \in \mathbb{R}^n_c$ where, by scalar multiplication definition, $2\otimes\mathbf{v} = \mathbf{v}\oplus\mathbf{v}$.

While it is clear how to define a right identity and a right inverse in a gyrogroup, the existence of such elements is not presumed. Indeed, the existence of a unique identity and a unique inverse, both left and right, is a consequence of the gyrogroup axioms, as the following theorem shows, along with other immediate results.

Theorem 1.8 (First Gyrogroup Properties) *Let $(G, \oplus)$ be a gyrogroup. For any elements $a, b, c, x \in G$ we have*:

1. *If $a\oplus b = a\oplus c$, then $b = c$ (general left cancellation law; see item* (9) *below).*
2. $\mathrm{gyr}[0, a] = I$ *for any left identity* 0 *in* G.
3. $\mathrm{gyr}[x, a] = I$ *for any left inverse x of a in G.*
4. $\mathrm{gyr}[a, a] = I$.
5. *There is a left identity which is a right identity.*
6. *There is only one left identity.*
7. *Every left inverse is a right inverse.*
8. *There is only one left inverse, $\ominus a$, of a, and $\ominus(\ominus a) = a$.*

9. *The Left Cancellation Law*:

$$\ominus a \oplus (a \oplus b) = b. \tag{1.38}$$

10. *The Gyrator Identity*:

$$\mathrm{gyr}[a,b]x = \ominus(a\oplus b)\oplus\{a\oplus(b\oplus x)\}. \tag{1.39}$$

11. $\mathrm{gyr}[a,b]0 = 0$.
12. $\mathrm{gyr}[a,b](\ominus x) = \ominus\,\mathrm{gyr}[a,b]x$.
13. $\mathrm{gyr}[a,0] = I$.

Proof

1. Let x be a left inverse of a corresponding to a left identity, 0, in G. We have $x\oplus(a\oplus b) = x\oplus(a\oplus c)$, implying $(x\oplus a)\oplus\mathrm{gyr}[x,a]b = (x\oplus a)\oplus\mathrm{gyr}[x,a]c$ by left gyroassociativity. Since 0 is a left identity, $\mathrm{gyr}[x,a]b = \mathrm{gyr}[x,a]c$. Since automorphisms are bijective, $b = c$.
2. By left gyroassociativity we have for any left identity 0 of G, $a\oplus x = 0\oplus(a\oplus x) = (0\oplus a)\oplus\mathrm{gyr}[0,a]x = a\oplus\mathrm{gyr}[0,a]x$. Hence, by Item 1 above we have $x = \mathrm{gyr}[0,a]x$ for all $x\in G$ so that $\mathrm{gyr}[0,a] = I$.
3. By the left loop property and by Item 2 above we have $\mathrm{gyr}[x,a] = \mathrm{gyr}[x\oplus a,a] = \mathrm{gyr}[0,a] = I$.
4. Follows from an application of the left loop property and Item 2 above.
5. Let x be a left inverse of a corresponding to a left identity, 0, of G. Then by left gyroassociativity and Item 3 above, $x\oplus(a\oplus 0) = (x\oplus a)\oplus\mathrm{gyr}[x,a]0 = 0\oplus 0 = 0 = x\oplus a$. Hence, by (1), $a\oplus 0 = a$ for all $a\in G$ so that 0 is a right identity.
6. Suppose 0 and 0^* are two left identities, one of which, say 0, is also a right identity. Then $0 = 0^*\oplus 0 = 0^*$.
7. Let x be a left inverse of a. Then $x\oplus(a\oplus x) = (x\oplus a)\oplus\mathrm{gyr}[x,a]x = 0\oplus x = x = x\oplus 0$, by left gyroassociativity, (G2) of Definition 1.5 and Items 3, 5, 6 above. By Item 1, we have $a\oplus x = 0$ so that x is a right inverse of a.
8. Suppose x and y are left inverses of a. By Item 7 above, they are also right inverses, so $a\oplus x = 0 = a\oplus y$. By Item 1, $x = y$. Let $\ominus a$ be the resulting unique inverse of a. Then $\ominus a\oplus a = 0$ so that the inverse $\ominus(\ominus a)$ of $\ominus a$ is a.
9. By left gyroassociativity and by Item 3, we have

$$\ominus a\oplus(a\oplus b) = (\ominus a\oplus a)\oplus\mathrm{gyr}[\ominus a,a]b = b \tag{1.40}$$

10. By an application of the left cancellation law in Item 9 to the left gyroassociative law (G3) in Definition 1.5, we obtain the result in Item 10.
11. We obtain Item 11 from Item 10 with $x = 0$.
12. Since $\mathrm{gyr}[a,b]$ is an automorphism of $(G,\oplus)$, we have from Item 11

$$\mathrm{gyr}[a,b](\ominus x)\oplus\mathrm{gyr}[a,b]x = \mathrm{gyr}[a,b](\ominus x\oplus x) = \mathrm{gyr}[a,b]0 = 0 \tag{1.41}$$

and hence the result.
13. We obtain Item 13 from Item 10 with $b = 0$, and a left cancellation, Item 9. □

1.7 Elements of Gyrogroup Theory

Einstein gyrogroups $(G, \oplus)$ possess the *gyroautomorphic inverse property*, according to which $\ominus(a\oplus b) = \ominus a\ominus b$ for all $a, b \in G$. In general, however, $\ominus(a\oplus b) \neq \ominus a\ominus b$ in other gyrogroups. Hence, the following theorem is interesting.

Theorem 1.9 (Gyrosum Inversion Law) *For any two elements a, b of a gyrogroup $(G, \oplus)$ we have the gyrosum inversion law*

$$\ominus(a\oplus b) = \mathrm{gyr}[a, b](\ominus b\ominus a) \tag{1.42}$$

Proof By the gyrator identity in Theorem 1.8(10) and a left cancellation, Theorem 1.8(9), we have

$$\begin{aligned}\mathrm{gyr}[a, b](\ominus b\ominus a) &= \ominus(a\oplus b)\oplus\big(a\oplus\big(b\oplus(\ominus b\ominus a)\big)\big)\\ &= \ominus(a\oplus b)\oplus(a\ominus a)\\ &= \ominus(a\oplus b)\end{aligned} \tag{1.43}$$

□

Theorem 1.10 *For any two elements, a and b, of a gyrogroup $(G, \oplus)$, we have*

$$\begin{aligned}\mathrm{gyr}[a, b]b &= \ominus\big\{\ominus(a\oplus b)\oplus a\big\}\\ \mathrm{gyr}[a, \ominus b]b &= \ominus(a\ominus b)\oplus a\end{aligned} \tag{1.44}$$

Proof The first identity in (1.44) follows from Theorem 1.8(10) with $x = \ominus b$, and Theorem 1.8(12), and the second part of Theorem 1.8(8). The second identity in (1.44) follows from the first one by replacing b by $\ominus b$. □

A nested gyroautomorphism is a gyration generated by points that depend on another gyration. Thus, for instance, some gyrations in (1.45)–(1.47) below are nested.

Theorem 1.11 *Any three elements a, b, c of a gyrogroup $(G, \oplus)$ satisfy the nested gyroautomorphism identities*

$$\mathrm{gyr}[a, b\oplus c]\,\mathrm{gyr}[b, c] = \mathrm{gyr}\big[a\oplus b, \mathrm{gyr}[a, b]c\big]\,\mathrm{gyr}[a, b] \tag{1.45}$$

$$\mathrm{gyr}\big[a\oplus b, \ominus\,\mathrm{gyr}[a, b]b\big]\,\mathrm{gyr}[a, b] = I \tag{1.46}$$

$$\mathrm{gyr}\big[a, \ominus\,\mathrm{gyr}[a, b]b\big]\,\mathrm{gyr}[a, b] = I \tag{1.47}$$

and the gyroautomorphism product identities

$$\mathrm{gyr}[\ominus a, a\oplus b]\,\mathrm{gyr}[a, b] = I \tag{1.48}$$

$$\mathrm{gyr}[b, a\oplus b]\,\mathrm{gyr}[a, b] = I \tag{1.49}$$

Proof By two successive applications of the left gyroassociative law in two different ways, we obtain the following two chains of equations for all $a, b, c, x \in G$,

$$\begin{aligned} a\oplus\big(b\oplus(c\oplus x)\big) &= a\oplus\big((b\oplus c)\oplus \operatorname{gyr}[b, c]x\big) \\ &= \big(a\oplus(b\oplus c)\big)\oplus \operatorname{gyr}[a, b\oplus c]\operatorname{gyr}[b, c]x \end{aligned} \tag{1.50}$$

and

$$\begin{aligned} a\oplus\big(b\oplus(c\oplus x)\big) &= (a\oplus b)\oplus \operatorname{gyr}[a, b](c\oplus x) \\ &= (a\oplus b)\oplus\big(\operatorname{gyr}[a, b]c\oplus \operatorname{gyr}[a, b]x\big) \\ &= \big((a\oplus b)\oplus \operatorname{gyr}[a, b]c\big)\oplus \operatorname{gyr}\big[a\oplus b, \operatorname{gyr}[a, b]c\big]\operatorname{gyr}[a, b]x \\ &= \big(a\oplus(b\oplus c)\big)\oplus \operatorname{gyr}\big[a\oplus b, \operatorname{gyr}[a, b]c\big]\operatorname{gyr}[a, b]x \end{aligned} \tag{1.51}$$

By comparing the extreme right-hand sides of these two chains of equations, and by employing the left cancellation law, Theorem 1.8(1), we obtain the identity

$$\operatorname{gyr}[a, b\oplus c]\operatorname{gyr}[b, c]x = \operatorname{gyr}\big[a\oplus b, \operatorname{gyr}[a, b]c\big]\operatorname{gyr}[a, b]x \tag{1.52}$$

for all $x \in G$, thus verifying (1.45).

In the special case when $c = \ominus b$, (1.45) reduces to (1.46), noting that the left-hand side of (1.45) becomes trivial owing to Items (2) and (3) of Theorem 1.8.

Identity (1.47) results from the following chain of equations, which are numbered for subsequent derivation:

$$\begin{aligned} I &\overbrace{=\!=}^{(1)} \operatorname{gyr}\big[a\oplus b, \ominus \operatorname{gyr}[a, b]b\big]\operatorname{gyr}[a, b] \\ &\overbrace{=\!=}^{(2)} \operatorname{gyr}\big[(a\oplus b)\ominus \operatorname{gyr}[a, b]b, \ominus \operatorname{gyr}[a, b]b\big]\operatorname{gyr}[a, b] \\ &\overbrace{=\!=}^{(3)} \operatorname{gyr}\big[a\oplus(b\ominus b), \ominus \operatorname{gyr}[a, b]b\big]\operatorname{gyr}[a, b] \\ &\overbrace{=\!=}^{(4)} \operatorname{gyr}\big[a, \ominus \operatorname{gyr}[a, b]b\big]\operatorname{gyr}[a, b] \end{aligned} \tag{1.53}$$

Derivation of the numbered equalities in (1.53) follows:

1. Follows from (1.46).
2. Follows from Item 1 by the left loop property.
3. Follows from Item 2 by the left gyroassociative law. Indeed, an application of the left gyroassociative law to the first entry of the left gyration in (3) gives the first entry of the left gyration in (2), that is, $a\oplus(b\ominus b) = (a\oplus b)\ominus \operatorname{gyr}[a, b]b$.
4. Follows from Item 3 immediately, since $b\ominus b = 0$.

To verify (1.48), we consider the special case of (1.45) when $b = \ominus a$, obtaining

$$\mathrm{gyr}[a, \ominus a \oplus c]\,\mathrm{gyr}[\ominus a, c] = \mathrm{gyr}\big[0, \mathrm{gyr}[a, \ominus a]c\big]\,\mathrm{gyr}[a, \ominus a] = I \tag{1.54}$$

where the second identity in (1.54) follows from Items (2) and (3) of Theorem 1.8. Replacing a by $\ominus a$ and c by b in (1.54), we obtain (1.48).

Finally, (1.49) is derived from (1.48) by an application of the left loop property to the first gyroautomorphism in (1.48) followed by a left cancellation, Theorem 1.8(9). Accordingly,

$$\begin{aligned} I &= \mathrm{gyr}[\ominus a, a \oplus b]\,\mathrm{gyr}[a, b] \\ &= \mathrm{gyr}\big[\ominus a \oplus (a \oplus b), a \oplus b\big]\,\mathrm{gyr}[a, b] \\ &= \mathrm{gyr}[b, a \oplus b]\,\mathrm{gyr}[a, b] \end{aligned} \tag{1.55}$$

□

The nested gyroautomorphism identity (1.47) in Theorem 1.11 allows the equation that defines the coaddition $\boxplus$ to be dualized with its corresponding equation in which the roles of the binary operations $\boxplus$ and $\oplus$ are interchanged, as shown in the following theorem:

Theorem 1.12 *Let* $(G, \oplus)$ *be a gyrogroup with cooperation* $\boxplus$ *given in Definition* 1.7, *p.* 13, *by the equation*

$$a \boxplus b = a \oplus \mathrm{gyr}[a, \ominus b]b \tag{1.56}$$

Then

$$a \oplus b = a \boxplus \mathrm{gyr}[a, b]b \tag{1.57}$$

Proof Let a and b be any two elements of G. By (1.56) and (1.47), we have

$$\begin{aligned} a \boxplus \mathrm{gyr}[a, b]b &= a \oplus \mathrm{gyr}\big[a, \ominus \mathrm{gyr}[a, b]b\big]\,\mathrm{gyr}[a, b]b \\ &= a \oplus b \end{aligned} \tag{1.58}$$

thus verifying (1.57). □

We naturally use the notation

$$a \boxminus b = a \boxplus (\ominus b) \tag{1.59}$$

in a gyrogroup $(G, \oplus)$, so that, by (1.59), (1.56) and Theorem 1.8(12),

$$\begin{aligned} a \boxminus b &= a \boxplus (\ominus b) \\ &= a \oplus \mathrm{gyr}[a, b](\ominus b) \\ &= a \ominus \mathrm{gyr}[a, b]b \end{aligned} \tag{1.60}$$

and hence

$$a \boxminus a = a \ominus a = 0 \tag{1.61}$$

as it should be. Identity (1.61), in turn, implies the equality between the inverses of $a \in G$ with respect to $\oplus$ and $\boxplus$,

$$\boxminus a = \ominus a \tag{1.62}$$

for all $a \in G$.

Theorem 1.13 *Let* $(G, \oplus)$ *be a gyrogroup. Then*

$$(\ominus a \oplus b) \oplus \mathrm{gyr}[\ominus a, b](\ominus b \oplus c) = \ominus a \oplus c \tag{1.63}$$

for all $a, b, c \in G$.

Proof By the left gyroassociative law and the left cancellation law, and using the notation $d = \ominus b \oplus c$, we have

$$\begin{aligned} (\ominus a \oplus b) \oplus \mathrm{gyr}[\ominus a, b](\ominus b \oplus c) &= (\ominus a \oplus b) \oplus \mathrm{gyr}[\ominus a, b]d \\ &= \ominus a \oplus (b \oplus d) \\ &= \ominus a \oplus \big(b \oplus (\ominus b \oplus c)\big) \\ &= \ominus a \oplus c \end{aligned} \tag{1.64}$$

□

Theorem 1.14 (The Gyrotranslation Theorem, I) *Let* $(G, \oplus)$ *be a gyrogroup. Then*

$$\ominus(\ominus a \oplus b) \oplus (\ominus a \oplus c) = \mathrm{gyr}[\ominus a, b](\ominus b \oplus c) \tag{1.65}$$

for all $a, b, c \in G$.

Proof Identity (1.65) is a rearrangement of Identity (1.63) obtained by a left cancellation. □

The importance of Identity (1.65) lies in the analogy it shares with its group counterpart, $-(-a + b) + (-a + c) = -b + c$ in any group $(\mathit{Group}, +)$.

The identity of Theorem 1.13 can readily be generalized to any number of terms, for instance,

$$(\ominus a \oplus b) \oplus \mathrm{gyr}[\ominus a, b]\big\{(\ominus b \oplus c) \oplus \mathrm{gyr}[\ominus b, c](\ominus c \oplus d)\big\} = \ominus a \oplus d \tag{1.66}$$

which generalizes the obvious group identity $(-a + b) + (-b + c) + (-c + d) = -a + d$ in any group $(\mathit{Group}, +)$.

1.8 The Two Basic Equations of Gyrogroups

The two basic equations of gyrogroup theory are

$$a \oplus x = b \tag{1.67}$$

and

$$x \oplus a = b \tag{1.68}$$

for $a, b, x \in G$, each for the unknown x in a gyrogroup $(G, \oplus)$.

Let x be a solution of the first basic equation, (1.67). Then we have by (1.67) and the left cancellation law, Theorem 1.8(9),

$$\ominus a \oplus b = \ominus a \oplus (a \oplus x) = x \tag{1.69}$$

Hence, if a solution x of (1.67) exists then it must be given by $x = \ominus a \oplus b$, as we see from (1.69).

Conversely, $x = \ominus a \oplus b$ is, indeed, a solution of (1.67) as wee see by substituting $x = \ominus a \oplus b$ into (1.67) and applying the left cancellation law in Theorem 1.8(9). Hence, the gyrogroup equation (1.67) possesses the unique solution $x = \ominus a \oplus b$.

The solution of the second basic gyrogroup equation, (1.68), is quiet different from that of the first, (1.67), owing to the noncommutativity of the gyrogroup operation. Let x be a solution of (1.68). Then we have the following chain of equations, which are numbered for subsequent derivation:

$$\begin{aligned}
x &\overbrace{=\!=\!=}^{(1)} x \oplus 0 \\
&\overbrace{=\!=\!=}^{(2)} x \oplus (a \ominus a) \\
&\overbrace{=\!=\!=}^{(3)} (x \oplus a) \oplus \mathrm{gyr}[x, a](\ominus a) \\
&\overbrace{=\!=\!=}^{(4)} (x \oplus a) \ominus \mathrm{gyr}[x, a]a \\
&\overbrace{=\!=\!=}^{(5)} (x \oplus a) \ominus \mathrm{gyr}[x \oplus a, a]a \\
&\overbrace{=\!=\!=}^{(6)} b \ominus \mathrm{gyr}[b, a]a \overbrace{=\!=\!=}^{(7)} b \boxminus a
\end{aligned} \tag{1.70}$$

Derivation of the numbered equalities in (1.70) follows:

1. Follows from the existence of a unique identity element, 0, in the gyrogroup $(G, \oplus)$ by Theorem 1.8.
2. Follows from the existence of a unique inverse element $\ominus a$ of a in the gyrogroup $(G, \oplus)$ by Theorem 1.8.
3. Follows from Item 2 by the left gyroassociative law in Axiom (G3) of gyrogroups in Definition 1.5, p. 12.

4. Follows from Item 3 by Theorem 1.8(12).
5. Follows from Item 4 by the left loop property (G5) of gyrogroups in Definition 1.5.
6. Follows from Item 5 by the assumption that x is a solution of (1.68).
7. Follows from Item 6 by (1.60).

Hence, if a solution x of (1.68) exists then it must be given by $x = b \boxminus a$, as we see from (1.70).

Conversely, $x = b \boxminus a$ is, indeed, a solution of (1.68), as we see from the following chain of equations:

$$\begin{aligned} x \oplus a &\overset{(1)}{=} (b \boxminus a) \oplus a \\ &\overset{(2)}{=} \big(b \ominus \mathrm{gyr}[b, a]a\big) \oplus a \\ &\overset{(3)}{=} \big(b \ominus \mathrm{gyr}[b, a]a\big) \oplus \mathrm{gyr}\big[b, \ominus \mathrm{gyr}[b, a]\big] \mathrm{gyr}[b, a]a \\ &\overset{(4)}{=} b \oplus \big(\ominus \mathrm{gyr}[b, a]a \oplus \mathrm{gyr}[b, a]a\big) \\ &\overset{(5)}{=} b \oplus 0 \\ &\overset{(6)}{=} b \end{aligned} \tag{1.71}$$

Derivation of the numbered equalities in (1.71) follows:

1. Follows from the assumption that $x = b \boxminus a$.
2. Follows from Item 1 by (1.60).
3. Follows from Item 2 by Identity (1.47) of Theorem 1.11, according to which the gyration product applied to a in (3) is trivial.
4. Follows from Item 3 by the left gyroassociative law. Indeed, an application of the left gyroassociative law to (4) results in (3).
5. Follows from Item 4 since $\ominus \mathrm{gyr}[b, a]a$ is the unique inverse of $\mathrm{gyr}[b, a]a$.
6. Follows from Item 5 since 0 is the unique identity element of the gyrogroup $(G, \oplus)$.

Formalizing the results of this section, we have the following theorem:

Theorem 1.15 (The Two Basic Gyrogroup Equations) *Let* $(G, \oplus)$ *be a gyrogroup, and let* $a, b \in G$. *The unique solution of the equation*

$$a \oplus x = b \tag{1.72}$$

in G *for the unknown* x *is*

$$x = \ominus a \oplus b \tag{1.73}$$

and the unique solution of the equation

$$x \oplus a = b \tag{1.74}$$

in G for the unknown x is

$$x = b \boxminus a \tag{1.75}$$

Let $(G, \oplus)$ be a gyrogroup, and let $a \in G$. The maps λ_a and ρ_a of G, given by

$$\begin{aligned} \lambda_a\colon \quad G \to G, \qquad \lambda_a\colon \quad g \mapsto a \oplus g \\ \rho_a\colon \quad G \to G, \qquad \rho_a\colon \quad g \mapsto g \oplus a \end{aligned} \tag{1.76}$$

are called, respectively, a *left gyrotranslation* of G by a and a *right gyrotranslation* of G by a. Theorem 1.15 asserts that each of these transformations of G is bijective, that is, it maps G onto itself in a one-to-one manner.

1.9 The Basic Gyrogroup Cancellation Laws

The basic cancellation laws of gyrogroup theory are obtained in this section from the basic equations of gyrogroups solved in Sect. 1.8. Substituting the solution (1.73) into its equation (1.72), we obtain the left cancellation law

$$a \oplus (\ominus a \oplus b) = b \tag{1.77}$$

for all $a, b \in G$, already verified in Theorem 1.8(9).

Similarly, substituting the solution (1.75) into its equation (1.74), we obtain the first right cancellation law

$$(b \boxminus a) \oplus a = b \tag{1.78}$$

for all $a, b \in G$. The latter can be dualized, obtaining the second right cancellation law

$$(b \ominus a) \boxplus a = b \tag{1.79}$$

for all $a, b \in G$. Indeed, (1.79) results from the following chain of equations

$$\begin{aligned} b &= b \oplus 0 \\ &= b \oplus (\ominus a \oplus a) \\ &= (b \ominus a) \oplus \mathrm{gyr}[b, \ominus a]a \\ &= (b \ominus a) \oplus \mathrm{gyr}[b \ominus a, \ominus a]a \\ &= (b \ominus a) \boxplus a \end{aligned} \tag{1.80}$$

where we employ the left gyroassociative law, the left loop property, and the definition of the gyrogroup cooperation. Identities (1.77)–(1.79) form the three basic cancellation laws of gyrogroup theory.

1.10 Commuting Automorphisms with Gyroautomorphisms

In this section, we will find that automorphisms of a gyrogroup commute with its gyroautomorphisms in a special, interesting way.

Theorem 1.16 *For any two elements a, b of a gyrogroup $(G, \oplus)$ and any automorphism A of $(G, \oplus)$, $A \in \mathrm{Aut}(G, \oplus)$,*

$$A\,\mathrm{gyr}[a, b] = \mathrm{gyr}[Aa, Ab]A \tag{1.81}$$

Proof For any three elements $a, b, x \in (G, \oplus)$ and any automorphism $A \in \mathrm{Aut}(G, \oplus)$, we have, by the left gyroassociative law,

$$\begin{aligned}(Aa \oplus Ab) \oplus A\,\mathrm{gyr}[a, b]x &= A\big((a \oplus b) \oplus \mathrm{gyr}[a, b]x\big)\\ &= A\big(a \oplus (b \oplus x)\big)\\ &= Aa \oplus (Ab \oplus Ax)\\ &= (Aa \oplus Ab) \oplus \mathrm{gyr}[Aa, Ab]Ax\end{aligned} \tag{1.82}$$

Hence, by a left cancellation, Theorem 1.8(1),

$$A\,\mathrm{gyr}[a, b]x = \mathrm{gyr}[Aa, Ab]Ax$$

for all $x \in G$, implying (1.81). □

Theorem 1.17 *Let a, b be any two elements of a gyrogroup $(G, \oplus)$ and let $A \in \mathrm{Aut}(G)$ be an automorphism of G. Then*

$$\mathrm{gyr}[a, b] = \mathrm{gyr}[Aa, Ab] \tag{1.83}$$

if and only if the automorphisms A and $\mathrm{gyr}[a, b]$ are commutative.

Proof If $\mathrm{gyr}[Aa, Ab] = \mathrm{gyr}[a, b]$ then by Theorem 1.16 the automorphisms $\mathrm{gyr}[a, b]$ and A commute, $A\,\mathrm{gyr}[a, b] = \mathrm{gyr}[Aa, Ab]A = \mathrm{gyr}[a, b]A$. Conversely, if $\mathrm{gyr}[a, b]$ and A commute then by Theorem 1.16 $\mathrm{gyr}[Aa, Ab] = A\,\mathrm{gyr}[a, b]A^{-1} = \mathrm{gyr}[a, b]AA^{-1} = \mathrm{gyr}[a, b]$. □

As a simple, but useful, consequence of Theorem 1.17 we note the elegant identity

$$\mathrm{gyr}\big[\mathrm{gyr}[a, b]a, \mathrm{gyr}[a, b]b\big] = \mathrm{gyr}[a, b] \tag{1.84}$$

1.11 The Gyrosemidirect Product

Definition 1.18 (Gyroautomorphism Groups, Gyrosemidirect Product) Let $G = (G, \oplus)$ be a gyrogroup, and let $\mathrm{Aut}(G) = \mathrm{Aut}(G, \oplus)$ be the automorphism group of G. A gyroautomorphism group, $\mathrm{Aut}_0(G)$, of G is any subgroup of $\mathrm{Aut}(G)$ containing all the gyroautomorphisms $\mathrm{gyr}[a, b]$ of G, $a, b \in G$. The *gyrosemidirect product group*

$$G \times \mathrm{Aut}_0(G) \tag{1.85}$$

of a gyrogroup G and any gyroautomorphism group, $\mathrm{Aut}_0(G)$ of G, is a group of pairs (x, X), where $x \in G$ and $X \in \mathrm{Aut}_0(G)$, with operation given by the *gyrosemidirect product*

$$(x, X)(y, Y) = \big(x \oplus Xy, \mathrm{gyr}[x, Xy]XY\big) \tag{1.86}$$

It is anticipated in Definition 1.18 that the gyrosemidirect product set (1.85) of a gyrogroup and any one of its gyroautomorphism groups is a set that forms a group with group operation given by the gyrosemidirect product (1.86). The following theorem shows that this is indeed the case.

Theorem 1.19 *Let* $(G, \oplus)$ *be a gyrogroup, and let* $\mathrm{Aut}_0(G, \oplus)$ *be a gyroautomorphism group of* G. *Then the gyrosemidirect product* $G \times \mathrm{Aut}_0(G)$ *is a group, with group operation given by the gyrosemidirect product* (1.86).

Proof We will show that the set $G \times \mathrm{Aut}_0(G)$ with its binary operation (1.86) satisfies the group axioms.

(i) Existence of a left identity: A left identity element of $G \times \mathrm{Aut}_0(G)$ is the pair $(0, I)$, where $0 \in G$ is the identity element of G, and $I \in \mathrm{Aut}_0(G)$ is the identity automorphism of G. Indeed,

$$(0, I)(a, A) = \big(0 \oplus Ia, \mathrm{gyr}[0, Ia]IA\big) = (a, A) \tag{1.87}$$

noting that the gyration in (1.87) is trivial by Theorem 1.8(2).

(ii) Existence of a left inverse: Let $A^{-1} \in \mathrm{Aut}_0(G)$ be the inverse automorphism of $A \in \mathrm{Aut}_0(G)$. Then, by the gyrosemidirect product (1.86) we have

$$\big(\ominus A^{-1}a, A^{-1}\big)(a, A) = \big(\ominus A^{-1}a \oplus A^{-1}a, \mathrm{gyr}\big[\ominus A^{-1}a, A^{-1}a\big]A^{-1}A\big) = (0, I) \tag{1.88}$$

Hence, a left inverse of $(a, A) \in G \times \mathrm{Aut}_0(G)$ is the pair $(\ominus A^{-1}a, A^{-1})$,

$$(a, A)^{-1} = \big(\ominus A^{-1}a, A^{-1}\big) \tag{1.89}$$

(iii) Validity of the associative law: We have to show that the successive products in (1.90) and in (1.91) below are equal.

On the one hand, we have

$$
\begin{aligned}
&(a_1, A_1)\big((a_2, A_2)(a_3, A_3)\big) \\
&\quad = (a_1, A_1)\big(a_2 \oplus A_2 a_3, \operatorname{gyr}[a_2, A_2 a_3] A_2 A_3\big) \\
&\quad = \big(a_1 \oplus A_1(a_2 \oplus A_2 a_3), \operatorname{gyr}\big[a_1, A_1(a_2 \oplus A_2 a_3)\big] A_1 \operatorname{gyr}[a_2, A_2 a_3] A_2 A_3\big) \\
&\quad = \big(a_1 \oplus (A_1 a_2 \oplus A_1 A_2 a_3), \\
&\qquad \operatorname{gyr}[a_1, A_1 a_2 \oplus A_1 A_2 a_3] \operatorname{gyr}[A_1 a_2, A_1 A_2 a_3] A_1 A_2 A_3\big)
\end{aligned}
\tag{1.90}
$$

where we employ the gyrosemidirect product (1.86) and the commuting law (1.81).

On the other hand, we have

$$
\begin{aligned}
&\big((a_1, A_1)(a_2, A_2)\big)(a_3, A_3) \\
&\quad = \big(a_1 \oplus A_1 a_2, \operatorname{gyr}[a_1, A_1 a_2] A_1 A_2\big)(a_3, A_3) \\
&\quad = \big((a_1 \oplus A_1 a_2) \oplus \operatorname{gyr}[a_1, A_1 a_2] A_1 A_2 a_3, \\
&\qquad \operatorname{gyr}\big[a_1 \oplus A_1 a_2, \operatorname{gyr}[a_1, A_1 a_2] A_1 A_2 a_3\big] \operatorname{gyr}[a_1, A_1 a_2] A_1 A_2 A_3\big)
\end{aligned}
\tag{1.91}
$$

where we employ (1.86).

In order to show that the gyrosemidirect products in (1.90) and (1.91) are equal, using the notation

$$
\begin{aligned}
a_1 &= a \\
A_1 a_2 &= b \\
A_1 A_2 A_3 &= c
\end{aligned}
\tag{1.92}
$$

we have to establish the identity

$$
\begin{aligned}
&\big(a \oplus (b \oplus c), \operatorname{gyr}[a, b \oplus c] \operatorname{gyr}[b, c] A_1 A_2 A_3\big) \\
&\quad = \big((a \oplus b) \oplus \operatorname{gyr}[a, b]c, \operatorname{gyr}\big[a \oplus b, \operatorname{gyr}[a, b]c\big] \operatorname{gyr}[a, b] A_1 A_2 A_3\big)
\end{aligned}
\tag{1.93}
$$

This identity between two pairs is equivalent to the two identities between their corresponding entries,

$$
\begin{aligned}
a \oplus (b \oplus c) &= (a \oplus b) \oplus \operatorname{gyr}[a, b]c \\
\operatorname{gyr}[a, b \oplus c] \operatorname{gyr}[b, c] &= \operatorname{gyr}\big[a \oplus b, \operatorname{gyr}[a, b]c\big] \operatorname{gyr}[a, b]
\end{aligned}
\tag{1.94}
$$

The first identity is valid, being the left gyroassociative law, and the second identity is valid by (1.45), p. 15. □

Instructively, a second proof of Theorem 1.19 is given below.

Proof A one-to-one map of a set Q_1 onto a set Q_2 is said to be bijective and, accordingly, the map is called a bijection. The set of all bijections of a set Q onto itself forms a group under bijection composition. Let S be the group of all bijections of the set G onto itself under bijection composition. Let each element

$$(a, A) \in S_0 := G \times \mathrm{Aut}_0(G) \tag{1.95}$$

act bijectively on the gyrogroup $(G, \oplus)$ according to the equation

$$(a, A)g = a \oplus Ag \tag{1.96}$$

the unique inverse of (a, A) in $S_0 = G \times \mathrm{Aut}_0(G)$ being, by (1.89),

$$(a, A)^{-1} = \left(\ominus A^{-1}a, A^{-1}\right) \tag{1.97}$$

Being a set of special bijections of G onto itself, given by (1.96), S_0 is a subset of the group S, $S_0 \subset S$. Employing the subgroup criterion in Theorem 1.4, p. 11, we will show that, under bijection composition, S_0 is a subgroup of the group S.

Two successive bijections $(a, A), (b, B) \in S_0$ of G are equivalent to a single bijection $(c, C) \in S_0$ according to the following chain of equations. Employing successively the bijection (1.96) along with the left gyroassociative law we have

$$\begin{aligned}
(a, A)(b, B)g &= (a, A)(b \oplus Bg) \\
&= a \oplus A(b \oplus Bg) \\
&= a \oplus (Ab \oplus ABg) \\
&= (a \oplus Ab) \oplus \mathrm{gyr}[a, Ab]ABg \\
&= \bigl(a \oplus Ab, \mathrm{gyr}[a, Ab]AB\bigr)g \\
&=: (c, C)g
\end{aligned} \tag{1.98}$$

for all $g \in G$, $(a, A), (b, B) \in S_0$.

It follows from (1.98) that bijection composition in S_0 is given by the gyrosemidirect product, (1.86),

$$(a, A)(b, B) = \bigl(a \oplus Ab, \mathrm{gyr}[a, Ab]AB\bigr) \tag{1.99}$$

Finally, for any $(a, A), (b, B) \in S_0$ we have by (1.97) and (1.99),

$$\begin{aligned}
(a, A)(b, B)^{-1} &= (a, A)\left(\ominus B^{-1}b, B^{-1}\right) \\
&= \left(a \ominus AB^{-1}b, \mathrm{gyr}\left[a, \ominus AB^{-1}b\right]AB^{-1}\right) \\
&\in S_0
\end{aligned} \tag{1.100}$$

Hence, by the subgroup criterion in Theorem 1.4, p. 11, the subset S_0 of the group S of all bijections of G onto itself is a subgroup under bijection composition. But, bijection composition in S_0 is given by the gyrosemidirect product (1.99). Hence, as desired, the set $S_0 = G \times \mathrm{Aut}_0(G)$ with composition given by the gyrosemidirect product (1.99) forms a group. □

The gyrosemidirect product group enables problems in gyrogroups to be converted to the group setting thus gaining access to the powerful group theoretic techniques. Illustrative examples for the use of gyrosemidirect product groups are provided by the proof of the following two Theorems 1.20 and 1.21.

Theorem 1.20 *Let* $(G, \oplus)$ *be a gyrogroup, let* $a, b \in G$ *be any two elements of* G*, and let* $Y \in \mathrm{Aut}(G)$ *be any automorphism of* $(G, \oplus)$*. Then, the unique solution of the automorphism equation*

$$Y = \ominus \mathrm{gyr}[b, Xa]X \tag{1.101}$$

for the unknown automorphism $X \in \mathrm{Aut}(G)$ *is*

$$X = \ominus \mathrm{gyr}[b, Ya]Y \tag{1.102}$$

Proof Let X be a solution of (1.101), and let $x \in G$ be given by the equation

$$x = b \boxminus Xa \tag{1.103}$$

so that, by a right cancellation, (1.78), $b = x \oplus Xa$.

Then we have the following gyrosemidirect product

$$\begin{aligned}(x, X)(a, I) &= \big(x \oplus Xa, \mathrm{gyr}[x, Xa]X\big) \\ &= \big(x \oplus Xa, \mathrm{gyr}[x \oplus Xa, Xa]X\big) \\ &= \big(b, \mathrm{gyr}[b, Xa]X\big) \\ &= (b, \ominus Y)\end{aligned} \tag{1.104}$$

so that

$$\begin{aligned}(x, X) &= (b, \ominus Y)(a, I)^{-1} \\ &= (b, \ominus Y)(\ominus a, I) \\ &= \big(b \oplus Ya, \ominus \mathrm{gyr}[b, Ya]Y\big)\end{aligned} \tag{1.105}$$

Comparing the second entries of the extreme sides of (1.105) we have

$$X = \ominus \mathrm{gyr}[b, Ya]Y \tag{1.106}$$

Hence, if a solution X of (1.101) exists, then it must be given by (1.102).

Conversely, the automorphism X in (1.106) is, indeed, a solution of (1.101) as we see by substituting X from (1.106) into the right-hand side of (1.101), and employing the nested gyration identity (1.47), p. 15. Indeed,

$$\ominus \operatorname{gyr}[b, Xa]X = \operatorname{gyr}\big[b, \ominus \operatorname{gyr}[b, Ya]Ya\big] \operatorname{gyr}[b, Ya]Y = Y \tag{1.107}$$

as desired. □

1.12 Basic Gyration Properties

The most important basic gyration properties that we establish in this section are the *gyration even property*

$$\operatorname{gyr}[\ominus a, \ominus b] = \operatorname{gyr}[a, b] \tag{1.108}$$

and the *gyroautomorphism inversion law*

$$\operatorname{gyr}^{-1}[a, b] = \operatorname{gyr}[b, a] \tag{1.109}$$

for any two elements a and b of a gyrogroup $(G, \oplus)$, where $\operatorname{gyr}^{-1}[a, b] = (\operatorname{gyr}[a, b])^{-1}$ is the inverse of the gyration $\operatorname{gyr}[a, b]$.

Theorem 1.21 (Gyrosum Inversion, Gyroautomorphism Inversion) *For any two elements a, b of a gyrogroup $(G, \oplus)$ we have the gyrosum inversion law*

$$\ominus(a \oplus b) = \operatorname{gyr}[a, b](\ominus b \ominus a) \tag{1.110}$$

and the gyroautomorphism inversion law

$$\operatorname{gyr}^{-1}[a, b] = \operatorname{gyr}[\ominus b, \ominus a] \tag{1.111}$$

Proof Let $\operatorname{Aut}_0(G)$ be any gyroautomorphism group of $(G, \oplus)$, and let $G \times \operatorname{Aut}_0(G)$ be the gyrosemidirect product of the gyrogroup G and the group $\operatorname{Aut}_0(G)$ according to Definition 1.18. Being a group, the product of two elements of the gyrosemidirect product group $G \times \operatorname{Aut}_0(G)$ has a unique inverse. This inverse can be calculated in two different ways.

On the one hand, the inverse of the left-hand side of the gyrosemidirect product

$$(a, I)(b, I) = \big(a \oplus b, \operatorname{gyr}[a, b]\big) \tag{1.112}$$

in $G \times \operatorname{Aut}_0(G)$ is

$$\begin{aligned}(b, I)^{-1}(a, I)^{-1} &= (\ominus b, I)(\ominus a, I) \\ &= \big(\ominus b \ominus a, \operatorname{gyr}[\ominus b, \ominus a]\big)\end{aligned} \tag{1.113}$$

On the other hand, the inverse of the right-hand side of the product (1.112) is, by (1.97),

$$\big(\ominus \mathrm{gyr}^{-1}[a,b](a\oplus b), \mathrm{gyr}^{-1}[a,b]\big) \tag{1.114}$$

for all $a, b \in G$. Comparing corresponding entries in (1.113) and (1.114), we have

$$\ominus b\ominus a = \ominus \mathrm{gyr}^{-1}[a,b](a\oplus b) \tag{1.115}$$

and

$$\mathrm{gyr}[\ominus b, \ominus a] = \mathrm{gyr}^{-1}[a,b] \tag{1.116}$$

Eliminating $\mathrm{gyr}^{-1}[a,b]$ between (1.115) and (1.116), we have

$$\ominus b\ominus a = \ominus \mathrm{gyr}[\ominus b, \ominus a](a\oplus b) \tag{1.117}$$

Replacing (a, b) by $(\ominus b, \ominus a)$, (1.117) becomes

$$a\oplus b = \ominus \mathrm{gyr}[a,b](\ominus b\ominus a) \tag{1.118}$$

Identities (1.118) and (1.116) complete the proof. □

Instructively, the gyrosum inversion law (1.110) is verified here as a by-product along with the gyroautomorphism inversion law (1.111) in Theorem 1.21 in terms of the gyrosemidirect product group. A direct proof of (1.110) is, however, simpler as we see in Theorem 1.9, p. 15.

Theorem 1.22 *Let* $(G, \oplus)$ *be a gyrogroup. Then for all* $a, b \in G$

$$\mathrm{gyr}^{-1}[a,b] = \mathrm{gyr}\big[a, \ominus \mathrm{gyr}[a,b]b\big] \tag{1.119}$$

$$\mathrm{gyr}^{-1}[a,b] = \mathrm{gyr}[\ominus a, a\oplus b] \tag{1.120}$$

$$\mathrm{gyr}^{-1}[a,b] = \mathrm{gyr}[b, a\oplus b] \tag{1.121}$$

$$\mathrm{gyr}[a,b] = \mathrm{gyr}[b, \ominus b\ominus a] \tag{1.122}$$

$$\mathrm{gyr}[a,b] = \mathrm{gyr}[\ominus a, \ominus b\ominus a] \tag{1.123}$$

$$\mathrm{gyr}[a,b] = \mathrm{gyr}\big[\ominus(a\oplus b), a\big] \tag{1.124}$$

Proof Identity (1.119) follows from (1.47).

Identity (1.120) follows from (1.48).

Identity (1.121) results from an application to (1.120) of the left loop property followed by a left cancellation.

Identity (1.122) follows from the gyroautomorphism inversion law (1.111) and from (1.120),

$$\mathrm{gyr}[a,b] = \mathrm{gyr}^{-1}[\ominus b, \ominus a] = \mathrm{gyr}[b, \ominus b\ominus a] \tag{1.125}$$

Identity (1.123) follows from an application, to the right-hand side of (1.122), of the left loop property followed by a left cancellation.

Identity (1.124) follows by inverting (1.120) by means of the gyroautomorphism inversion law (1.111). □

Theorem 1.23 (The Gyration Inversion Law; The Gyration Even Property) *The gyroautomorphisms of any gyrogroup $(G,\oplus)$ obey the gyration inversion law*

$$\mathrm{gyr}^{-1}[a,b]=\mathrm{gyr}[b,a] \tag{1.126}$$

and possess the gyration even property

$$\mathrm{gyr}[\ominus a,\ominus b]=\mathrm{gyr}[a,b] \tag{1.127}$$

satisfying the four mutually equivalent nested gyroautomorphism identities

$$\begin{aligned}
\mathrm{gyr}\big[b,\ominus\mathrm{gyr}[b,a]a\big]&=\mathrm{gyr}[a,b]\\
\mathrm{gyr}\big[b,\mathrm{gyr}[b,\ominus a]a\big]&=\mathrm{gyr}[a,\ominus b]\\
\mathrm{gyr}\big[\ominus\mathrm{gyr}[a,b]b,a\big]&=\mathrm{gyr}[a,b]\\
\mathrm{gyr}\big[\mathrm{gyr}[a,\ominus b]b,a\big]&=\mathrm{gyr}[a,\ominus b]
\end{aligned} \tag{1.128}$$

for all $a,b\in G$.

Proof By the left loop property and (1.121), we have

$$\begin{aligned}
\mathrm{gyr}^{-1}[a\oplus b,b]&=\mathrm{gyr}^{-1}[a,b]\\
&=\mathrm{gyr}[b,a\oplus b]
\end{aligned} \tag{1.129}$$

for all $a,b\in G$. Let us substitute $a=c\boxminus b$ into (1.129), so that by a right cancellation $a\oplus b=c$, obtaining the identity

$$\mathrm{gyr}^{-1}[c,b]=\mathrm{gyr}[b,c] \tag{1.130}$$

for all $c,b\in G$. Renaming c in (1.130) as a, we obtain (1.126), as desired.

Identity (1.127) results from (1.111) and (1.126),

$$\begin{aligned}
\mathrm{gyr}[\ominus a,\ominus b]&=\mathrm{gyr}^{-1}[b,a]\\
&=\mathrm{gyr}[a,b]
\end{aligned} \tag{1.131}$$

Finally, the first identity in (1.128) follows from (1.119) and (1.126).

By means of the gyroautomorphism inversion law (1.126), the third identity in (1.128) is equivalent to the first one.

The second (fourth) identity in (1.128) follows from the first (third) by replacing a by $\ominus a$ (or, alternatively, by replacing b by $\ominus b$), noting that gyrations are even by (1.127). □

The left gyroassociative law and the left loop property of gyrogroups admit right counterparts, as we see from the following theorem.

Theorem 1.24 *For any three elements* a, b, *and* c *of a gyrogroup* $(G, \oplus)$ *we have*

(i) $(a \oplus b) \oplus c = a \oplus (b \oplus \mathrm{gyr}[b, a]c)$ (*Right Gyroassociative Law*).
(ii) $\mathrm{gyr}[a, b] = \mathrm{gyr}[a, b \oplus a]$ (*Right Loop Property*).

Proof The right gyroassociative law follows from the left gyroassociative law and the gyration inversion law (1.126) of gyroautomorphisms,

$$\begin{aligned} a \oplus \bigl(b \oplus \mathrm{gyr}[b, a]c\bigr) &= (a \oplus b) \oplus \mathrm{gyr}[a, b]\,\mathrm{gyr}[b, a]c \\ &= (a \oplus b) \oplus c \end{aligned} \tag{1.132}$$

The right loop property results from (1.121) and the gyration inversion law (1.126),

$$\begin{aligned} \mathrm{gyr}[b, a \oplus b] &= \mathrm{gyr}^{-1}[a, b] \\ &= \mathrm{gyr}[b, a] \end{aligned} \tag{1.133}$$

□

The right cancellation law allows the loop property to be dualized in the following theorem.

Theorem 1.25 (The Coloop Property—Left and Right) *Let* $(G, \oplus)$ *be a gyrogroup. Then*

$$\begin{aligned} \mathrm{gyr}[a, b] &= \mathrm{gyr}[a \boxminus b, b] \quad (\textit{Left Coloop Property}) \\ \mathrm{gyr}[a, b] &= \mathrm{gyr}[a, b \boxminus a] \quad (\textit{Right Coloop Property}) \end{aligned}$$

for all $a, b \in G$.

Proof The proof follows from an application of the left and the right loop property followed by a right cancellation,

$$\begin{aligned} \mathrm{gyr}[a \boxminus b, b] &= \mathrm{gyr}\bigl[(a \boxminus b) \oplus b, b\bigr] = \mathrm{gyr}[a, b] \\ \mathrm{gyr}[a, b \boxminus a] &= \mathrm{gyr}\bigl[a, (b \boxminus a) \oplus a\bigr] = \mathrm{gyr}[a, b] \end{aligned} \tag{1.134}$$

□

A right and a left loop give rise to the identities in the following theorem.

Theorem 1.26 *Let* $(G, \oplus)$ *be a gyrogroup. Then*

$$\begin{aligned} \mathrm{gyr}[a \oplus b, \ominus a] &= \mathrm{gyr}[a, b] \\ \mathrm{gyr}[\ominus a, a \oplus b] &= \mathrm{gyr}[b, a] \end{aligned} \tag{1.135}$$

for all $a, b \in G$.

Proof By a right loop, a left cancellation and a left loop we have

$$\begin{aligned}\mathrm{gyr}[a\oplus b,\ominus a] &= \mathrm{gyr}\big[a\oplus b,\ominus a\oplus(a\oplus b)\big] \\ &= \mathrm{gyr}[a\oplus b,b] \\ &= \mathrm{gyr}[a,b]\end{aligned} \tag{1.136}$$

thus verifying the first identity in (1.135). The second identity in (1.135) follows from the first one by gyroautomorphism inversion, (1.126). □

In general, $\ominus(a\oplus b)\neq\ominus a\ominus b$ in a gyrogroup $(G,\oplus)$. In fact, we have $\ominus(a\oplus b)=\ominus a\ominus b$ for all $a,b\in G$ if and only if the gyrogroup $(G,\oplus)$ is gyrocommutative, as we see from Theorem 1.32, p. 34, of Chap. 2 on gyrocommutative gyrogroups. In this sense, the gyrogroup cooperation $\boxplus$ conducts itself more properly than its associated gyrogroup operation, as we see from the following theorem.

Theorem 1.27 (The Cogyroautomorphic Inverse Property) *Any gyrogroup $(G,\oplus)$ possesses the cogyroautomorphic inverse property,*

$$\ominus(a\boxplus b)=(\ominus b)\boxplus(\ominus a) \tag{1.137}$$

for any $a,b\in G$.

Proof We verify (1.137) in the following chain of equations, which are numbered for subsequent derivation:

$$\begin{aligned}a\boxplus b &\overset{(1)}{=\!=\!=} a\oplus\mathrm{gyr}[a,\ominus b]b \\ &\overset{(2)}{=\!=\!=} \ominus\mathrm{gyr}\big[a,\mathrm{gyr}[a,\ominus b]b\big]\big\{\ominus\mathrm{gyr}[a,\ominus b]b\ominus a\big\} \\ &\overset{(3)}{=\!=\!=} \mathrm{gyr}\big[a,\mathrm{gyr}[a,\ominus b]b\big]\big\{\ominus\big(\ominus\mathrm{gyr}[a,\ominus b]b\ominus a\big)\big\} \\ &\overset{(4)}{=\!=\!=} \mathrm{gyr}\big[a,\ominus\mathrm{gyr}[a,\ominus b](\ominus b)\big]\big\{\ominus\big(\ominus\mathrm{gyr}[a,\ominus b]b\ominus a\big)\big\} \\ &\overset{(5)}{=\!=\!=} \mathrm{gyr}^{-1}[a,\ominus b]\big\{\ominus\big(\ominus\mathrm{gyr}[a,\ominus b]b\ominus a\big)\big\} \\ &\overset{(6)}{=\!=\!=} \ominus\big(\ominus b\ominus\mathrm{gyr}^{-1}[a,\ominus b]a\big) \\ &\overset{(7)}{=\!=\!=} \ominus\big\{\ominus b\ominus\mathrm{gyr}[b,\ominus a]a\big\} \\ &\overset{(8)}{=\!=\!=} \ominus\big\{(\ominus b)\boxplus(\ominus a)\big\}\end{aligned} \tag{1.138}$$

Inverting both extreme sides of (1.138) we obtain the desired identity (1.137).

Derivation of the numbered equalities in (1.138) follows:

1. Follows from Definition 1.7, p. 13, of the gyrogroup cooperation $\boxplus$.
2. Follows from Item 1 by the gyrosum inversion, (1.110).
3. Follows from Item 2 by Theorem 1.8(12) applied to the term $\{\dots\}$ in (2).
4. Follows from Item 3 by Theorem 1.8(12) applied to b, that is, $\mathrm{gyr}[a,\ominus b]b = \ominus\,\mathrm{gyr}[a,\ominus b](\ominus b)$.
5. Follows from Item 4 by Identity (1.119) of Theorem 1.22.
6. Follows from Item 5 by distributing the gyroautomorphism $\mathrm{gyr}^{-1}[a,\ominus b]$ over each of the two terms in $\{\dots\}$.
7. Follows from Item 6 by the gyroautomorphism inversion law (1.111).
8. Follows from Item 7 by Definition 1.7, p. 13, of the gyrogroup cooperation $\boxplus$.

□

Theorem 1.28 *Let* $(G,\oplus)$ *be a gyrogroup. Then*

$$a\oplus\big\{(\ominus a\oplus b)\oplus a\big\} = b\boxplus a \tag{1.139}$$

for all $a,b\in G$.

Proof The proof rests on the following chain of equations, which are numbered for subsequent explanation:

$$\begin{aligned} a\oplus\big\{(\ominus a\oplus b)\oplus a\big\} &\overset{(1)}{=\!=\!=} \big\{a\oplus(\ominus a\oplus b)\big\}\oplus\mathrm{gyr}[a,\ominus a\oplus b]a \\ &\overset{(2)}{=\!=\!=} b\oplus\mathrm{gyr}[b,\ominus a\oplus b]a \\ &\overset{(3)}{=\!=\!=} b\oplus\mathrm{gyr}[b,\ominus a]a \\ &\overset{(4)}{=\!=\!=} b\boxplus a \end{aligned} \tag{1.140}$$

The derivation of the equalities in (1.140) follows.

1. Follows from the left gyroassociative law.
2. Follows from Item 1 by a left cancellation, and by a left loop followed by a left cancellation.
3. Follows from Item 2 by a right loop, that is, an application of the right loop property to (3) gives (2).
4. Follows from Item 3 by Definition 1.7, p. 13, of the gyrogroup cooperation $\boxplus$.

□

1.13 An Advanced Gyrogroup Equation

As an example, we present in Theorem 1.29 below an advanced gyrogroup equation and its unique solution. The equation is advanced in the sense that its unknown appears in the equation both directly, and indirectly in the argument of a gyration.

Theorem 1.29 *Let*

$$c = \mathrm{gyr}[b, \ominus x]x \tag{1.141}$$

be an equation for the unknown x in a gyrogroup $(G, \oplus)$. The unique solution of (1.141) *is*

$$x = \ominus\big(\ominus b \ominus (c \boxminus b)\big) \tag{1.142}$$

Proof If a solution x to the gyrogroup equation (1.141) exists then, by (1.141) and by the second identity in (1.44), p. 15,

$$c = \mathrm{gyr}[b, \ominus x]x = \ominus(b \ominus x) \oplus b \tag{1.143}$$

Applying a right cancellation to (1.143), we obtain

$$\ominus(b \ominus x) = c \boxminus b \tag{1.144}$$

or, equivalently, by gyro-sign inversion,

$$b \ominus x = \ominus(c \boxminus b) \tag{1.145}$$

so that, by a left cancellation

$$\ominus x = \ominus b \ominus (c \boxminus b) \tag{1.146}$$

implying, by gyro-sign inversion,

$$x = \ominus\big(\ominus b \ominus (c \boxminus b)\big) \tag{1.147}$$

Hence, if a solution x to (1.141) exists, it must be given by (1.147).

Conversely, x given by (1.147) is a solution. Indeed, the substitution of x of (1.147) into the right-hand side of (1.143) results in c, as we see in the following chain of equations, which are numbered for subsequent derivation:

$$\begin{aligned}
\mathrm{gyr}[b, \ominus x]x &\overbrace{=\!=\!=}^{(1)} \ominus \mathrm{gyr}\big[b, \ominus b \ominus (c \boxminus b)\big]\big\{\ominus b \ominus (c \boxminus b)\big\} \\
&\overbrace{=\!=\!=}^{(2)} \ominus\big\{b \oplus \big(\ominus b \ominus (c \boxminus b)\big)\big\} \oplus b \\
&\overbrace{=\!=\!=}^{(3)} \ominus \mathrm{gyr}\big[\ominus\big\{b \oplus \big(\ominus b \ominus (c \boxminus b)\big)\big\}, b\big]\big(\ominus b \oplus \big\{b \oplus \big(\ominus b \ominus (c \boxminus b)\big)\big\}\big)
\end{aligned}$$

$$\overset{(4)}{=\!=} \ominus \operatorname{gyr}\big[b\oplus\big(\ominus b\ominus(c \boxminus b)\big), \ominus b\big]\big\{\ominus b\ominus(c \boxminus b)\big\}$$

$$\overset{(5)}{=\!=} \ominus \operatorname{gyr}\big[b, \ominus b\ominus(c \boxminus b)\big]\big\{\ominus b\ominus(c \boxminus b)\big\}$$

$$\overset{(6)}{=\!=} \ominus \operatorname{gyr}\big[\ominus(c \boxminus b), \ominus b\big]\big\{\ominus b\ominus(c \boxminus b)\big\}$$

$$\overset{(7)}{=\!=} \ominus \operatorname{gyr}[c \boxminus b, b]\big\{\ominus b\ominus(c \boxminus b)\big\}$$

$$\overset{(8)}{=\!=} (c \boxminus b)\oplus b$$

$$\overset{(9)}{=\!=} c \tag{1.148}$$

Derivation of the numbered equalities in (1.148) follows:

1. Follows from the substitution of x from (1.147), and from Theorem 1.8(12).
2. Follows from Item 1 by the first identity in (1.44), p. 15.
3. Follows from Item 2 by the gyrosum inversion law (1.42), p. 15.
4. Follows from Item 3 by the gyration even property and by a left cancellation.
5. Follows from Item 4 by the first identity in (1.135).
6. Follows from Item 5 by the second identity in (1.135).
7. Follows from Item 6 by the gyration even property.
8. Follows from Item 7 by the gyrosum inversion law (1.42), p. 15.
9. Follows from Item 8 by a right cancellation. □

Corollary 1.30 *Let* $(G, \oplus)$ *be a gyrogroup. The map*

$$a \mapsto c = \operatorname{gyr}[b, \ominus a]a \tag{1.149}$$

of G *into itself is bijective so that when* a *runs over all the elements of* G *its image,* c*, runs over all the elements of* G *as well, for any given element* $b \in G$.

Proof It follows immediately from Theorem 1.29 that, for any given $b \in G$, the map (1.149) is bijective and, hence, the result of the corollary. □

1.14 Gyrocommutative Gyrogroups

Definition 1.31 (Gyroautomorphic Inverse Property) A gyrogroup $(G, \oplus)$ possesses the *gyroautomorphic inverse property* if for all $a, b \in G$,

$$\ominus(a\oplus b) = \ominus a\ominus b \tag{1.150}$$

Theorem 1.32 (The Gyroautomorphic Inverse Property) *A gyrogroup is gyrocommutative if and only if it possesses the gyroautomorphic inverse property.*

Proof Let $(G, \oplus)$ be a gyrogroup possessing the gyroautomorphic inverse property. Then the gyrosum inversion law (1.42), p. 15, specializes, by means of Theorem 1.8(12), p. 13, to the gyrocommutative law $(G6)$ in Definition 1.6, p. 12,

$$\begin{aligned} a \oplus b &= \ominus \operatorname{gyr}[a, b](\ominus b \ominus a) \\ &= \operatorname{gyr}[a, b]\{\ominus(\ominus b \ominus a)\} \\ &= \operatorname{gyr}[a, b](b \oplus a) \end{aligned} \tag{1.151}$$

for all $a, b \in G$.

Conversely, if the gyrocommutative law is valid then by Theorem 1.8(12) and the gyrosum inversion law, (1.42), p. 15, we have

$$\operatorname{gyr}[a, b]\{\ominus(\ominus b \ominus a)\} = \ominus \operatorname{gyr}[a, b](\ominus b \ominus a) = a \oplus b = \operatorname{gyr}[a, b](b \oplus a) \tag{1.152}$$

so that, by eliminating the gyroautomorphism $\operatorname{gyr}[a, b]$ on both extreme sides of (1.152) and inverting the gyro-sign, we recover the gyroautomorphic inverse property,

$$\ominus(b \oplus a) = \ominus b \ominus a \tag{1.153}$$

for all $a, b \in G$. □

Theorem 1.33 *The gyrogroup cooperation* $\boxplus$ *of a gyrogroup* $(G, \oplus)$ *is commutative if and only if the gyrogroup* $(G, \oplus)$ *is gyrocommutative.*

Proof For any $a, b \in G$ we have, by Equality (7) of the chain of equations (1.138) in the proof of Theorem 1.27, p. 31,

$$a \boxplus b = \ominus\big(\ominus b \ominus \operatorname{gyr}[b, \ominus a]a\big) \tag{1.154}$$

But by definition,

$$b \boxplus a = b \oplus \operatorname{gyr}[b, \ominus a]a \tag{1.155}$$

Hence

$$a \boxplus b = b \boxplus a \tag{1.156}$$

for all $a, b \in G$ if and only if

$$\ominus(\ominus b \ominus c) = b \oplus c \tag{1.157}$$

for all $a, b \in G$, where

$$c = \operatorname{gyr}[b, \ominus a]a \tag{1.158}$$

as we see from (1.154) and (1.155). But the self-map of G that takes a to c in (1.158),

$$a \mapsto \operatorname{gyr}[b, \ominus a]a = c \tag{1.159}$$

for any given $b \in G$ is bijective, by Corollary 1.30, p. 34. Hence, the commutative relation (1.156) for $\boxplus$ holds for all $a, b \in G$ if and only if (1.157) holds for all $b, c \in G$. The latter, in turn, is the gyroautomorphic inverse property that, by Theorem 1.32, is equivalent to the gyrocommutativity of the gyrogroup $(G, \oplus)$. Hence, (1.156) holds for all $a, b \in G$ if and only if the gyrogroup $(G, \oplus)$ is gyrocommutative. □

Theorem 1.34 *Let $(G, \oplus)$ be a gyrocommutative gyrogroup. Then*

$$\mathrm{gyr}[a, b]\,\mathrm{gyr}[b\oplus a, c] = \mathrm{gyr}[a, b\oplus c]\,\mathrm{gyr}[b, c] \tag{1.160}$$

for all $a, b, c \in G$.

Proof Using the notation $g_{a,b} = \mathrm{gyr}[a, b]$ whenever convenient, we have by Theorem 1.16, p. 22, by the gyrocommutative law, and by Identity (1.45), p. 15,

$$\begin{aligned}
\mathrm{gyr}[a, b]\,\mathrm{gyr}[b\oplus a, c] &= \mathrm{gyr}\big[g_{a,b}(b\oplus a), g_{a,b}c\big]\,\mathrm{gyr}[a, b] \\
&= \mathrm{gyr}\big[a\oplus b, \mathrm{gyr}[a, b]c\big]\,\mathrm{gyr}[a, b] \\
&= \mathrm{gyr}[a, b\oplus c]\,\mathrm{gyr}[b, c]
\end{aligned} \tag{1.161}$$

□

Theorem 1.35 *Let $a, b, c \in G$ be any three elements of a gyrocommutative gyrogroup $(G, \oplus)$, and let $d \in G$ be determined by the "gyroparallelogram condition"*

$$d = (b \boxplus c)\ominus a \tag{1.162}$$

Then, the elements a, b, c and d satisfy the telescopic gyration identity

$$\mathrm{gyr}[a, \ominus b]\,\mathrm{gyr}[b, \ominus c]\,\mathrm{gyr}[c, \ominus d] = \mathrm{gyr}[a, \ominus d] \tag{1.163}$$

for all $a, b, c \in G$.

Proof By Identity (1.160), along with an application of the right and the left loop property, we have

$$\mathrm{gyr}\big[a', b'\oplus a'\big]\,\mathrm{gyr}\big[b'\oplus a', c'\big] = \mathrm{gyr}\big[a', b'\oplus c'\big]\,\mathrm{gyr}\big[b'\oplus c', c'\big] \tag{1.164}$$

Let

$$\begin{aligned}
a &= \ominus c' \\
c &= \ominus a' \\
b &= b'\oplus a'
\end{aligned} \tag{1.165}$$

so that, by the third equation in (1.165) and by (1.162), we have

$$\begin{aligned} b'\oplus c' &= \bigl(b \boxminus a'\bigr)\oplus c' \\ &= (b \boxplus c)\ominus a \\ &= d \end{aligned} \tag{1.166}$$

Then (1.164), expressed in terms of a, b, c, d in (1.165)–(1.166), takes the form

$$\mathrm{gyr}[\ominus c, b]\,\mathrm{gyr}[b, \ominus a] = \mathrm{gyr}[\ominus c, d]\,\mathrm{gyr}[d, \ominus a] \tag{1.167}$$

Inverting both sides of (1.167) by means of the gyration inversion law (1.126), p. 29, and the gyration even property (1.127), p. 29, we obtain the identity

$$\mathrm{gyr}[a, \ominus b]\,\mathrm{gyr}[b, \ominus c] = \mathrm{gyr}[a, \ominus d]\,\mathrm{gyr}[d, \ominus c] \tag{1.168}$$

from which the telescopic gyration identity (1.163) follows immediately, by a gyration inversion and the gyration even property. □

Theorem 1.36 *The gyroparallelogram condition* (1.162),

$$d = (b \boxplus c)\ominus a \tag{1.169}$$

in a gyrocommutative gyrogroup $(G, \oplus)$ *is equivalent to the identity*

$$\ominus c\oplus d = \mathrm{gyr}[c, \ominus b](b\ominus a) \tag{1.170}$$

Proof In a gyrocommutative gyrogroup $(G, \oplus)$, the gyroparallelogram condition (1.162) implies the following chain of equations, which are numbered for subsequent derivation.

$$\begin{aligned} d &\overset{(1)}{=\!=\!=} (b \boxplus c)\ominus a \\ &\overset{(2)}{=\!=\!=} (c \boxplus b)\ominus \mathrm{gyr}[b, \ominus c]\,\mathrm{gyr}[c, \ominus b]a \\ &\overset{(3)}{=\!=\!=} (c \boxplus b)\ominus \mathrm{gyr}\bigl[c, \mathrm{gyr}[c, \ominus b]b\bigr]\,\mathrm{gyr}[c, \ominus b]a \\ &\overset{(4)}{=\!=\!=} \bigl(c\oplus \mathrm{gyr}[c, \ominus b]b\bigr)\ominus \mathrm{gyr}\bigl[c, \mathrm{gyr}[c, \ominus b]b\bigr]\,\mathrm{gyr}[c, \ominus b]a \\ &\overset{(5)}{=\!=\!=} c\oplus\bigl(\mathrm{gyr}[c, \ominus b]b\ominus \mathrm{gyr}[c, \ominus b]a\bigr) \\ &\overset{(6)}{=\!=\!=} c\oplus \mathrm{gyr}[c, \ominus b](b\ominus a) \end{aligned} \tag{1.171}$$

for all $a, b, c \in G$. Derivation of the numbered equalities in (1.171) follows:

1. This is the gyroparallelogram condition (1.162).
2. Follows from Item 1 (i) since the gyrogroup cooperation $\boxplus$ is commutative, by Theorem 1.33, p. 35, and (ii) since, by gyration inversion along with the gyration even property in (1.126)–(1.127), p. 29, the gyration product applied to a in (2) is trivial.
3. Follows from Item 2 by the second nested gyration identity in (1.128), p. 29.
4. Follows from Item 3 by Definition 1.7, p. 13, of the gyrogroup cooperation $\boxplus$.
5. Follows from Item 4 by the left gyroassociative law. Indeed, an application of the left gyroassociative law to (5) results in (4).
6. Follows from Item 5 since gyroautomorphisms respect their gyrogroup operation.

Finally, (1.170) follows from (1.171) by a left cancellation, moving c from the extreme right-hand side of (1.171) to its extreme left-hand side. □

Theorem 1.37 *Let* $(G, \oplus)$ *be a gyrocommutative gyrogroup. Then*

$$\mathrm{gyr}[a,b]\{b\oplus(a\oplus c)\} = (a\oplus b)\oplus c \tag{1.172}$$

for all $a, b, c \in G$.

Proof By the left gyroassociative law and by the gyrocommutative law we have the chain of equations

$$\begin{aligned} b\oplus(a\oplus c) &= (b\oplus a)\oplus \mathrm{gyr}[b,a]c \\ &= \mathrm{gyr}[b,a](a\oplus b)\oplus \mathrm{gyr}[b,a]c \\ &= \mathrm{gyr}[b,a]\{(a\oplus b)\oplus c\} \end{aligned} \tag{1.173}$$

from which (1.172) is derived by the gyroautomorphism inversion law (1.111), p. 27. □

The special case of Theorem 1.37 corresponding to $c = \ominus a$ gives rise to a new cancellation law in gyrocommutative gyrogroups, called the *left-right cancellation law*.

Theorem 1.38 (The Left-Right Cancellation Law) *Let* $(G, \oplus)$ *be a gyrocommutative gyrogroup. Then*

$$(a\oplus b)\ominus a = \mathrm{gyr}[a,b]b \tag{1.174}$$

for all $a, b, c \in G$.

Proof Identity (1.174) follows from (1.44), p. 15, and the gyroautomorphic inverse property (1.150), p. 34. Alternatively, Identity (1.174) is equivalent to the special case of (1.173) when $c = \ominus a$. □

The left-right cancellation law (1.174) is not a complete cancellation since the echo of the "canceled" a remains in the argument of the involved gyroautomorphism.

Theorem 1.39 *Let* $(G, \oplus)$ *be a gyrocommutative gyrogroup. Then*

$$a\oplus\{(\ominus a\oplus b)\oplus a\} = a \boxplus b \tag{1.175}$$

for all $a, b \in G$.

Proof By Theorem 1.28, p. 32, and Theorem 1.33,

$$a\oplus\{(\ominus a\oplus b)\oplus a\} = b \boxplus a = a \boxplus b \tag{1.176}$$

□

Theorem 1.40 *Let* $(G, \oplus)$ *be a gyrocommutative gyrogroup. Then*

$$a \boxplus (a\oplus b) = a\oplus(b\oplus a) \tag{1.177}$$

for all $a, b \in G$.

Proof By a left cancellation and Theorem 1.39, we have

$$a\oplus(b\oplus a) = a\oplus(\{\ominus a\oplus(a\oplus b)\}\oplus a) = a \boxplus (a\oplus b) \tag{1.178}$$

□

Theorem 1.41 (The Gyrotranslation Theorem, II) *Let* $(G, \oplus)$ *be a gyrocommutative gyrogroup. For all* $a, b, c \in G$,

$$\begin{aligned} \ominus(a\oplus b)\oplus(a\oplus c) &= \mathrm{gyr}[a, b](\ominus b\oplus c) \\ (a\oplus b)\ominus(a\oplus c) &= \mathrm{gyr}[a, b](b\ominus c) \end{aligned} \tag{1.179}$$

Proof The first identity in (1.179) follows from the Gyrotranslation Theorem 1.14, p. 18, with a replaced by $\ominus a$. Hence, it is valid in nongyrocommutative gyrogroups as well. The second identity in (1.179) follows from the first by the gyroautomorphic inverse property of gyrocommutative gyrogroups, Theorem 1.32, p. 34. Hence, it is valid in gyrocommutative gyrogroups. □

The following theorem gives an elegant gyration identity in which the product of three telescopic gyrations is equivalent to a single gyration.

Theorem 1.42 *Let* $a, b, c \in G$ *be any three elements of a gyrocommutative gyrogroup* $(G, \oplus)$. *Then,*

$$\mathrm{gyr}[\ominus a\oplus b, a\ominus c] = \mathrm{gyr}[a, \ominus b]\,\mathrm{gyr}[b, \ominus c]\,\mathrm{gyr}[c, \ominus a] \tag{1.180}$$

Proof By Theorem 1.16, p. 22, and by the gyrocommutative law, we have

$$\begin{aligned}\mathrm{gyr}[a,b]\,\mathrm{gyr}[b\oplus a,c] &= \mathrm{gyr}\big[\mathrm{gyr}[a,b](b\oplus a),\mathrm{gyr}[a,b]c\big]\,\mathrm{gyr}[a,b]\\ &= \mathrm{gyr}\big[a\oplus b,\mathrm{gyr}[a,b]c\big]\,\mathrm{gyr}[a,b]\end{aligned}\tag{1.181}$$

Hence, Identity (1.45) in Theorem 1.11, p. 15, can be written as

$$\mathrm{gyr}[a,b\oplus c]\,\mathrm{gyr}[b,c] = \mathrm{gyr}[a,b]\,\mathrm{gyr}[b\oplus a,c]\tag{1.182}$$

By gyroautomorphism inversion, the latter can be written as

$$\mathrm{gyr}[a,b\oplus c] = \mathrm{gyr}[a,b]\,\mathrm{gyr}[b\oplus a,c]\,\mathrm{gyr}[c,b]\tag{1.183}$$

Using the notation $b\oplus a = d$, which implies $a = \ominus b\oplus d$, Identity (1.183) becomes, by means of Theorem 1.26, p. 30,

$$\begin{aligned}\mathrm{gyr}[\ominus b\oplus d,b\oplus c] &= \mathrm{gyr}[\ominus b\oplus d,b]\,\mathrm{gyr}[d,c]\,\mathrm{gyr}[c,b]\\ &= \mathrm{gyr}[\ominus b,d]\,\mathrm{gyr}[d,c]\,\mathrm{gyr}[c,b]\end{aligned}\tag{1.184}$$

Renaming the elements $b,c,d\in G$, $(b,c,d)\to(\ominus a,c,\ominus b)$, (1.184) becomes

$$\mathrm{gyr}[a\ominus b,\ominus a\oplus c] = \mathrm{gyr}[a,\ominus b]\,\mathrm{gyr}[\ominus b,c]\,\mathrm{gyr}[c,\ominus a]\tag{1.185}$$

By means of the gyroautomorphic inverse property, Theorem 1.32, p. 34, and the gyration even property (1.127) in Theorem 1.23, p. 29, Identity (1.185) can be written, finally, in the desired form (1.180). □

The special case of Theorem 1.42 when $c = \ominus b$ is interesting, giving rise to the following theorem.

Theorem 1.43 *Let* $(G,\oplus)$ *be a gyrocommutative gyrogroup. Then*

$$\mathrm{gyr}[a,\ominus b] = \mathrm{gyr}[\ominus a\oplus b,a\oplus b]\,\mathrm{gyr}[a,b]\tag{1.186}$$

Proof Owing to the gyration inversion law in Theorem 1.23, p. 29, Identity (1.180) can be written as

$$\mathrm{gyr}[\ominus a\oplus b,a\ominus c]\,\mathrm{gyr}[\ominus a,c] = \mathrm{gyr}[a,\ominus b]\,\mathrm{gyr}[b,\ominus c]\tag{1.187}$$

from which the result (1.186) of the theorem follows in the spacial case when $c = \ominus b$ by applying the gyration even property. □

Identity (1.186) is interesting since it relates the gyration $\mathrm{gyr}[a,\ominus b]$ to the gyration $\mathrm{gyr}[a,b]$. Furthermore, it gives rise, by gyration inversion and the gyration even property, to the following elegant gyration identity,

$$\mathrm{gyr}[\ominus a,b]\,\mathrm{gyr}[b,a] = \mathrm{gyr}[\ominus a\oplus b,a\oplus b]\tag{1.188}$$

in which the product of two gyrations is equivalent to a single gyration.

As an application of Theorem 1.42, and for later reference, we present the following Theorem 1.44. This theorem, the proof of which involves a long chain of gyrocommutative gyrogroup identities, will prove crucially important for the introduction of hyperbolic vectors, called gyrovectors, into hyperbolic geometry in Chap. 3.

Theorem 1.44 *Let* $(G, \oplus)$ *be a gyrocommutative gyrogroup. Then*

$$(x\oplus a) \boxplus (x\oplus b) = x\oplus\big\{(a \boxplus b)\oplus x\big\} \tag{1.189}$$

for all $a, b, x \in G$.

Before presenting the proof of Theorem 1.44, it would be instructive to present an immediate corollary of the theorem. Renaming the elements $x, a, b \in G$ of the identity, (1.189), of Theorem 1.44 as $\ominus a, b, c$, respectively, we obtain the *parallelogram addition law* as a corollary of Theorem 1.44:

Corollary 1.45 (The Gyroparallelogram Addition Law) *Let* a, b, c *be any three elements of a gyrocommutative gyrogroup* $(G, \oplus)$, *and let* $d \in G$ *be given by the gyroparallelogram condition*, (1.162),

$$d = (b \boxplus c)\ominus a \tag{1.190}$$

Then, the points $a, b, d, c \in G$ *satisfy the gyroparallelogram addition law*

$$(\ominus a\oplus b) \boxplus (\ominus a\oplus c) = (\ominus a\oplus d) \tag{1.191}$$

The term *gyroparallelogram addition law* in Corollary 1.45 will be justified in Sect. 5.4, p. 123, in terms of analogies it shares with the common parallelogram addition law of Euclidean geometry.

Proof The proof of Theorem 1.44 is given by the following chain of equations, which are numbered for subsequent derivation.

$$\begin{aligned}
(x\oplus a) \boxplus (x\oplus b) &\overset{(1)}{\overbrace{=\!=\!=}} (x\oplus a)\oplus \mathrm{gyr}\big[x\oplus a, \ominus(x\oplus b)\big](x\oplus b) \\
&\overset{(2)}{\overbrace{=\!=\!=}} (x\oplus a)\oplus \mathrm{gyr}[x\oplus a, \ominus x\ominus b](x\oplus b) \\
&\overset{(3)}{\overbrace{=\!=\!=}} x\oplus\big\{a\oplus \mathrm{gyr}[a, x]\,\mathrm{gyr}[x\oplus a, \ominus x\ominus b](x\oplus b)\big\} \\
&\overset{(4)}{\overbrace{=\!=\!=}} x\oplus\big\{a\oplus \mathrm{gyr}[a, x]\,\mathrm{gyr}[\ominus x, \ominus a]\,\mathrm{gyr}[a, \ominus b]\,\mathrm{gyr}[b, x](x\oplus b)\big\} \\
&\overset{(5)}{\overbrace{=\!=\!=}} x\oplus\big\{a\oplus \mathrm{gyr}[a, x]\,\mathrm{gyr}[x, a]\,\mathrm{gyr}[a, \ominus b](b\oplus x)\big\}
\end{aligned}$$

$$\overset{(6)}{=\!=} x\oplus\{a\oplus\mathrm{gyr}[a,\ominus b](b\oplus x)\}$$

$$\overset{(7)}{=\!=} x\oplus\{a\oplus(\mathrm{gyr}[a,\ominus b]b\oplus\mathrm{gyr}[a,\ominus b]x)\}$$

$$\overset{(8)}{=\!=} x\oplus\{(a\oplus\mathrm{gyr}[a,\ominus b]b)\oplus\mathrm{gyr}[a,\mathrm{gyr}[a,\ominus b]b]\,\mathrm{gyr}[a,\ominus b]x\}$$

$$\overset{(9)}{=\!=} x\oplus\{(a\oplus\mathrm{gyr}[a,\ominus b]b)\oplus\mathrm{gyr}[b,\ominus a]\,\mathrm{gyr}[a,\ominus b]x\}$$

$$\overset{(10)}{=\!=} x\oplus\{(a\oplus\mathrm{gyr}[a,\ominus b]b)\oplus x\}$$

$$\overset{(11)}{=\!=} x\oplus\{(a\boxplus b)\oplus x\} \tag{1.192}$$

Derivation of the numbered equalities in (1.192) follows:

1. Follows from Definition 1.7, p. 13, of the gyrogroup cooperation $\boxplus$.
2. Follows from Item 1 by the gyroautomorphic inverse property, Theorem 1.32, p. 34.
3. Follows from Item 2 by the right gyroassociative law.
4. Follows from Item 3 by Identity (1.180) of Theorem 1.42, thus providing an elegant example for an application of that theorem.
5. Follows from Item 4 by the gyration even property, and by the gyrocommutative law.
6. Follows from item 5 by the gyration inversion law (1.126), p. 29.
7. Follows from Item 6 by expanding the gyration application term by term.
8. Follows from Item 7 by the left gyroassociative law.
9. Follows from item 8 by the second identity in (1.128), p. 29.
10. Follows from Item 9 by the gyration even property and the gyration inversion law in Theorem 1.23, p. 29, implying $\mathrm{gyr}[b,\ominus a]\,\mathrm{gyr}[a,\ominus b]=I$.
11. Follows from Item 10 by Definition 1.7, p. 13, of the gyrogroup cooperation $\boxplus$. □

1.15 Problems

Problem 1.1 The Automorphic Inverse Property:
Verify the automorphic inverse property (1.4), p. 4, of Einstein addition.

Problem 1.2 The Left Cancellation Law:
Verify the left cancellation law of Einstein addition (1.5), p. 5. You may use a computer software for symbolic manipulation, like Mathematica or Maple.

Problem 1.3 The Gamma Identity:
Verify the gamma identity (1.7), p. 5. You may use a computer software for symbolic manipulation, like Mathematica or Maple.

Problem 1.4 The Algebra of Einstein Addition:
Verify the algebraic laws of Einstein addition in (1.25), p. 9. You may use a computer software for symbolic manipulation, like Mathematica or Maple.

Remark Note that in the proof of gyration identities, gyrations must be applied to a generic element. Thus, for instance, in order to verify the gyration left loop property $\mathrm{gyr}[\mathbf{u}\oplus\mathbf{v}, \mathbf{v}] = \mathrm{gyr}[\mathbf{u}, \mathbf{v}]$ for all $\mathbf{u}, \mathbf{v} \in \mathbb{R}_c^n$ one has to show that $\mathrm{gyr}[\mathbf{u}\oplus\mathbf{v}, \mathbf{v}]\mathbf{w} = \mathrm{gyr}[\mathbf{u}, \mathbf{v}]\mathbf{w}$ for all $\mathbf{u}, \mathbf{v}, \mathbf{w} \in \mathbb{R}_c^n$.

Problem 1.5 Gyrations in Explicit Form:
Verify identity (1.27), p. 9. You may use a computer software for symbolic manipulation, like Mathematica or Maple.

Problem 1.6 Einstein Coaddition:
Prove that the gyrogroup coaddition definition (1.35), p. 13, specializes to Einstein coaddition (1.37).

Chapter 2
Einstein Gyrovector Spaces

Abstract Einstein addition admits scalar multiplication between any real number and any relativistically admissible velocity vector, giving rise to the Einstein gyrovector spaces. As an example, Einstein scalar multiplication enables hyperbolic lines to be calculated with respect to Cartesian coordinates just as Euclidean lines are calculated with respect to Cartesian coordinates. Along with remarkable analogies that Einstein scalar multiplication shares with the common scalar multiplication in vector spaces there is a striking disanalogy. Einstein scalar multiplication does not distribute over Einstein addition. However, a weaker law, called the *monodistributive law*, remains valid. It is shown in this chapter that Einstein gyrovector spaces form the setting for the Cartesian–Beltrami–Klein ball model of hyperbolic geometry just as vector spaces form the setting for the standard Cartesian model of Euclidean geometry.

2.1 Einstein Scalar Multiplication

The rich structure of Einstein addition is not limited to its gyrocommutative gyrogroup structure. Indeed, Einstein addition admits scalar multiplication, giving rise to the Einstein gyrovector space. Remarkably, the resulting Einstein gyrovector spaces form the setting for the Cartesian–Beltrami–Klein ball model of hyperbolic geometry just as vector spaces form the setting for the standard Cartesian model of Euclidean geometry, as we will see in this book.

Let $k \otimes \mathbf{v}$ be the Einstein addition of k copies of $\mathbf{v} \in \mathbb{R}_c^n$, that is $k \otimes \mathbf{v} = \mathbf{v} \oplus \mathbf{v} \cdots \oplus \mathbf{v}$ (k terms). Then,

$$k \otimes \mathbf{v} = c \frac{(1 + \frac{\|\mathbf{v}\|}{c})^k - (1 - \frac{\|\mathbf{v}\|}{c})^k}{(1 + \frac{\|\mathbf{v}\|}{c})^k + (1 - \frac{\|\mathbf{v}\|}{c})^k} \frac{\mathbf{v}}{\|\mathbf{v}\|}. \tag{2.1}$$

The definition of scalar multiplication in an Einstein gyrovector space requires analytically continuing k off the positive integers, thus obtaining the following definition:

A.A. Ungar, *Hyperbolic Triangle Centers*,
Fundamental Theories of Physics 166,
DOI 10.1007/978-90-481-8637-2_2, © Springer Science+Business Media B.V. 2010

Definition 2.1 (Einstein Scalar Multiplication; Einstein Gyrovector Spaces) An Einstein gyrovector space $(\mathbb{R}^n_s, \oplus, \otimes)$ is an Einstein gyrogroup $(\mathbb{R}^n_s, \oplus)$ with scalar multiplication $\otimes$ given by

$$r\otimes\mathbf{v} = s\frac{(1+\frac{\|\mathbf{v}\|}{s})^r - (1-\frac{\|\mathbf{v}\|}{s})^r}{(1+\frac{\|\mathbf{v}\|}{s})^r + (1-\frac{\|\mathbf{v}\|}{s})^r}\frac{\mathbf{v}}{\|\mathbf{v}\|} = s\tanh\left(r\tanh^{-1}\frac{\|\mathbf{v}\|}{s}\right)\frac{\mathbf{v}}{\|\mathbf{v}\|}, \tag{2.2}$$

where r is any real number, $r \in \mathbb{R}$, $\mathbf{v} \in \mathbb{R}^n_s$, $\mathbf{v} \neq \mathbf{0}$, and $r\otimes\mathbf{0} = \mathbf{0}$, and with which we use the notation $\mathbf{v}\otimes r = r\otimes\mathbf{v}$.

Example 2.2 (The Einstein Half) In the special case when $r = 1/2$, (2.2) reduces to

$$\frac{1}{2}\otimes\mathbf{v} = \frac{\gamma_{\mathbf{v}}}{1+\gamma_{\mathbf{v}}}\mathbf{v} \tag{2.3}$$

so that

$$\frac{\gamma_{\mathbf{v}}}{1+\gamma_{\mathbf{v}}}\mathbf{v}\oplus\frac{\gamma_{\mathbf{v}}}{1+\gamma_{\mathbf{v}}}\mathbf{v} = \mathbf{v}. \tag{2.4}$$

Einstein gyrovector spaces are studied in [63, Sect. 6.18]. Einstein scalar multiplication does not distribute with Einstein addition, but it possesses other properties of vector spaces. For any positive integer k, and for all real numbers $r, r_1, r_2 \in \mathbb{R}$ and $\mathbf{v} \in \mathbb{R}^n_s$, we have

$$\begin{aligned} k\otimes\mathbf{v} &= \mathbf{v}\oplus\cdots\oplus\mathbf{v}, && (k \text{ terms}) \\ (r_1+r_2)\otimes\mathbf{v} &= r_1\otimes\mathbf{v}\oplus r_2\otimes\mathbf{v}, && \text{(Scalar Distributive Law)} \\ (r_1 r_2)\otimes\mathbf{v} &= r_1\otimes(r_2\otimes\mathbf{v}) && \text{(Scalar Associative Law)} \end{aligned} \tag{2.5}$$

in any Einstein gyrovector space $(\mathbb{R}^n_s, \oplus, \otimes)$.

Additionally, Einstein gyrovector spaces possess the *scaling property*

$$\frac{|r|\otimes\mathbf{a}}{\|r\otimes\mathbf{a}\|} = \frac{\mathbf{a}}{\|\mathbf{a}\|} \tag{2.6}$$

for $\mathbf{a} \in \mathbb{R}^n_s$, $\mathbf{a} \neq \mathbf{0}$, $r \in \mathbb{R}$, $r \neq 0$, the *gyroautomorphism property*

$$\mathrm{gyr}[\mathbf{u}, \mathbf{v}](r\otimes\mathbf{a}) = r\otimes\mathrm{gyr}[\mathbf{u}, \mathbf{v}]\mathbf{a} \tag{2.7}$$

for $\mathbf{a}, \mathbf{u}, \mathbf{v} \in \mathbb{R}^n_s$, $r \in \mathbb{R}$, and the identity gyroautomorphism

$$\mathrm{gyr}[r_1\otimes\mathbf{v}, r_2\otimes\mathbf{v}] = I \tag{2.8}$$

for $r_1, r_2 \in \mathbb{R}$, $\mathbf{v} \in \mathbb{R}^n_s$.

Any Einstein gyrovector space $(\mathbb{R}^n_s, \oplus, \otimes)$ inherits an inner product and a norm from its vector space $\mathbb{R}^n$. These turn out to be invariant under gyrations, that is,

$$\begin{aligned} \text{gyr}[\mathbf{a}, \mathbf{b}]\mathbf{u}\cdot \text{gyr}[\mathbf{a}, \mathbf{b}]\mathbf{v} &= \mathbf{u}\cdot\mathbf{v}, \\ \big\|\text{gyr}[\mathbf{a}, \mathbf{b}]\mathbf{v}\big\| &= \|\mathbf{v}\| \end{aligned} \tag{2.9}$$

for all $\mathbf{a}, \mathbf{b}, \mathbf{u}, \mathbf{v} \in \mathbb{R}^n_s$.

Unlike vector spaces, Einstein gyrovector spaces $(\mathbb{R}^n_s, \oplus, \otimes)$ do not possess the distributive law since, in general,

$$r\otimes(\mathbf{u}\oplus\mathbf{v}) \neq r\otimes\mathbf{u}\oplus r\otimes\mathbf{v} \tag{2.10}$$

for $r \in \mathbb{R}$ and $\mathbf{u}, \mathbf{v} \in \mathbb{R}^n_s$. However, a weak form of the distributive law does exist, as we see from the following theorem:

Theorem 2.3 (The Monodistributive Law) *Let* $(\mathbb{R}^n_s, \oplus, \otimes)$ *be an Einstein gyrovector space. Then,*

$$r\otimes(r_1\otimes\mathbf{v}\oplus r_2\otimes\mathbf{v}) = r\otimes(r_1\otimes\mathbf{v})\oplus r\otimes(r_2\otimes\mathbf{v}) \tag{2.11}$$

for all $r, r_1, r_2 \in \mathbb{R}$ *and* $\mathbf{v} \in \mathbb{R}^n_s$.

Proof By the scalar distributive and associative laws, (2.5), we have

$$\begin{aligned} r\otimes(r_1\otimes\mathbf{v}\oplus r_2\otimes\mathbf{v}) &= r\otimes\big\{(r_1 + r_2)\otimes\mathbf{v}\big\} \\ &= \big(r(r_1 + r_2)\big)\otimes\mathbf{v} \\ &= (rr_1 + rr_2)\otimes\mathbf{v} \\ &= (rr_1)\otimes\mathbf{v}\oplus(rr_2)\otimes\mathbf{v} \\ &= r\otimes(r_1\otimes\mathbf{v})\oplus r\otimes(r_2\otimes\mathbf{v}), \end{aligned} \tag{2.12}$$

as desired. □

Since scalar multiplication in Einstein gyrovector spaces does not distribute with Einstein addition, the following theorem is interesting.

Theorem 2.4 (The Two-Sum Identity) *Let* $\mathbf{u}, \mathbf{v}$ *be any two points of an Einstein gyrovector space* $(\mathbb{R}^n_s, \oplus, \otimes)$. *Then*

$$2\otimes(\mathbf{u}\oplus\mathbf{v}) = \mathbf{u}\oplus(2\otimes\mathbf{v}\oplus\mathbf{u}). \tag{2.13}$$

Proof Employing the right gyroassociative law in (1.25), the identity $\text{gyr}[\mathbf{v}, \mathbf{v}] = I$, Theorem 1.8(4), the left gyroassociative law, and the gyrocommutative law in (1.25) we have the following chain of equations that gives (2.13),

$$\begin{aligned}
\mathbf{u}\oplus(2\otimes\mathbf{v}\oplus\mathbf{u}) &= \mathbf{u}\oplus\big((\mathbf{v}\oplus\mathbf{v})\oplus\mathbf{u}\big) \\
&= \mathbf{u}\oplus\big(\mathbf{v}\oplus\big(\mathbf{v}\oplus \operatorname{gyr}[\mathbf{v},\mathbf{v}]\mathbf{u}\big)\big) \\
&= \mathbf{u}\oplus\big(\mathbf{v}\oplus(\mathbf{v}\oplus\mathbf{u})\big) \\
&= (\mathbf{u}\oplus\mathbf{v})\oplus \operatorname{gyr}[\mathbf{u},\mathbf{v}](\mathbf{v}\oplus\mathbf{u}) \\
&= (\mathbf{u}\oplus\mathbf{v})\oplus(\mathbf{u}\oplus\mathbf{v}) \\
&= 2\otimes(\mathbf{u}\oplus\mathbf{v}).
\end{aligned} \tag{2.14}$$

□

As an application of Theorem 2.4, we prove the following theorem:

Theorem 2.5 *Let* $\mathbf{u}, \mathbf{v}$ *be any two points of an Einstein gyrovector space* $(\mathbb{R}^n_s, \oplus, \otimes)$. *Then*

$$\mathbf{u}\oplus(\ominus\mathbf{u}\oplus\mathbf{v})\otimes\frac{1}{2} = \frac{1}{2}\otimes(\mathbf{u}\boxplus\mathbf{v}). \tag{2.15}$$

Proof The proof is given by the following chain of equations, which are numbered for subsequent derivation:

$$\begin{aligned}
2\otimes\left\{\mathbf{u}\oplus(\ominus\mathbf{u}\oplus\mathbf{v})\otimes\frac{1}{2}\right\} &\overbrace{=\!=\!=}^{(1)} \mathbf{u}\oplus\big\{(\ominus\mathbf{u}\oplus\mathbf{v})\oplus\mathbf{u}\big\} \\
&\overbrace{=\!=\!=}^{(2)} \big\{\mathbf{u}\oplus(\ominus\mathbf{u}\oplus\mathbf{v})\big\}\oplus \operatorname{gyr}[\mathbf{u},\ominus\mathbf{u}\oplus\mathbf{v}]\mathbf{u} \\
&\overbrace{=\!=\!=}^{(3)} \mathbf{v}\oplus \operatorname{gyr}[\mathbf{u},\ominus\mathbf{u}\oplus\mathbf{v}]\mathbf{u} \\
&\overbrace{=\!=\!=}^{(4)} \mathbf{v}\oplus \operatorname{gyr}[\mathbf{v},\ominus\mathbf{u}]\mathbf{u} \\
&\overbrace{=\!=\!=}^{(5)} \mathbf{v}\boxplus\mathbf{u} \\
&\overbrace{=\!=\!=}^{(6)} \mathbf{u}\boxplus\mathbf{v},
\end{aligned} \tag{2.16}$$

implying (2.15) by the scalar associative law in (2.5). Derivation of the numbered equalities in (2.16) follows.

1. Follows from the Two-Sum Identity in Theorem 2.4 and the scalar associative law in (2.5).
2. Follows from Item 1 by the left gyroassociative law.
3. Follows from Item 2 by a left cancellation.
4. Follows from Item 3 by applying successively the left loop property and the right loop property.

5. Follows from Item 4 by Definition 1.7, p. 13, of the gyrogroup cooperation $\boxplus$.
6. Follows from Item 5 by the commutativity of Einstein coaddition $\boxplus$ according to (1.37), p. 13. □

2.2 Linking Einstein Addition to Hyperbolic Geometry

The Einstein distance function, $d(\mathbf{u},\mathbf{v})$, in an Einstein gyrovector space $(\mathbb{R}^n_c, \oplus, \otimes)$ is given by the equation

$$d(\mathbf{u},\mathbf{v}) = \|\mathbf{u}\ominus\mathbf{v}\| \tag{2.17}$$

for $\mathbf{u},\mathbf{v}\in\mathbb{R}^n_c$. We call it a *gyrodistance function* in order to emphasize the analogies it shares with its Euclidean counterpart, the distance function $\|\mathbf{u}-\mathbf{v}\|$ in $\mathbb{R}^n$. Among these analogies is the gyrotriangle inequality according to which

$$\|\mathbf{u}\oplus\mathbf{v}\| \le \|\mathbf{u}\|\oplus\|\mathbf{v}\| \tag{2.18}$$

for all $\mathbf{u},\mathbf{v}\in\mathbb{R}^n_c$. For this and other analogies that distance and gyrodistance functions share, see [60, 63].

In a two dimensional Einstein gyrovector space $(\mathbb{R}^2_c, \oplus, \otimes)$, the squared gyrodistance between a point $\mathbf{x}\in\mathbb{R}^2_c$ and an infinitesimally nearby point $\mathbf{x}+d\mathbf{x}\in\mathbb{R}^2_c$, $d\mathbf{x}=(dx_1, dx_2)$, is defined by the equation [63, Sect. 7.5], [60, Sect. 7.5]

$$\begin{aligned} ds^2 &= \left\|(\mathbf{x}+d\mathbf{x})\ominus\mathbf{x}\right\|^2 \\ &= E\,dx_1^2 + 2F\,dx_1\,dx_2 + G\,dx_2^2 + \cdots, \end{aligned} \tag{2.19}$$

where, if we use the notation $r^2 = x_1^2 + x_2^2$, we have

$$\begin{aligned} E &= c^2\frac{c^2 - x_2^2}{(c^2-r^2)^2}, \\ F &= c^2\frac{x_1x_2}{(c^2-r^2)^2}, \\ G &= c^2\frac{c^2 - x_1^2}{(c^2-r^2)^2}. \end{aligned} \tag{2.20}$$

The triple $(g_{11}, g_{12}, g_{22}) = (E, F, G)$ along with $g_{21} = g_{12}$ is known in differential geometry as the metric tensor g_{ij} [31]. It turns out to be the metric tensor of the Beltrami–Klein disc model of hyperbolic geometry [37, p. 220]. Hence, ds^2 in (2.19)–(2.20) is the Riemannian line element of the Beltrami–Klein disc model of hyperbolic geometry, linked to Einstein velocity addition (1.2), p. 4, and to Einstein gyrodistance function (2.17) [61].

The link between Einstein gyrovector spaces and the Beltrami–Klein ball model of hyperbolic geometry, already noted by Fock [18, p. 39], has thus been established

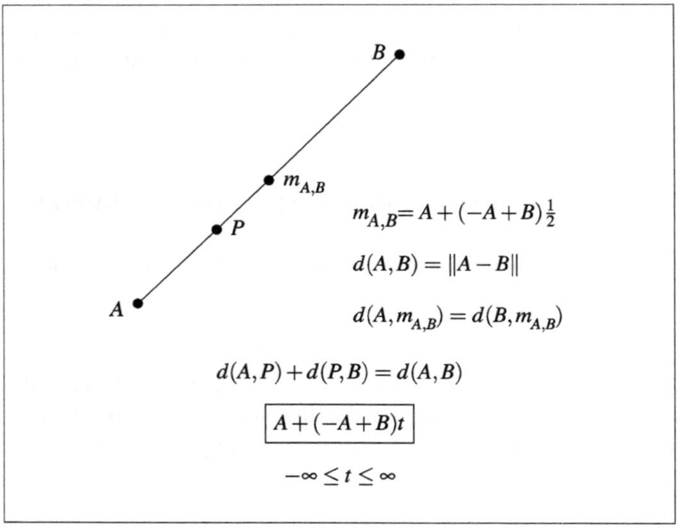

Fig. 2.1 The Euclidean line. The line $A + (-A + B)t$, $t \in \mathbb{R}$, in a Euclidean plane is shown. The points A and B correspond to $t = 0$ and $t = 1$, respectively. The point P is a generic point on the line through the points A and B lying between these points. The Einstein sum, $+$, of the distance from A to P and from P to B equals the distance from A to B. The point $m_{A,B}$ is the midpoint of the points A and B, corresponding to $t = 1/2$

in (2.17)–(2.20) in two dimensions. The extension of the link to higher dimensions is presented in [58, Sect. 9, Chap. 3], [63, Sect. 7.5], [60, Sect. 7.5], and [61]. For a brief account of the history of linking Einstein's velocity addition law with hyperbolic geometry, see [44, p. 943].

2.3 The Euclidean Line

We introduce Cartesian coordinates into $\mathbb{R}^n$ in the usual way in order to specify uniquely each point P of the Euclidean n-space $\mathbb{R}^n$ by an n-tuple of real numbers, called the coordinates, or components, of P. Cartesian coordinates provide a method of indicating the position of points and rendering graphs on a two-dimensional Euclidean plane $\mathbb{R}^2$ and in a three-dimensional Euclidean space $\mathbb{R}^3$.

As an example, Fig. 2.1 presents a Euclidean plane $\mathbb{R}^2$ equipped with an unseen Cartesian coordinate system Σ. The position of points A and B and their midpoint m_{AB} with respect to Σ are shown. The missing Cartesian coordinates in Fig. 2.1 are shown in Fig. 2.3.

The set of all points

$$A + (-A + B)t, \tag{2.21}$$

$t \in \mathbb{R}$, forms a Euclidean line. The segment of this line, corresponding to $1 \leq t \leq 1$, and a generic point P on the segment, are shown in Fig. 2.1. Being collinear, the

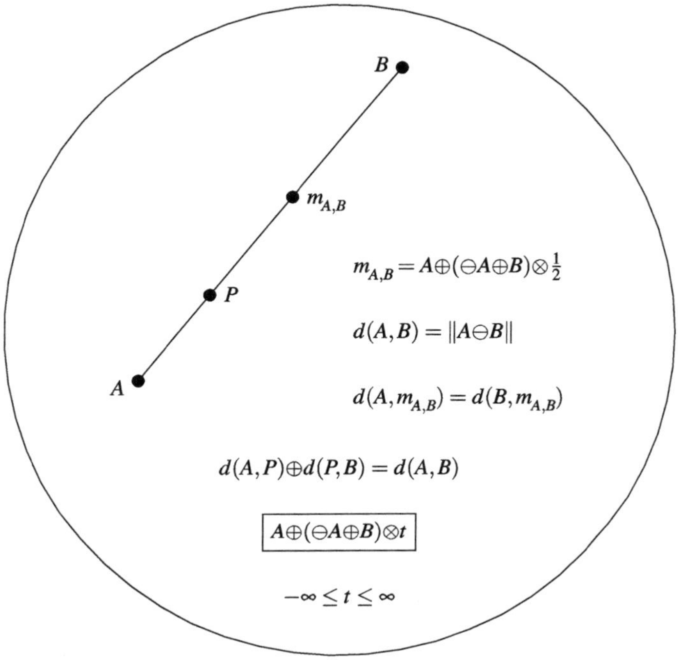

Fig. 2.2 Gyroline, the hyperbolic line. The gyroline $A\oplus(\ominus A\oplus B)\otimes t$, $t\in\mathbb{R}$, in an Einstein gyrovector space $(\mathbb{R}_s^n,\oplus,\otimes)$ is a geodesic line in the Beltrami–Klein ball model of hyperbolic geometry, fully analogous to the straight line $A+(-A+B)t$, $t\in\mathbb{R}$, in the Euclidean geometry of $\mathbb{R}^n$. The points A and B correspond to $t=0$ and $t=1$, respectively. The point P is a generic point on the gyroline through the points A and B lying between these points. The Einstein sum, $\oplus$, of the gyrodistance from A to P and from P to B equals the gyrodistance from A to B. The point $m_{A,B}$ is the gyromidpoint of the points A and B, corresponding to $t=1/2$. The analogies between lines and gyrolines, as illustrated in Figs. 2.1 and 2.2, are obvious

points A, P and B obey the triangle equality $d(A,P)+d(P,B)=d(A,B)$, where $d(A,B)=\|-A+B\|$ is the Euclidean distance function in $\mathbb{R}^n$.

Figure 2.1 demonstrates the use of the standard Cartesian model of Euclidean geometry for graphical presentations. In a fully analogous way, Fig. 2.2 demonstrates the use of the Cartesian–Beltrami–Klein model of hyperbolic geometry, as we will see in Sects. 2.4 and 2.5.

2.4 Gyrolines—the Hyperbolic Lines

In the study of triangles and gyrotriangles, we use extensively the letters a, b, c to denote triangle side-lengths and gyrotriangle side-gyrolengths. Hence, it is convenient in applications to geometry to replace the notation $\mathbb{R}_c^n$ for the c-ball of an Einstein gyrovector space by the s-ball, $\mathbb{R}_s^n$. Moreover, it is understood in this book that $n\geq 2$ is any integer greater than 2, unless specified otherwise.

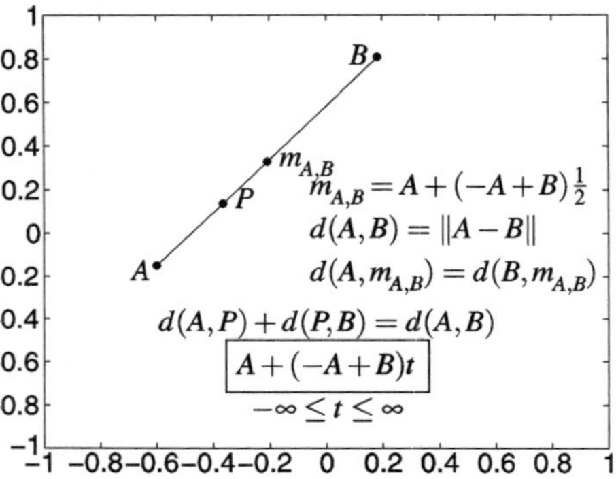

Fig. 2.3 The Cartesian coordinates for the Euclidean plane $\mathbb{R}^2$, (x_1, x_2), $x_1^2 + x_2^2 < \infty$, unseen in Fig. 2.1, are shown here. The points A and B are given, with respect to these Cartesian coordinates by $A = (-0.60, -0.15)$ and $B = (0.18, 0.80)$

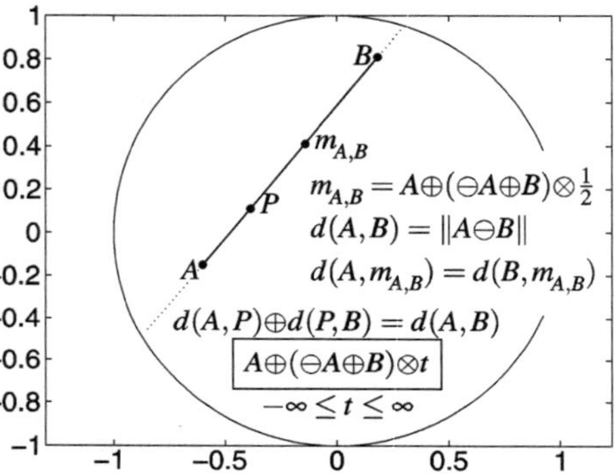

Fig. 2.4 The Cartesian coordinates for the unit disc in the Euclidean plane $\mathbb{R}^2$, (x_1, x_2), $x_1^2 + x_2^2 < 1$, unseen in Fig. 2.2, are shown here. The points A and B are given, with respect to these Cartesian coordinates by $A = (-0.60, -0.15)$ and $B = (0.18, 0.80)$

Let $A, B \in \mathbb{R}_s^n$ be two distinct points of the Einstein gyrovector space $(\mathbb{R}_s^n, \oplus, \otimes)$, and let $t \in \mathbb{R}$ be a real parameter. Then, in full analogy with the Euclidean line (2.21), the graph of the set of all points, Fig. 2.2,

$$A \oplus (\ominus A \oplus B) \otimes t \tag{2.22}$$

for $t \in \mathbb{R}$, in the Einstein gyrovector space $(\mathbb{R}_s^n, \oplus, \otimes)$ is a chord of the ball $\mathbb{R}_s^n$. As such, it is a geodesic line of the Cartesian–Beltrami–Klein ball model of hyperbolic geometry, shown in Fig. 2.2 for $n = 2$. The geodesic line (2.22) is the unique geodesic passing through the points A and B. It passes through the point A when $t = 0$ and, owing to the left cancellation law, (1.38), it passes through the point B when $t = 1$. Furthermore, it passes through the midpoint $m_{A,B}$ of A and B when $t = 1/2$. Accordingly, the *gyrosegment* that joins the points A and B in Fig. 2.2 is obtained from gyroline (2.22) with $0 \le t \le 1$.

Each point of (2.22) with $0 < t < 1$ is said to lie *between* A and B. Thus, for instance, the point P in Fig. 2.2 lies between the points A and B. As such, the points A, P and B obey the *gyrotriangle equality* according to which $d(A, P) \oplus d(P, B) = d(A, B)$, in full analogy with Euclidean geometry. The points in Fig. 2.2 are drawn with respect to an unseen Cartesian coordinate system. The missing Cartesian coordinates for the hyperbolic disc in Fig. 2.2 are shown in Fig. 2.4.

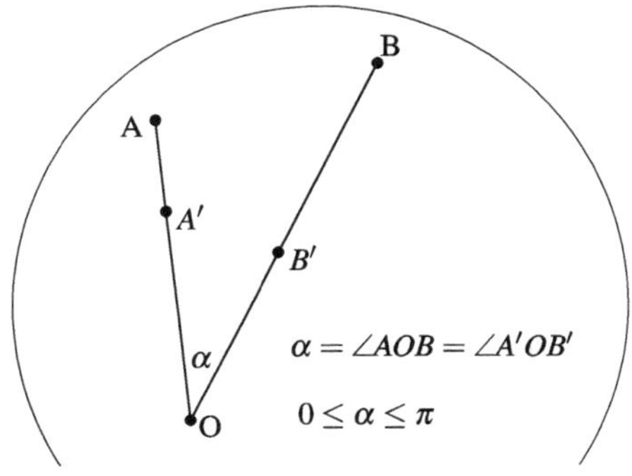

$$\cos\alpha := \frac{\ominus O\oplus A}{\|\ominus O\oplus A\|}\cdot\frac{\ominus O\oplus B}{\|\ominus O\oplus B\|} = \frac{\ominus O\oplus A'}{\|\ominus O\oplus A'\|}\cdot\frac{\ominus O\oplus B'}{\|\ominus O\oplus B'\|}$$

Fig. 2.5 Gyroangles share remarkable analogies with angles, allowing the use of the elementary trigonometric functions cos, sin, etc., in gyrotrigonometry as well. Let A' and B' be points different from O, lying arbitrarily on the gyrosegments OA and OB, respectively, that emanate from a common point O in an Einstein gyrovector space $(\mathbb{R}_s^n, \oplus, \otimes)$ as shown here for $n = 2$. The measure of the gyroangle α formed by the two gyrosegments OA and OB or, equivalently, formed by the two gyrosegments OA' and OB', is given by $\cos\alpha$, as shown here. In full analogy with angles, the measure of gyroangle α is independent of the choice of A' and B'

2.5 The Cartesian Model of Euclidean and Hyperbolic Geometry

The introduction of Cartesian coordinates $(x_1, x_2, \ldots, x_n)$, $x_1^2 + x_2^2 + \cdots + x_n^2 < \infty$, (1.14), p. 6, into the Euclidean n-space $\mathbb{R}^n$, along with the common vector addition in Cartesian coordinates, results in the Cartesian model of Euclidean geometry. The latter, in turn, enables Euclidean geometry to be studied analytically.

In full analogy, the introduction of Cartesian coordinates $(x_1, x_2, \ldots, x_n)$, $x_1^2 + x_2^2 + \cdots + x_n^2 < s^2$, (1.15), p. 6, into the s-ball $\mathbb{R}_s^n$ of the Euclidean n-space $\mathbb{R}^n$, along with the common Einstein addition in Cartesian coordinates, presented in Sect. 1.3, p. 6, results in the Cartesian model of hyperbolic geometry. The latter, in turn, enables hyperbolic geometry to be studied analytically. Indeed, Figs. 2.3 and 2.4 of Sect. 2.4 and Figs. 2.5 and 2.6 of Sect. 2.6 indicate the way we study analytic hyperbolic geometry, guided by analogies with analytic Euclidean geometry.

2.6 Gyroangles—the Hyperbolic Angles

The analogies between lines and gyrolines suggest corresponding analogies between angles and gyroangles. Let O, A and B be any three distinct points in an Einstein gyrovector space $(\mathbb{R}_s^n, \oplus, \otimes)$. The resulting gyrosegments OA and OB that emanate from the point O form a gyroangle $\alpha = \angle AOB$ with vertex O, as shown in Fig. 2.5

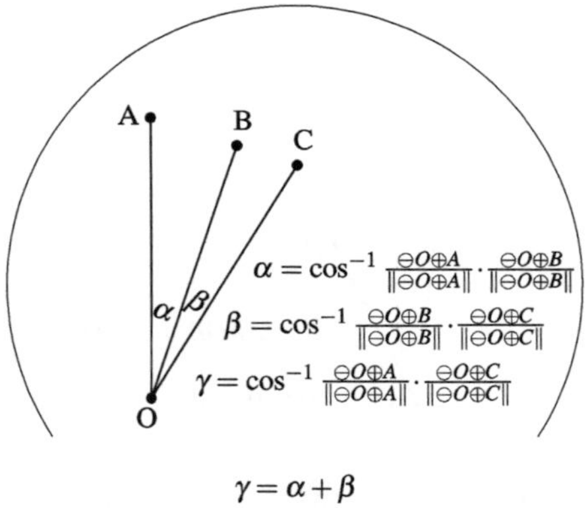

Fig. 2.6 Let A and C be two distinct points, let O be a point not on gyroline AC, and let B be a point between A and C in an Einstein gyrovector space $(\mathbb{R}_s^n, \oplus, \otimes)$. Furthermore, let $\alpha = \angle AOB$ and $\beta = \angle BOC$ be the two adjacent gyroangles that the three gyrosegments OA, OB and OC form, and let γ be their composite gyroangle, formed by gyrosegments OA and OC. Then, $\gamma = \alpha + \beta$, demonstrating that, like angles, gyroangles are additive. We call $(\ominus O \oplus A)/\|\ominus O \oplus A\|$ a unit gyrovector. When applied to an inner product of unit gyrovectors, the common cosine function of trigonometry becomes the gyrocosine function of gyrotrigonometry

for $n = 2$. Following the analogies between gyrolines and lines, the radian measure of gyroangle α is, suggestively, given by the equation

$$\cos\alpha = \frac{\ominus O \oplus A}{\|\ominus O \oplus A\|} \cdot \frac{\ominus O \oplus B}{\|\ominus O \oplus B\|}. \tag{2.23}$$

Here, $(\ominus O \oplus A)/\|\ominus O \oplus A\|$ and $(\ominus O \oplus B)/\|\ominus O \oplus B\|$ are unit *gyrovectors*, and cos is the common cosine function of trigonometry, which we apply to the inner product between unit gyrovectors rather than unit vectors. Accordingly, in the context of gyrovector spaces rather than vector spaces, we refer the function "cosine" of trigonometry to as the function "gyrocosine" of gyrotrigonometry. Similarly, all the other elementary trigonometric functions and their interrelationships survive unimpaired in their transition from the common trigonometry in Euclidean spaces $\mathbb{R}^n$ to a corresponding gyrotrigonometry in Einstein gyrovector space $\mathbb{R}_s^n$, as we will see in Chap. 6.

The center $\mathbf{0} = (0, \dots, 0) \in \mathbb{R}_s^n$ of the ball $\mathbb{R}_s^n = (\mathbb{R}_s^n, \oplus, \otimes)$ is conformal (to Euclidean geometry) in the sense that the measure of any gyroangle with vertex $\mathbf{0}$ is equal to the measure of its Euclidean counterpart. Indeed, if $O = \mathbf{0}$ then (2.23) reduces to

$$\cos\alpha = \frac{A}{\|A\|} \cdot \frac{B}{\|B\|}, \tag{2.24}$$

which is indistinguishable from its Euclidean counterpart.

2.7 The Euclidean Group of Motions

The Euclidean group of motions of $\mathbb{R}^n$ consists of the commutative group of all translations of $\mathbb{R}^n$ and the group of all rotations of $\mathbb{R}^n$ about its origin.

For any $\mathbf{x} \in \mathbb{R}^n$, a translation of $\mathbb{R}^n$ by $\mathbf{x} \in \mathbb{R}^n$ is the map $L_\mathbf{x} : \mathbb{R}^n \to \mathbb{R}^n$ given by

$$L_\mathbf{x}\mathbf{v} = \mathbf{x} + \mathbf{v} \tag{2.25a}$$

for all $\mathbf{v} \in \mathbb{R}^n$.

A rotation R of $\mathbb{R}^n$ about its origin is an element of the group $SO(n)$ of all $n \times n$ orthogonal matrices with determinant 1. The rotation of $\mathbf{v} \in \mathbb{R}^n$ by $R \in SO(n)$ is given by $R\mathbf{v}$. The map $R \in SO(n)$ is a linear map of $\mathbb{R}^n$ that keeps the inner product invariant, that is,

$$\begin{aligned} R(\mathbf{u} + \mathbf{v}) &= R\mathbf{u} + R\mathbf{v}, \\ R\mathbf{u}\cdot R\mathbf{v} &= \mathbf{u}\cdot\mathbf{v} \end{aligned} \tag{2.25b}$$

for all $R \in \mathbb{R}^n$ and $\mathbf{u}, \mathbf{v} \in \mathbb{R}^n$.

The *Euclidean group of motions* is the *semidirect product group*

$$\mathbb{R}^n \times SO(n) \tag{2.26}$$

of the Euclidean commutative group $\mathbb{R}^n = (\mathbb{R}^n, +)$ and the rotation group $SO(n)$. It is a group of pairs $(\mathbf{x}, R)$, $\mathbf{x} \in (\mathbb{R}^n, +)$, $R \in SO(n)$, acting on elements $\mathbf{v} \in \mathbb{R}^n$ according to the equation

$$(\mathbf{x}, R)\mathbf{v} = \mathbf{x} + R\mathbf{v}. \tag{2.27}$$

The group operation of the semidirect product group (2.26) is given by action composition. The latter, in turn, is determined by the following chain of equations, in which we employ the associative law:

$$\begin{aligned} (\mathbf{x}_1, R_1)(\mathbf{x}_2, R_2)\mathbf{v} &= (\mathbf{x}_1, R_1)(\mathbf{x}_2 + R_2\mathbf{v}) \\ &= \mathbf{x}_1 + R_1(\mathbf{x}_2 + R_2\mathbf{v}) \\ &= \mathbf{x}_1 + (R_1\mathbf{x}_2 + R_1R_2\mathbf{v}) \\ &= (\mathbf{x}_1 + R_1\mathbf{x}_2) + R_1R_2\mathbf{v} \\ &= (\mathbf{x}_1 + R_1\mathbf{x}_2, R_1R_2)\mathbf{v} \end{aligned} \tag{2.28}$$

for all $\mathbf{v} \in \mathbb{R}^n$.

Hence, by (2.28), the group operation of the semidirect product group (2.26) is given by the *semidirect product*

$$(\mathbf{x}_1, R_1)(\mathbf{x}_2, R_2) = (\mathbf{x}_1 + R_1\mathbf{x}_2, R_1R_2) \tag{2.29}$$

for any $(\mathbf{x}_1, R_1), (\mathbf{x}_2, R_2) \in \mathbb{R}^n \times SO(n)$.

Definition 2.6 (Covariance) An identity in $\mathbb{R}^n$ that remains invariant in form under the action of the Euclidean group of motions of $\mathbb{R}^n$ is said to be covariant.

We will see in Chap. 4 that Euclidean barycentric coordinate representations of points of $\mathbb{R}^n$ are covariant, by Theorem 4.3, p. 87.

2.8 The Hyperbolic Group of Motions

The hyperbolic group of motions of $\mathbb{R}^n_s$ consists of the gyrocommutative gyrogroup of all left gyrotranslations of $\mathbb{R}^n_s$ and the group of all rotations of $\mathbb{R}^n_s$ about its center.

For any $\mathbf{x} \in \mathbb{R}^n_s$, a left gyrotranslation of $\mathbb{R}^n_s$ by $\mathbf{x} \in \mathbb{R}^n_s$ is the map $L_\mathbf{x} : \mathbb{R}^n_s \to \mathbb{R}^n_s$ given by

$$L_\mathbf{x}\mathbf{v} = \mathbf{x} \oplus \mathbf{v} \tag{2.30a}$$

for all $\mathbf{v} \in \mathbb{R}^n_s$.

The group of all rotations of the ball $\mathbb{R}^n_s$ about its center is $SO(n)$. Following (2.25b), we have

$$\begin{aligned} R(\mathbf{u} \oplus \mathbf{v}) &= R\mathbf{u} \oplus R\mathbf{v}, \\ R\mathbf{u} \cdot R\mathbf{v} &= \mathbf{u} \cdot \mathbf{v} \end{aligned} \tag{2.30b}$$

for all $R \in SO(n)$ and $\mathbf{u}, \mathbf{v} \in \mathbb{R}^n$.

The *hyperbolic group of motions* is the *gyrosemidirect product group*

$$\mathbb{R}^n_s \times SO(n) \tag{2.31}$$

of the Einsteinian gyrocommutative gyrogroup $\mathbb{R}^n_s = (\mathbb{R}^n_s, \oplus)$ and the rotation group $SO(n)$. It is a group of pairs $(\mathbf{x}, R)$, $\mathbf{x} \in (\mathbb{R}^n_s, \oplus)$, $R \in SO(n)$, acting on elements $\mathbf{v} \in \mathbb{R}^n_s$ according to the equation

$$(\mathbf{x}, R)\mathbf{v} = \mathbf{x} \oplus R\mathbf{v}. \tag{2.32}$$

The group operation of the gyrosemidirect product group (2.31) is given by action composition. The latter, in turn, is determined by the following chain of equations, in which we employ the left gyroassociative law:

$$\begin{aligned} (\mathbf{x}_1, R_1)(\mathbf{x}_2, R_2)\mathbf{v} &= (\mathbf{x}_1, R_1)(\mathbf{x}_2 \oplus R_2\mathbf{v}) \\ &= \mathbf{x}_1 \oplus R_1(\mathbf{x}_2 \oplus R_2\mathbf{v}) \\ &= \mathbf{x}_1 \oplus (R_1\mathbf{x}_2 \oplus R_1R_2\mathbf{v}) \\ &= (\mathbf{x}_1 \oplus R_1\mathbf{x}_2) \oplus \mathrm{gyr}[\mathbf{x}, R_1\mathbf{x}_2]R_1R_2\mathbf{v} \\ &= \big(\mathbf{x}_1 \oplus R_1\mathbf{x}_2, \mathrm{gyr}[\mathbf{x}, R_1\mathbf{x}_2]R_1R_2\big)\mathbf{v} \end{aligned} \tag{2.33}$$

for all $\mathbf{v} \in \mathbb{R}^n_s$.

Hence, by (2.33), the group operation of the gyrosemidirect product group (2.31) is given by the *gyrosemidirect product*

$$(\mathbf{x}_1, R_1)(\mathbf{x}_2, R_2) = \big(\mathbf{x}_1 \oplus R_1\mathbf{x}_2, \mathrm{gyr}[\mathbf{x}, R_1\mathbf{x}_2]R_1R_2\big) \qquad (2.34)$$

for any $(\mathbf{x}_1, R_1), (\mathbf{x}_2, R_2) \in \mathbb{R}_s^n \times SO(n)$. Indeed, the gyrosemidirect product is a group operation, as demonstrated in Sect. 1.11, p. 23.

Definition 2.7 (Gyrocovariance) An identity in $\mathbb{R}_s^n$ that remains invariant in form under the action of the hyperbolic group of motions of $\mathbb{R}_s^n$ is said to be gyrocovariant.

We will see in Chap. 4 that hyperbolic barycentric (that is, gyrobarycentric) coordinate representations of points of $\mathbb{R}_s^n$ are gyrocovariant, by the Gyrobarycentric Coordinate Representation Gyrocovariance Theorem 4.6, p. 90.

2.9 Problems

Problem 2.1 Einstein Scalar Multiplication:
Show that $k\otimes\mathbf{v} := \mathbf{v}\oplus\cdots\oplus\mathbf{v}$ (k terms) is given by (2.1), p. 45.

Problem 2.2 Einstein Scalar Multiplication:
Prove the second equation in (2.2), p. 46.

Problem 2.3 The Einstein Half:
Prove the Einstein-half identities (2.3)–(2.4), p. 46.

Problem 2.4 Einstein Scalar Distributive Law:
Prove the scalar distributive law in (2.5), p. 46.

Problem 2.5 Einstein Scalar Associative Law:
Prove the scalar associative law in (2.5), p. 46.

Problem 2.6 Scaling Property:
Prove the scaling property (2.6), p. 46.

Problem 2.7 A Gyroautomorphism Property:
Prove the gyroautomorphism property (2.7), p. 46.

Problem 2.8 Inner Product Invariance Under Gyrations:
Prove the identities in (2.9), p. 47.

Problem 2.9 Rotations Respect Einstein Addition:
Show that the first identity in (2.30b), p. 56, follows from (2.25b), p. 55.

Chapter 3
When Einstein Meets Minkowski

Abstract Einstein's addition law of three-dimensional relativistically admissible velocities is the corner stone of Einstein's three-vector formalism of the special theory of relativity that he founded in 1905. Soon later, in 1908, special relativity was reformulated by Minkowski who introduced his elegant four-vector formalism. In this chapter, we present the harmonious interplay between Einstein's three-dimensional velocity addition and the Minkowskian four-vector formalism of Einstein's special theory of relativity, along with its relevant consequences to the study of hyperbolic geometry in Part II and hyperbolic triangle centers in Part III of the book.

3.1 Introduction

Einstein's addition law of three-dimensional relativistically admissible velocities, studied in Chap. 1, appears in [58, 60, 63, 64] as the corner stone of Einstein's three-vector formalism of the special theory of relativity that he founded in 1905 [12]. In 1908, special relativity was reformulated by Minkowski who introduced his elegant four-vector formalism [25, 59, 73]. The elegance and usefulness of the Minkowskian formalism posed an annoying problem: the concept of the relativistic mass, according to which mass is velocity dependent, seemed too wild, defying attempts to place it under the umbrella of the Minkowskian four-vector formalism of special relativity [2, 4, 42]. This intriguing puzzle challenges our mind in this chapter as well, in our mission to capture analogies that Euclidean and hyperbolic triangle centers share.

The study of Euclidean triangle centers can be approached by considering, in classical mechanics, the center of momentum of an isolated system of three non-interacting, uniformly moving massive particles with *Newtonian masses*, and with velocities in the Euclidean 3-space $\mathbb{R}^3$ of *Newtonian velocities*. A related example is found, for instance, in [24, p. 3], in which a classical mechanical interpretation of the Euclidean triangle centroid is presented.

In full analogy, the mission of this book is to approach the study of hyperbolic triangle centers by considering, in relativistic mechanics, the center of momentum

A.A. Ungar, *Hyperbolic Triangle Centers*,
Fundamental Theories of Physics 166,
DOI 10.1007/978-90-481-8637-2_3,

of an isolated system of three noninteracting, uniformly moving massive particles with *relativistic masses*, and with relativistically admissible velocities in the ball $\mathbb{R}_c^3 \subset \mathbb{R}^3$ of *Einsteinian velocities*. But, in order to harness the relativistic mass for our mission, we must tame it by placing it under the umbrella of the Minkowskian four-vector formalism of special relativity.

Accordingly, the mission of this chapter is to show how the relativistic mass is tamed when Einstein, with his three-dimensional vector formalism, and Minkowski, with his four-dimensional vector formalism, meet.

3.2 Lorentz Transformation and Minkowski's Four-Velocity

Einstein addition underlies the Lorentz transformation group of special relativity. A Lorentz transformation is a linear transformation of spacetime coordinates that fixes the spacetime origin. A Lorentz boost, $L(\mathbf{v})$, is a Lorentz transformation without rotation, parametrized by a velocity parameter $\mathbf{v} = (v_1, v_2, v_3) \in \mathbb{R}_c^3$.

Being linear, the Lorentz boost has a matrix representation $L_m(\mathbf{v})$, which turns out to be [40]

$$L_m(\mathbf{v}) = \begin{pmatrix} \gamma_{\mathbf{v}} & c^{-2}\gamma_{\mathbf{v}} v_1 & c^{-2}\gamma_{\mathbf{v}} v_2 & c^{-2}\gamma_{\mathbf{v}} v_3 \\ \gamma_{\mathbf{v}} v_1 & 1 + c^{-2}\frac{\gamma_{\mathbf{v}}^2}{\gamma_{\mathbf{v}}+1} v_1^2 & c^{-2}\frac{\gamma_{\mathbf{v}}^2}{\gamma_{\mathbf{v}}+1} v_1 v_2 & c^{-2}\frac{\gamma_{\mathbf{v}}^2}{\gamma_{\mathbf{v}}+1} v_1 v_3 \\ \gamma_{\mathbf{v}} v_2 & c^{-2}\frac{\gamma_{\mathbf{v}}^2}{\gamma_{\mathbf{v}}+1} v_1 v_2 & 1 + c^{-2}\frac{\gamma_{\mathbf{v}}^2}{\gamma_{\mathbf{v}}+1} v_2^2 & c^{-2}\frac{\gamma_{\mathbf{v}}^2}{\gamma_{\mathbf{v}}+1} v_2 v_3 \\ \gamma_{\mathbf{v}} v_3 & c^{-2}\frac{\gamma_{\mathbf{v}}^2}{\gamma_{\mathbf{v}}+1} v_1 v_3 & c^{-2}\frac{\gamma_{\mathbf{v}}^2}{\gamma_{\mathbf{v}}+1} v_2 v_3 & 1 + c^{-2}\frac{\gamma_{\mathbf{v}}^2}{\gamma_{\mathbf{v}}+1} v_3^2 \end{pmatrix}. \tag{3.1}$$

Employing the matrix representation (3.1) of the Lorentz transformation boost, the Lorentz boost application to spacetime coordinates takes the form

$$L(\mathbf{v})\begin{pmatrix} t \\ \mathbf{x} \end{pmatrix} = L_m(\mathbf{v})\begin{pmatrix} t \\ x_1 \\ x_2 \\ x_3 \end{pmatrix} =: \begin{pmatrix} t' \\ x_1' \\ x_2' \\ x_3' \end{pmatrix} = \begin{pmatrix} t' \\ \mathbf{x}' \end{pmatrix}, \tag{3.2}$$

where $\mathbf{v} = (v_1, v_2, v_3)^t \in \mathbb{R}_c^3$, $\mathbf{x} = (x_1, x_2, x_3)^t \in \mathbb{R}^3$, $\mathbf{x}' = (x_1', x_2', x_3')^t \in \mathbb{R}^3$, and $t, t' \in \mathbb{R}$, where exponent t denotes transposition.

In our approach to special relativity, analogies with classical results form the right tool. Hence, we emphasize that in the Newtonian limit of large vacuum speed of light c, $c \to \infty$, the Lorentz boost $L(\mathbf{v})$, (3.1)–(3.2), reduces to the Galilei boost $G(\mathbf{v})$, $\mathbf{v} = (v_1, v_2, v_3) \in \mathbb{R}^3$,

$$G(\mathbf{v})\begin{pmatrix} t \\ \mathbf{x} \end{pmatrix} = \lim_{c\to\infty} L(\mathbf{v})\begin{pmatrix} t \\ \mathbf{x} \end{pmatrix}$$

$$
= \begin{pmatrix} 1 & 0 & 0 & 0 \\ v_1 & 1 & 0 & 0 \\ v_2 & 0 & 1 & 0 \\ v_3 & 0 & 0 & 1 \end{pmatrix} \begin{pmatrix} t \\ x_1 \\ x_2 \\ x_3 \end{pmatrix}
$$

$$
= \begin{pmatrix} t \\ x_1 + v_1 t \\ x_2 + v_2 t \\ x_3 + v_3 t \end{pmatrix} = \begin{pmatrix} t \\ \mathbf{x} + \mathbf{v}t \end{pmatrix}, \tag{3.3}
$$

where $\mathbf{x} = (x_1, x_2, x_3)^t \in \mathbb{R}^3$ and $t \in \mathbb{R}$.

As we see from (3.2)–(3.3), our spacetime coordinates are $(t, \mathbf{x})^t$ and, as a result, the Lorentz boost matrix representation $L_m(\mathbf{v})$ in (3.1) is *non-symmetric* for $c \neq 1$. In contrast, some authors present spacetime coordinates as $(ct, \mathbf{x})^t$, resulting in a *symmetric* Lorentz boost matrix representation found, for instance, in [27, (11.98), p. 541].

Since in our approach to special relativity analogies with classical results form the right tool, the representation of spacetime coordinates as $(t, \mathbf{x})^t$ is more advantageous than its representation as $(ct, \mathbf{x})^t$. Indeed, unlike the latter representation, the former representation of spacetime coordinates allows one to recover the Galilei boost from the Lorentz boost by taking the Newtonian limit of large speed of light c, as shown in the transition from (3.2) to (3.3).

As a result of adopting $(t, \mathbf{x})^t$ rather than $(ct, \mathbf{x})^t$ as our four-vector that represents four-position, our four-velocity is given by $(\gamma_{\mathbf{v}}, \gamma_{\mathbf{v}} \mathbf{v})$ rather than $(\gamma_{\mathbf{v}} c, \gamma_{\mathbf{v}} \mathbf{v})$, $\mathbf{v} \in \mathbb{R}_c^3$. Similarly, our four-momentum is given by

$$
\begin{pmatrix} p_0 \\ \mathbf{p} \end{pmatrix} = \begin{pmatrix} \frac{E}{c^2} \\ \mathbf{p} \end{pmatrix} = m \begin{pmatrix} \gamma_{\mathbf{v}} \\ \gamma_{\mathbf{v}} \mathbf{v} \end{pmatrix} \tag{3.4}
$$

rather than the standard four-momentum, which is given by $(p_0, \mathbf{p})^t = (E/c, \mathbf{p})^t = (m\gamma_{\mathbf{v}} c, m\gamma_{\mathbf{v}} \mathbf{v})^t$, as found in most relativity physics books. According to (3.4), the relativistically invariant mass (that is, rest mass) m of a particle is the ratio of the particle's four-momentum $(p_0, \mathbf{p})^t$ to its four-velocity $(\gamma_{\mathbf{v}}, \gamma_{\mathbf{v}} \mathbf{v})^t$.

For the sake of simplicity, and without loss of generality, some authors normalize the vacuum speed of light to $c = 1$ as, for instance, in [17]. We, however, prefer to leave c as a free positive parameter, enabling related modern results to be reduced to classical ones under the limit of large c, $c \to \infty$, as, for instance, in the transition from a Lorentz boost into a corresponding Galilei boost in (3.1)–(3.3).

The Lorentz boost (3.1)–(3.2) can be written vectorially in the form

$$
L(\mathbf{u}) \begin{pmatrix} t \\ \mathbf{x} \end{pmatrix} = \begin{pmatrix} \gamma_{\mathbf{u}} (t + \frac{1}{c^2} \mathbf{u} \cdot \mathbf{x}) \\ \gamma_{\mathbf{u}} \mathbf{u} t + \mathbf{x} + \frac{1}{c^2} \frac{\gamma_{\mathbf{u}}^2}{1 + \gamma_{\mathbf{u}}} (\mathbf{u} \cdot \mathbf{x}) \mathbf{u} \end{pmatrix}. \tag{3.5}
$$

Being written in a vector form, the Lorentz boost in (3.5) survives unimpaired in higher dimensions. Rewriting (3.5) in higher dimensional spaces, with $\mathbf{x} = \mathbf{v}t$, $\mathbf{u}, \mathbf{v} \in \mathbb{R}^n_c \subset \mathbb{R}^n$, we have

$$
\begin{aligned}
L(\mathbf{u}) \begin{pmatrix} t \\ \mathbf{v}t \end{pmatrix} &= \begin{pmatrix} \gamma_{\mathbf{u}} (t + \frac{1}{c^2} \mathbf{u}\cdot\mathbf{v}t) \\ \gamma_{\mathbf{u}} \mathbf{u}t + \mathbf{v}t + \frac{1}{c^2} \frac{\gamma_{\mathbf{u}}^2}{1+\gamma_{\mathbf{u}}} (\mathbf{u}\cdot\mathbf{v}t)\mathbf{u} \end{pmatrix} \\
&= \begin{pmatrix} \frac{\gamma_{\mathbf{u}\oplus\mathbf{v}}}{\gamma_{\mathbf{v}}} t \\ \frac{\gamma_{\mathbf{u}\oplus\mathbf{v}}}{\gamma_{\mathbf{v}}} (\mathbf{u}\oplus\mathbf{v})t \end{pmatrix}.
\end{aligned}
\tag{3.6}
$$

Equation (3.6) reveals explicitly the way Einstein velocity addition underlies the Lorentz boost. The second equation in (3.6) follows from the first by (1.7), p. 5, and (1.2), p. 4.

The special case of $t = \gamma_{\mathbf{v}}$ in (3.6) proves useful, giving rise to the elegant identity

$$
L(\mathbf{u}) \begin{pmatrix} \gamma_{\mathbf{v}} \\ \gamma_{\mathbf{v}} \mathbf{v} \end{pmatrix} = \begin{pmatrix} \gamma_{\mathbf{u}\oplus\mathbf{v}} \\ \gamma_{\mathbf{u}\oplus\mathbf{v}} (\mathbf{u}\oplus\mathbf{v}) \end{pmatrix}
\tag{3.7}
$$

of the Lorentz boost of four-velocities, $\mathbf{u}, \mathbf{v} \in \mathbb{R}^n_c$. Since in physical applications $n = 3$, in the context of n-dimensional special relativity we call $\mathbf{v}$ a three-vector and $(\gamma_{\mathbf{v}}, \gamma_{\mathbf{v}} \mathbf{v})^t$ a four-vector, etc.

The four-vector $m(\gamma_{\mathbf{v}}, \gamma_{\mathbf{v}} \mathbf{v})^t$ is the four-momentum of a particle with invariant mass (or, rest mass) m and velocity $\mathbf{v}$ relative to a given inertial rest frame $\Sigma_{\mathbf{0}}$. Let $\Sigma_{\ominus\mathbf{u}}$ be an inertial frame that moves with velocity $\ominus\mathbf{u} = -\mathbf{u}$ relative to the rest frame $\Sigma_{\mathbf{0}}$, $\mathbf{u}, \mathbf{v} \in \mathbb{R}^n_c$. Then, the particle with velocity $\mathbf{v}$ relative to $\Sigma_{\mathbf{0}}$ has velocity $\mathbf{u}\oplus\mathbf{v}$ relative to the frame $\Sigma_{\ominus\mathbf{u}}$. In full agreement and, owing to the linearity of the Lorentz boost, it follows from (3.7) that the four-momentum of the particle relative to the frame $\Sigma_{\ominus\mathbf{u}}$ is

$$
\begin{aligned}
L(\mathbf{u}) m \begin{pmatrix} \gamma_{\mathbf{v}} \\ \gamma_{\mathbf{v}} \mathbf{v} \end{pmatrix} &= m L(\mathbf{u}) \begin{pmatrix} \gamma_{\mathbf{v}} \\ \gamma_{\mathbf{v}} \mathbf{v} \end{pmatrix} \\
&= m \begin{pmatrix} \gamma_{\mathbf{u}\oplus\mathbf{v}} \\ \gamma_{\mathbf{u}\oplus\mathbf{v}} (\mathbf{u}\oplus\mathbf{v}) \end{pmatrix}.
\end{aligned}
\tag{3.8}
$$

It follows from the linearity of the Lorentz boost and from (3.7) that

$$
\begin{aligned}
L(\mathbf{w}) \sum_{k=1}^{N} m_k \begin{pmatrix} \gamma_{\mathbf{v}_k} \\ \gamma_{\mathbf{v}_k} \mathbf{v}_k \end{pmatrix} &= \sum_{k=1}^{N} m_k L(\mathbf{w}) \begin{pmatrix} \gamma_{\mathbf{v}_k} \\ \gamma_{\mathbf{v}_k} \mathbf{v}_k \end{pmatrix} \\
&= \sum_{k=1}^{N} m_k \begin{pmatrix} \gamma_{\mathbf{w}\oplus\mathbf{v}_k} \\ \gamma_{\mathbf{w}\oplus\mathbf{v}_k} (\mathbf{w}\oplus\mathbf{v}_k) \end{pmatrix}
\end{aligned}
$$

$$= \begin{pmatrix} \sum_{k=1}^{N} m_k \gamma_{\mathbf{w}\oplus\mathbf{v}_k} \\ \sum_{k=1}^{N} m_k \gamma_{\mathbf{w}\oplus\mathbf{v}_k} (\mathbf{w}\oplus\mathbf{v}_k) \end{pmatrix}. \tag{3.9}$$

The chain of equations (3.9) reveals the interplay of Einstein addition, $\oplus$, in $\mathbb{R}_c^n$ and vector addition, $+$, in $\mathbb{R}^n$ that appears implicitly in the sigma-notation for scalar and vector addition. This harmonious interplay between $\oplus$ and $+$, which is crucially important in our mission to determine hyperbolic triangle centers, reveals itself in (3.9) where Einstein's three-vector formalism of special relativity meets Minkowski's four-vector formalism of special relativity.

The (Minkowski) norm of a four-vector is Lorentz transformation invariant. The norm of the four-position $(t, \mathbf{x})^t$ is

$$\left\| \begin{pmatrix} t \\ \mathbf{x} \end{pmatrix} \right\| = \sqrt{t^2 - \frac{\|\mathbf{x}\|^2}{c^2}} \tag{3.10}$$

and, accordingly, the norm of the four-velocity $(\gamma_{\mathbf{v}}, \gamma_{\mathbf{v}} \mathbf{v})^t$ is

$$\left\| \begin{pmatrix} \gamma_{\mathbf{v}} \\ \gamma_{\mathbf{v}} \mathbf{v} \end{pmatrix} \right\| = \gamma_{\mathbf{v}} \left\| \begin{pmatrix} 1 \\ \mathbf{v} \end{pmatrix} \right\| = \gamma_{\mathbf{v}} \sqrt{1 - \frac{\|\mathbf{v}\|^2}{c^2}} = 1. \tag{3.11}$$

3.3 Invariant Mass of a System of Particles

In obtaining the result in (3.8), we exploit the linearity of the Lorentz boost. We will now further exploit that linearity, demonstrated in (3.9), to obtain the relativistically invariant mass of a system of particles. Being observer's invariant, we refer the Newtonian, rest mass, m, to as the (relativistically) invariant mass, as opposed to the common relativistic mass, $m\gamma_{\mathbf{v}}$, which is observer's dependent.

Let

$$S = S(m_k, \mathbf{v}_k, \Sigma_{\mathbf{0}}, k = 1, \ldots, N) \tag{3.12}$$

be an isolated system of N noninteracting material particles the kth particle of which has invariant mass $m_k > 0$ and velocity $\mathbf{v}_k \in \mathbb{R}_c^n$ relative to an inertial frame $\Sigma_{\mathbf{0}}$, $k = 1, \ldots, N$.

Classically, the Newtonian mass m_{newton} of the system S equals the sum of the Newtonian masses of its constituent particles, that is,

$$m_{newton} = \sum_{k=1}^{N} m_k, \tag{3.13}$$

and it forms the total mass of the system. Relativistically, however, this need not be the case since dark matter may emerge, as we will see in Theorem 3.2 of Sect. 3.4.

Accordingly, we wish to determine the relativistically invariant mass m_0 of the system S, and the velocity $\mathbf{v}_0$ relative to $\Sigma_\mathbf{0}$ of a fictitious inertial frame, called the center of momentum frame, relative to which the three-momentum of S vanishes.

Assuming that the four-momentum is additive, the sum of the four-momenta of the N particles of the system S gives the four-momentum $(m_0\gamma_{\mathbf{v}_0}, m_0\gamma_{\mathbf{v}_0}\mathbf{v}_0)^t$ of S. Accordingly,

$$\sum_{k=1}^{N} m_k \begin{pmatrix} \gamma_{\mathbf{v}_k} \\ \gamma_{\mathbf{v}_k}\mathbf{v}_k \end{pmatrix} = m_0 \begin{pmatrix} \gamma_{\mathbf{v}_0} \\ \gamma_{\mathbf{v}_0}\mathbf{v}_0 \end{pmatrix}, \tag{3.14}$$

where

(i) The invariant masses $m_k > 0$ and the velocities $\mathbf{v}_k \in \mathbb{R}_c^n$, $k = 1, \ldots, N$, relative to $\Sigma_\mathbf{0}$ of the constituent particles of S are given, while
(ii) The invariant mass m_0 of S and the velocity $\mathbf{v}_0$ of the center of momentum frame of S relative to $\Sigma_\mathbf{0}$ are to be determined uniquely by the *Resultant Relativistically Invariant Mass Theorem*, which is Theorem 3.2 in Sect. 3.4

If $m_0 > 0$ and $\mathbf{v}_0 \in \mathbb{R}_c^n$ that satisfy (3.14) exist then, as anticipated, the three-momentum of the system S relative to its center of momentum frame vanishes since, by (3.8) and (3.14), the four-momentum of S relative to its center of momentum frame is given by

$$\begin{aligned} L(\ominus\mathbf{v}_0)\sum_{k=1}^{N} m_k \begin{pmatrix} \gamma_{\mathbf{v}_k} \\ \gamma_{\mathbf{v}_k}\mathbf{v}_k \end{pmatrix} &= L(\ominus\mathbf{v}_0) m_0 \begin{pmatrix} \gamma_{\mathbf{v}_0} \\ \gamma_{\mathbf{v}_0}\mathbf{v}_0 \end{pmatrix} \\ &= m_0 \begin{pmatrix} \gamma_{\ominus\mathbf{v}_0\oplus\mathbf{v}_0} \\ \gamma_{\ominus\mathbf{v}_0\oplus\mathbf{v}_0}(\ominus\mathbf{v}_0\oplus\mathbf{v}_0) \end{pmatrix} = m_0 \begin{pmatrix} 1 \\ \mathbf{0} \end{pmatrix}, \end{aligned} \tag{3.15}$$

noting that $\gamma_{\ominus\mathbf{v}_0\oplus\mathbf{v}_0} = \gamma_\mathbf{0} = 1$.

3.4 The Resultant Relativistically Invariant Mass Theorem

Lemma 3.1 below presents an identity that we need for the proof of the *Resultant Relativistically Invariant Mass Theorem* 3.2.

Lemma 3.1 *Let N be any positive integer, and let $m_k \in \mathbb{R}$ and $\mathbf{v}_k \in \mathbb{R}_c^n$, $k = 1, \ldots, N$, be N real numbers and N points of an Einstein gyrogroup $\mathbb{R}_c^n = (\mathbb{R}_c^n, \oplus)$. Then*

$$\left(\sum_{k=1}^{N} m_k\gamma_{\mathbf{v}_k}\frac{\mathbf{v}_k}{c}\right)^2 = \left(\sum_{k=1}^{N} m_k\gamma_{\mathbf{v}_k}\right)^2 - \left\{\left(\sum_{k=1}^{N} m_k\right)^2 + 2\sum_{\substack{j,k=1\\ j<k}}^{h} m_j m_k(\gamma_{\ominus\mathbf{v}_j\oplus\mathbf{v}_k} - 1)\right\}. \tag{3.16}$$

Proof The proof is given by the following chain of equations, which are numbered for subsequent explanation:

$$
\begin{aligned}
&\left(\sum_{k=1}^{N} m_k \gamma_{\mathbf{v}_k} \frac{\mathbf{v}_k}{c}\right)^2 \\
&\overset{(1)}{=\!=\!=} \sum_{k=1}^{N} m_k^2 \gamma_{\mathbf{v}_k}^2 \frac{\mathbf{v}_k^2}{c^2} + 2 \sum_{\substack{j,k=1\\ j<k}}^{h} m_j m_k \gamma_{\mathbf{v}_j} \gamma_{\mathbf{v}_k} \frac{\mathbf{v}_j \cdot \mathbf{v}_k}{c^2} \\
&\overset{(2)}{=\!=\!=} \sum_{k=1}^{N} m_k^2 (\gamma_{\mathbf{v}_k}^2 - 1) + 2 \sum_{\substack{j,k=1\\ j<k}}^{h} m_j m_k (\gamma_{\mathbf{v}_j} \gamma_{\mathbf{v}_k} - \gamma_{\ominus\mathbf{v}_j \oplus \mathbf{v}_k}) \\
&\overset{(3)}{=\!=\!=} \sum_{k=1}^{N} m_k^2 \gamma_{\mathbf{v}_k}^2 - \sum_{k=1}^{N} m_k^2 + 2 \sum_{\substack{j,k=1\\ j<k}}^{h} m_j m_k \gamma_{\mathbf{v}_j} \gamma_{\mathbf{v}_k} \\
&\qquad - 2 \sum_{\substack{j,k=1\\ j<k}}^{h} m_j m_k \gamma_{\ominus\mathbf{v}_j \oplus \mathbf{v}_k} \\
&\overset{(4)}{=\!=\!=} \left(\sum_{k=1}^{N} m_k \gamma_{\mathbf{v}_k}\right)^2 - \left\{\sum_{k=1}^{N} m_k^2 + 2 \sum_{\substack{j,k=1\\ j<k}}^{h} m_j m_k \gamma_{\ominus\mathbf{v}_j \oplus \mathbf{v}_k}\right\} \\
&\overset{(5)}{=\!=\!=} \left(\sum_{k=1}^{N} m_k \gamma_{\mathbf{v}_k}\right)^2 - \left\{\left(\sum_{k=1}^{N} m_k\right)^2 + 2 \sum_{\substack{j,k=1\\ j<k}}^{h} m_j m_k (\gamma_{\ominus\mathbf{v}_j \oplus \mathbf{v}_k} - 1)\right\}.
\end{aligned}
\tag{3.17}
$$

The assumption $\mathbf{v}_k \in \mathbb{R}_c^n$ implies, by (1.3), p. 4, that all gamma factors in (3.16)–(3.17) are real and greater than 1. Derivation of the numbered equalities in (3.17) follows:

1. This equation is obtained by an expansion of the square of a sum of vectors in $\mathbb{R}^n$.
2. Follows from Item 1 by (1.9)–(1.10), p. 5.
3. Follows from Item 2 by an obvious expansion.
4. Follows from Item 3 by an expansion of the square of a sum of real numbers.
5. Follows from Item 4 by an expansion of another square of a sum of real numbers. □

Einstein velocity addition law (1.2), p. 4, admits the following theorem:

Theorem 3.2 (Resultant Relativistically Invariant Mass Theorem) *Let* $(\mathbb{R}_c^n, \oplus)$ *be an Einstein gyrogroup, and let* $m_k \in \mathbb{R}$ *and* $\mathbf{v}_k \in \mathbb{R}_c^n$, $k = 1, 2, \ldots, N$, *be N real* *numbers and N elements of* $\mathbb{R}^n_c$ *satisfying*

$$\sum_{k=1}^{N} m_k \gamma_{\mathbf{v}_k} \neq 0. \tag{3.18}$$

Furthermore, let

$$\sum_{k=1}^{N} m_k \begin{pmatrix} \gamma_{\mathbf{v}_k} \\ \gamma_{\mathbf{v}_k} \mathbf{v}_k \end{pmatrix} = m_0 \begin{pmatrix} \gamma_{\mathbf{v}_0} \\ \gamma_{\mathbf{v}_0} \mathbf{v}_0 \end{pmatrix} \tag{3.19}$$

be an $(n+1)$*-vector equation for the two unknowns* $m_0 \in \mathbb{R}$ *and* $\mathbf{v}_0 \in \mathbb{R}^n$.

Then (3.19) *possesses a unique solution* $(m_0, \mathbf{v}_0)$, $m_0 \neq 0$, $\mathbf{v}_0 \in \mathbb{R}^n_c$, *satisfying the following three identities for all* $\mathbf{w} \in \mathbb{R}^n_c$ *(including, in particular, the interesting special case of* $\mathbf{w} = \mathbf{0}$*):*

$$\mathbf{w} \oplus \mathbf{v}_0 = \frac{\sum_{k=1}^{N} m_k \gamma_{\mathbf{w} \oplus \mathbf{v}_k} (\mathbf{w} \oplus \mathbf{v}_k)}{\sum_{k=1}^{N} m_k \gamma_{\mathbf{w} \oplus \mathbf{v}_k}}, \tag{3.20}$$

$$\gamma_{\mathbf{w} \oplus \mathbf{v}_0} = \frac{\sum_{k=1}^{N} m_k \gamma_{\mathbf{w} \oplus \mathbf{v}_k}}{m_0}, \tag{3.21}$$

$$\gamma_{\mathbf{w} \oplus \mathbf{v}_0} (\mathbf{w} \oplus \mathbf{v}_0) = \frac{\sum_{k=1}^{N} m_k \gamma_{\mathbf{w} \oplus \mathbf{v}_k} (\mathbf{w} \oplus \mathbf{v}_0)}{m_0}, \tag{3.22}$$

where

$$m_0 = \sqrt{\left(\sum_{k=1}^{N} m_k\right)^2 + 2 \sum_{\substack{j,k=1 \\ j<k}}^{N} m_j m_k (\gamma_{\ominus(\mathbf{w} \oplus \mathbf{v}_j) \oplus (\mathbf{w} \oplus \mathbf{v}_k)} - 1)}. \tag{3.23}$$

Proof Following (3.18), we assume, without loss of generality, that

$$\sum_{k=1}^{N} m_k \gamma_{\mathbf{v}_k} > 0 \tag{3.24}$$

(otherwise, we replace each m_k by $-m_k$, resulting in the replacement of m_0 by $-m_0$). Let us consider the following four equations, (3.25)–(3.28), which are specialized from (3.20)–(3.23) by taking $\mathbf{w} = \mathbf{0}$:

$$\mathbf{v}_0 = \frac{\sum_{k=1}^{N} m_k \gamma_{\mathbf{v}_k} \mathbf{v}_k}{\sum_{k=1}^{N} m_k \gamma_{\mathbf{v}_k}}, \tag{3.25}$$

$$\gamma_{\mathbf{v}_0} = \frac{\sum_{k=1}^{N} m_k \gamma_{\mathbf{v}_k}}{m_0}, \tag{3.26}$$

$$\gamma_{\mathbf{v}_0}\mathbf{v}_0 = \frac{\sum_{k=1}^{N} m_k \gamma_{\mathbf{v}_k} \mathbf{v}_0}{m_0}, \tag{3.27}$$

where

$$m_0 = \sqrt{\left(\sum_{k=1}^{N} m_k\right)^2 + 2\sum_{\substack{j,k=1\\ j<k}}^{N} m_j m_k (\gamma_{\ominus \mathbf{v}_j \oplus \mathbf{v}_k} - 1)}. \tag{3.28}$$

The proof of this theorem consists of two parts. In the first part of the proof, we show that if (3.19) for the unknowns $\mathbf{v}_0 \in \mathbb{R}^n$ and $m_0 \in \mathbb{R}$ possesses a solution, then the solution must be given uniquely by $\mathbf{v}_0$ of (3.25) and m_0 of (3.28), with $\mathbf{v}_0 \in \mathbb{R}^n_c$ and $m_0 > 0$, satisfying (3.26)–(3.27).

In the second part of the proof, we show that $\mathbf{v}_0$ of (3.25) and m_0 of (3.28), indeed, form a solution of (3.19) for the unknowns $\mathbf{v}_0 \in \mathbb{R}^n_c$ and $m_0 > 0$, and that the solution satisfies (3.20)–(3.23).

Part I In this part of the proof, we assume that there exist $m_0 \in \mathbb{R}$ and $\mathbf{v}_0 \in \mathbb{R}^n$ that satisfy (3.19). Then, the norms of the two sides of (3.19) are equal while, by (3.11), the norm of the right-hand side of (3.19) is m_0. Hence, the norm of the left-hand side of (3.19) equals m_0 as well, obtaining the following chain of equations, which are numbered for subsequent explanation:

$$\begin{aligned}
m_0^2 &\overset{(1)}{=\!=} \left\| \sum_{k=1}^{N} m_k \begin{pmatrix} \gamma_{\mathbf{v}_k} \\ \gamma_{\mathbf{v}_k}\mathbf{v}_k \end{pmatrix} \right\|^2 \\
&\overset{(2)}{=\!=} \left\| \begin{pmatrix} \sum_{k=1}^{N} m_k \gamma_{\mathbf{v}_k} \\ \sum_{k=1}^{N} m_k \gamma_{\mathbf{v}_k}\mathbf{v}_k \end{pmatrix} \right\|^2 \\
&\overset{(3)}{=\!=} \left(\sum_{k=1}^{N} m_k \gamma_{\mathbf{v}_k}\right)^2 - \left(\sum_{k=1}^{N} m_k \gamma_{\mathbf{v}_k} \frac{\mathbf{v}_k}{c}\right)^2 \\
&\overset{(4)}{=\!=} \left(\sum_{k=1}^{N} m_k\right)^2 + 2\sum_{\substack{j,k=1\\ j<k}}^{h} m_j m_k \gamma_{\ominus \mathbf{v}_j \oplus \mathbf{v}_k} - 1.
\end{aligned} \tag{3.29}$$

Derivation of the numbered equalities in (3.29) follows:

1. This equation follows from the result that the norm of the left-hand side of (3.19) equals the norm of the right-hand side of (3.19), the latter being m_0 by (3.11).
2. Follows from Item 1 by the common "four-vector" addition of $(n+1)$-vectors (where $n = 3$ in physical applications).
3. Follows from Item 2 by (3.10).
4. Follows from Item 3 by Identity (3.16) of Lemma 3.1.

It follows from the upper entry of (3.19), along with Assumption (3.24) that

$$m_0 > 0. \tag{3.30}$$

We thus obtained in (3.29) the desired equation, (3.28), for m_0.

Hence, if $m_0 \in \mathbb{R}$ and $\mathbf{v}_0 \in \mathbb{R}^n$ that satisfy (3.19) exist, m_0 is positive and must be given by (3.28). Clearly, $\mathbf{v}_0 \in \mathbb{R}^n_c$ since $\gamma_{\mathbf{v}_0}$ is real, as we see from (3.19).

By assumption, $\mathbf{v}_0$ satisfies (3.19). Equation (3.19) is equivalent to two equations, formed by the upper entry and by the lower entry of (3.19). Dividing the lower entry of (3.19) by its upper entry, noting that $m_0 \neq 0$ by (3.30), we obtain (3.25).

Similarly, dividing the upper entry of (3.19) by $m_0 > 0$ we obtain (3.26), and dividing the lower entry of (3.19) by $m_0 > 0$ we obtain (3.27).

Hence, if $m_0 \in \mathbb{R}$ and $\mathbf{v}_0 \in \mathbb{R}^n$ that satisfy (3.19) exist, then $m_0 > 0$, $\mathbf{v}_0 \in \mathbb{R}^n_c$, and they must be given by (3.28) and (3.25), and satisfy (3.26)–(3.27).

Part II In Part I, we have shown that if (3.19) possesses a solution for the unknowns $\mathbf{v}_0 \in \mathbb{R}^n$ and $m_0 \in \mathbb{R}$, then $\mathbf{v}_0 \in \mathbb{R}^n_c$ is given uniquely by (3.25) and $m_0 > 0$ is given uniquely by (3.28), satisfying (3.26)–(3.27). We will now show that, indeed, $\mathbf{v}_0 \in \mathbb{R}^n_c$, given by (3.25), and $m_0 > 0$, given by (3.28), form a solution of (3.19), and that the solution satisfies (3.20)–(3.23). Accordingly, in this second part of the proof we assume that $\mathbf{v}_0 \in \mathbb{R}^n_c$ and $m_0 > 0$ are given by (3.25) and (3.28), and we will prove that the pair $(m_0, \mathbf{v}_0)$ forms a solution of (3.19).

It follows from Identity (3.16) of Lemma 3.1, along with m_0 of (3.28) that

$$\left(\sum_{k=1}^{N} m_k \gamma_{\mathbf{v}_k} \frac{\mathbf{v}_k}{c}\right)^2 = \left(\sum_{k=1}^{N} m_k \gamma_{\mathbf{v}_k}\right)^2 - m_0^2. \tag{3.31}$$

Hence, by (3.25) and (3.31), we have the following chain of equations, which are numbered for subsequent explanation:

$$\begin{aligned} \frac{\mathbf{v}_0^2}{c^2} &\overbrace{=\!=}^{(1)} \frac{(\sum_{k=1}^{N} m_k \gamma_{\mathbf{v}_k} \frac{\mathbf{v}_k}{c})^2}{(\sum_{k=1}^{N} m_k \gamma_{\mathbf{v}_k})^2} \\ &\overbrace{=\!=}^{(2)} \frac{(\sum_{k=1}^{N} m_k \gamma_{\mathbf{v}_k})^2 - m_0^2}{(\sum_{k=1}^{N} m_k \gamma_{\mathbf{v}_k})^2} \\ &=\!= 1 - \frac{m_0^2}{(\sum_{k=1}^{N} m_k \gamma_{\mathbf{v}_k})^2}. \end{aligned} \tag{3.32}$$

Derivation of the numbered equalities in (3.32) follows:

1. This equation is given by Assumption (3.25).
2. Follows from Item 1 by (3.31).

It follows from (3.32) that

$$\gamma_{\mathbf{v}_0} = \frac{1}{\sqrt{1-\frac{\mathbf{v}_0^2}{c^2}}} = \frac{\sum_{k=1}^{N} m_k \gamma_{\mathbf{v}_k}}{m_0}, \tag{3.33}$$

thus verifying (3.26).

Following (3.26) and (3.25), we have

$$\gamma_{\mathbf{v}_0} \mathbf{v}_0 = \frac{\sum_{k=1}^{N} m_k \gamma_{\mathbf{v}_k} \mathbf{v}_k}{m_0}, \tag{3.34}$$

thus verifying (3.27).

Finally, (3.33) implies that m_0 and $\mathbf{v}_0$ satisfy the upper entry of (3.19) and, similarly, (3.34) implies that m_0 and $\mathbf{v}_0$ satisfy the lower entry of (3.19). Hence, the pair consisting of m_0 and $\mathbf{v}_0$ forms a solution of (3.19). We have thus shown that $\mathbf{v}_0 \in \mathbb{R}_c^n$ and $m_0 > 0$ given by (3.25) and (3.28) form a solution of (3.19), and that this solution satisfies (3.26)–(3.28).

To complete the proof it remains to show that the pair $(m_0, \mathbf{v}_0)$ satisfies (3.20)–(3.23) as well.

Let us first show that m_0 given by (3.28) is given by (3.23) as well. Indeed, following (1.3), p. 4, (1.33), p. 10, and (1.65), p. 18, we have

$$\gamma_{\ominus(\mathbf{w}\oplus\mathbf{v}_j)\oplus(\mathbf{w}\oplus\mathbf{v}_k)} = \gamma_{\ominus\mathbf{v}_j\oplus\mathbf{v}_k}, \tag{3.35}$$

implying that the right-hand sides of (3.23) and (3.28) are equal, so that m_0 is independent of $\mathbf{w} \in \mathbb{R}_c^n$, as desired. As such, m_0 is given by each of (3.23) and (3.28).

We have thus shown that the unique solution of (3.19) is formed by $\mathbf{v}_0 \in \mathbb{R}_c^n$ and $m_0 > 0$ that are given by (3.25) and (3.28), and that the solution satisfies (3.26)–(3.27). It, therefore, remains to show that the solution satisfies (3.20)–(3.22) as well.

Applying the Lorentz boost $L(\mathbf{w})$, $\mathbf{w} \in \mathbb{R}_c^n$, to each side of (3.19), we have

$$L(\mathbf{w}) \sum_{k=1}^{N} m_k \begin{pmatrix} \gamma_{\mathbf{v}_k} \\ \gamma_{\mathbf{v}_k} \mathbf{v}_k \end{pmatrix} = L(\mathbf{w}) m_0 \begin{pmatrix} \gamma_{\mathbf{v}_0} \\ \gamma_{\mathbf{v}_0} \mathbf{v}_0 \end{pmatrix}. \tag{3.36}$$

Following the linearity of the Lorentz boost, illustrated in (3.8) and (3.9), (3.36) can be written as

$$\sum_{k=1}^{N} m_k \begin{pmatrix} \gamma_{\mathbf{w}\oplus\mathbf{v}_k} \\ \gamma_{\mathbf{w}\oplus\mathbf{v}_k} (\mathbf{w}\oplus\mathbf{v}_k) \end{pmatrix} = m_0 \begin{pmatrix} \gamma_{\mathbf{w}\oplus\mathbf{v}_0} \\ \gamma_{\mathbf{w}\oplus\mathbf{v}_0} (\mathbf{w}\oplus\mathbf{v}_0) \end{pmatrix}. \tag{3.37}$$

Equation (3.37) is identical with (3.19) in which $\mathbf{v}_k \in \mathbb{R}_c^n$ is replaced by $\mathbf{w}\oplus\mathbf{v}_k \in \mathbb{R}_c^n$, $k = 0, 1, \ldots, N$.

But, the unique solution of (3.19) is the pair $(m_0 > 0, \mathbf{v}_0 \in \mathbb{R}_c^n)$ that satisfies (3.25)–(3.28). Hence, the unique solution of (3.37) is the pair $(m_0 > 0, \mathbf{w}\oplus\mathbf{v}_0 \in \mathbb{R}_c^n)$

that satisfies (3.20)–(3.23). Hence, the unique solution ($m_0 > 0, \mathbf{v}_0 \in \mathbb{R}^n_c$) of (3.19) satisfies not only (3.25)–(3.28) but, more generally, (3.20)–(3.23), and the proof is complete. □

In physical applications to particle systems, the real numbers m_k in Theorem 3.2 represent particle masses. As such, they are positive so that Assumption (3.24) is satisfied. However, anticipating applications of Theorem 3.2 to barycentric coordinates in hyperbolic geometry, in Chap. 4, we need the validity of Theorem 3.2 for real numbers m_k that need not be positive as well.

We have thus established in Theorem 3.2 the following four results concerning an isolated system S, (3.12),

$$S = S(m_k, \mathbf{v}_k, \Sigma_{\mathbf{0}}, k = 1, \dots, N) \tag{3.38}$$

of N noninteracting material particles the kth particle of which has invariant mass $m_k > 0$ and velocity $\mathbf{v}_k \in \mathbb{R}^n_c$ relative to an inertial frame $\Sigma_{\mathbf{0}}$, $k = 1, \dots, N$:

1. The relativistically invariant (or, rest) mass m_0 of the system S is given by

$$m_0 = \sqrt{\left(\sum_{k=1}^{N} m_k\right)^2 + 2\sum_{\substack{j,k=1\\ j<k}}^{N} m_j m_k (\gamma_{\ominus\mathbf{v}_j\oplus\mathbf{v}_k} - 1)} \tag{3.39}$$

   according to (3.23) with $\mathbf{w} = \mathbf{0}$.
2. The relativistic mass of the system S is

$$m_0\gamma_{\mathbf{v}_0} \tag{3.40}$$

   relative to the rest frame $\Sigma_{\mathbf{0}}$, where $\mathbf{v}_0$ is the velocity of the center of momentum frame of S relative to $\Sigma_{\mathbf{0}}$, given by

$$\mathbf{v}_0 = \frac{\sum_{k=1}^{N} m_k\gamma_{\mathbf{v}_k}\mathbf{v}_k}{\sum_{k=1}^{N} m_k\gamma_{\mathbf{v}_k}} \tag{3.41}$$

   according to (3.20) with $\mathbf{w} = \mathbf{0}$.
3. Like energy and momentum, the relativistic mass is additive, that is, in particular for the system S relative to the rest frame $\Sigma_{\mathbf{0}}$, by (3.21) with $\mathbf{w} = \mathbf{0}$,

$$m_0\gamma_{\mathbf{v}_0} = \sum_{k=1}^{N} m_k\gamma_{\mathbf{v}_k}. \tag{3.42}$$

4. The relativistic mass $m_0\gamma_{\mathbf{v}_0}$ of a system meshes extraordinarily well with the Minkowskian four-vector formalism of special relativity. In particular, for the system S relative to the rest frame $\Sigma_{\mathbf{0}}$, we have, by (3.19),

$$\sum_{k=1}^{N} \begin{pmatrix} m_k\gamma_{\mathbf{v}_k} \\ m_k\gamma_{\mathbf{v}_k}\mathbf{v}_k \end{pmatrix} = \begin{pmatrix} m_0\gamma_{\mathbf{v}_0} \\ m_0\gamma_{\mathbf{v}_0}\mathbf{v}_0 \end{pmatrix}, \tag{3.43}$$

where m_0 and $\mathbf{v}_0$ are given uniquely by (3.39) and (3.41).

Thus, the relativistically invariant mass m_0 of a particle system S in (3.39) gives rise to its associated relativistic mass $m_0\gamma_{\mathbf{v}_0}$ relative to the rest frame $\Sigma_\mathbf{0}$. The latter, in turn, brings in (3.43) the concept of the relativistic mass into conformity with the Minkowskian four-vector formalism of special relativity. Moreover, we will see in Sects. 3.7 and 3.8 that the relativistically invariant mass m_0 of a particle system S provides a natural interpretation of observations in astrophysics and in particle physics.

To appreciate the power and elegance of Theorem 3.2 in relativistic mechanics in terms of novel analogies that it shares with familiar results in classical mechanics, we present below the classical counterpart, Theorem 3.3, of Theorem 3.2. The latter is obtained from the former by approaching the Newtonian limit when c tends to infinity. The resulting Theorem 3.3 is immediate, and its importance in classical mechanics is well-known.

Theorem 3.3 (Resultant Newtonian Invariant Mass Theorem) *Let* $(\mathbb{R}^n, +)$ *be a Euclidean n-space, and let* $m_k \in \mathbb{R}$ *and* $\mathbf{v}_k \in \mathbb{R}^n$, $k = 1, 2, \dots, N$, *be N real numbers and N elements of* $\mathbb{R}^n$ *satisfying*

$$\sum_{k=1}^{N} m_k \neq 0. \tag{3.44}$$

Furthermore, let

$$\sum_{k=1}^{N} m_k \begin{pmatrix} 1 \\ \mathbf{v}_k \end{pmatrix} = m_0 \begin{pmatrix} 1 \\ \mathbf{v}_0 \end{pmatrix} \tag{3.45}$$

be an $(n+1)$*-vector equation for the two unknowns* $m_0 \in \mathbb{R}$ *and* $\mathbf{v}_0 \in \mathbb{R}^n$.

Then (3.45) *possesses a unique solution* $(m_0, \mathbf{v}_0)$, $m_0 \neq 0$, *satisfying the following equations for all* $\mathbf{w} \in \mathbb{R}^n$ *(including, in particular, the interesting special case of* $\mathbf{w} = \mathbf{0}$*):*

$$\mathbf{w} + \mathbf{v}_0 = \frac{\sum_{k=1}^{N} m_k(\mathbf{w} + \mathbf{v}_k)}{\sum_{k=1}^{N} m_k} \tag{3.46}$$

and

$$m_0 = \sum_{k=1}^{N} m_k. \tag{3.47}$$

Proof While the proof of Theorem 3.3 is trivial, our point is to present a proof that emphasizes how Theorem 3.3 is derived from Theorem 3.2. Indeed, in the limit as $c \to \infty$, the results of Theorem 3.2 tend to corresponding results of Theorem 3.3, noting that in this limit gamma factors tend to 1. Accordingly, Theorem 3.3 is a special case of Theorem 3.2 corresponding to $c = \infty$. □

In physical applications to particle systems, the real numbers m_k in Theorem 3.3 represent particle masses and, hence, they are positive. However, anticipating applications of Theorem 3.3 to barycentric coordinates in Euclidean geometry, in Chap. 4, we need the validity of Theorem 3.3 for real numbers m_k that need not be positive as well.

Identity (3.46) of Theorem 3.3 is immediate. Yet, it is geometrically important. The geometric importance of the validity of (3.46) for all $\mathbf{w} \in \mathbb{R}^n$ lies on its implication that the velocity $\mathbf{v}_0$ of the center of momentum frame of a particle system relative to a given inertial rest frame in classical mechanics is independent of the choice of the origin of the classical velocity space $\mathbb{R}^n$ with its underlying standard model of Euclidean geometry.

Unlike Identity (3.46) of Theorem 3.3, which is immediate, its counterpart in Theorem 3.2, Identity (3.20), is not immediate. Yet, in full analogy with Theorem 3.3, the validity of Identity (3.20) in Theorem 3.2 for all $\mathbf{w} \in \mathbb{R}^n_c$ is geometrically important. This geometric importance of Identity (3.20) lies on its implication that the velocity $\mathbf{v}_0$ of the center of momentum frame of a particle system relative to a given inertial rest frame in relativistic mechanics is independent of the choice of the origin of the relativistic velocity space $\mathbb{R}^n_c$ with its underlying Cartesian–Beltrami–Klein ball model of hyperbolic geometry.

3.5 Mass and Velocity of Particle Systems

In this section, we emphasize the analogies that the classical mass and velocity of a particle system share with their relativistic counterparts. Let

$$\begin{aligned} S_{newton} &= S_{newton}\big(m_k, \mathbf{v}_k \in \mathbb{R}^n, \Sigma_{\mathbf{0}}, k = 1, \dots, N\big), \\ S_{einstein} &= S_{einstein}\big(m_k, \mathbf{v}_k \in \mathbb{R}^n_c, \Sigma_{\mathbf{0}}, k = 1, \dots, N\big) \end{aligned} \tag{3.48}$$

be a Newtonian particle system and its corresponding Einsteinian particle system. Each of these is an isolated system of N noninteracting material particles the kth particle of which has invariant mass $m_k > 0$ and velocity $\mathbf{v}_k$ relative to a rest frame $\Sigma_{\mathbf{0}}$. These velocities are Newtonian, $\mathbf{v}_k \in \mathbb{R}^n$, for the Newtonian system S_{newton} and Einsteinian, $\mathbf{v}_k \in \mathbb{R}^n_c$, for the Einsteinian system $S_{einstein}$.

The mass m_0 in Identity (3.45) of Theorem 3.3 is the Newtonian mass of the Newtonian particle system S_{newton}. It is given by, (3.47),

$$m_0 = \sum_{k=1}^{N} m_k. \tag{3.49}$$

The velocity $\mathbf{v}_0 \in \mathbb{R}^n$ in Identity (3.45) is the Newtonian velocity of the center of momentum frame of the system S_{newton} relative to $\Sigma_{\mathbf{0}}$. It is given by (3.46) with $\mathbf{w} = \mathbf{0}$,

$$\mathbf{v}_0 = \frac{\sum_{k=1}^{N} m_k \mathbf{v}_k}{\sum_{k=1}^{N} m_k} \tag{3.50}$$

satisfying (3.46), i.e.,

$$\mathbf{w}+\mathbf{v}_0=\frac{\sum_{k=1}^{N} m_k(\mathbf{w}+\mathbf{v}_k)}{\sum_{k=1}^{N} m_k} \tag{3.51}$$

for all $\mathbf{w}\in\mathbb{R}^n$.

In full analogy, the mass m_0 in Identity (3.19), p. 66, of Theorem 3.2 is the Einsteinian rest mass of the Einsteinian particle system $S_{einstein}$. It is given by (3.28) as

$$m_0 = \sqrt{\left(\sum_{k=1}^{N} m_k\right)^2 + 2\sum_{\substack{j,k=1\\ j<k}}^{N} m_j m_k(\gamma_{\ominus\mathbf{v}_j\oplus\mathbf{v}_k}-1)} \tag{3.52}$$

satisfying (3.23), i.e.,

$$m_0 = \sqrt{\left(\sum_{k=1}^{N} m_k\right)^2 + 2\sum_{\substack{j,k=1\\ j<k}}^{N} m_j m_k(\gamma_{\ominus(\mathbf{w}\oplus\mathbf{v}_j)\oplus(\mathbf{w}\oplus\mathbf{v}_k)}-1)} \tag{3.53}$$

for all $\mathbf{w}\in\mathbb{R}_c^n$.

The velocity $\mathbf{v}_0\in\mathbb{R}_c^n$ in Identity (3.19) is the Einsteinian velocity of the center of momentum frame of the system $S_{einstein}$ relative to $\Sigma_\mathbf{0}$. It is given by (3.25) as

$$\mathbf{v}_0=\frac{\sum_{k=1}^{N} m_k\gamma_{\mathbf{v}_k}\mathbf{v}_k}{\sum_{k=1}^{N} m_k\gamma_{\mathbf{v}_k}} \tag{3.54}$$

satisfying (3.20), i.e.,

$$\mathbf{w}\oplus\mathbf{v}_0=\frac{\sum_{k=1}^{N} m_k\gamma_{\mathbf{w}\oplus\mathbf{v}_k}(\mathbf{w}\oplus\mathbf{v}_k)}{\sum_{k=1}^{N} m_k\gamma_{\mathbf{w}\oplus\mathbf{v}_k}} \tag{3.55}$$

for all $\mathbf{w}\in\mathbb{R}_c^n$.

Accordingly, the relativistic mass of the system $S_{einstein}$ is $m_0\gamma_{\mathbf{v}_0}$.

In the Newtonian limit of large c, $c\to\infty$, gamma factors tend to 1. Hence, in that limit the relativistically invariant rest mass m_0 in (3.52)–(3.53) tends to its Newtonian counterpart m_0 in (3.49), and the relativistic center of momentum velocity $\mathbf{v}_0\in\mathbb{R}_c^n$ in (3.54)–(3.55) tends to its corresponding classical center of momentum velocity $\mathbf{v}_0\in\mathbb{R}^n$ in (3.50)–(3.51).

3.6 The Relativistic Mass is Additive

Suppose that the system S, (3.38), is made up of M subsystems each itself a system of particles. Let $m_{0,p}$ be the relativistically invariant mass and $\mathbf{v}_{0,p}$ the center of momentum frame velocity of the pth subsystem, $p = 1, \ldots, M$, so that the relativistic mass of the pth subsystem is $m_{0,p}\gamma_{\mathbf{v}_{0,p}}$.

Then, the relativistic mass $m_0\gamma_{\mathbf{v}_0}$ of the system S, given by (3.39)–(3.41), is additive, that is, it possesses the relativistic mass additivity property

$$m_0\gamma_{\mathbf{v}_0} = \sum_{p=1}^{M} m_{0,p}\gamma_{\mathbf{v}_{0,p}}. \tag{3.56}$$

For simplicity, we prove the relativistic mass additivity property (3.56) for the case of $M = 2$ subsystems, the proof for any $M > 2$ being similar.

Let us, therefore, view the system S of N particles, (3.38)–(3.41), with $N \geq 3$, as a system of the two subsystems S_1 and S_2,

$$\begin{aligned} S_1 &= S_1(m_k, \mathbf{v}_k, \Sigma_\mathbf{0}, k = 1, \ldots, N_1), \\ S_2 &= S_2(m_k, \mathbf{v}_k, \Sigma_\mathbf{0}, k = N_1 + 1, \ldots, N) \end{aligned} \tag{3.57}$$

for any fixed N_1, $1 < N_1 < N$.

Then, the relativistically invariant masses $m_{0,1}$ and $m_{0,2}$ of the subsystems S_1 and S_2 and their center of momentum frame velocities, $\mathbf{v}_{0,1}$ and $\mathbf{v}_{0,2}$ relative to $\Sigma_\mathbf{0}$, respectively, are

$$\begin{aligned} m_{0,1} &= \sqrt{\left(\sum_{k=1}^{N_1} m_k\right)^2 + 2\sum_{\substack{j,k=1\\ j<k}}^{N_1} m_j m_k(\gamma_{\ominus\mathbf{v}_j\oplus\mathbf{v}_k} - 1)}, \\ m_{0,2} &= \sqrt{\left(\sum_{k=N_1+1}^{N} m_k\right)^2 + 2\sum_{\substack{j,k=N_1+1\\ j<k}}^{N} m_j m_k(\gamma_{\ominus\mathbf{v}_j\oplus\mathbf{v}_k} - 1)}, \end{aligned} \tag{3.58}$$

and

$$\begin{aligned} \mathbf{v}_{0,1} &= \frac{\sum_{k=1}^{N_1} m_k\gamma_{\mathbf{v}_k}\mathbf{v}_k}{\sum_{k=1}^{N_1} m_k\gamma_{\mathbf{v}_k}}, \\ \mathbf{v}_{0,2} &= \frac{\sum_{k=N_1+1}^{N} m_k\gamma_{\mathbf{v}_k}\mathbf{v}_k}{\sum_{k=N_1+1}^{N} m_k\gamma_{\mathbf{v}_k}}, \end{aligned} \tag{3.59}$$

possessing the relativistic mass additivity property

$$m_{0,1}\gamma_{\mathbf{v}_{0,1}} + m_{0,2}\gamma_{\mathbf{v}_{0,2}} = m_0\gamma_{\mathbf{v}_0}. \tag{3.60}$$

The proof of the additivity property (3.60) follows from (3.42) immediately. Indeed, by applying the identity in (3.42) to each of the particle systems S_1, S_2 and S, we have the chain of equations

$$\begin{aligned} m_{0,1}\gamma_{\mathbf{v}_{0,1}} + m_{0,2}\gamma_{\mathbf{v}_{0,2}} &= \sum_{k=1}^{N_1} m_k\gamma_{\mathbf{v}_k} + \sum_{k=N_1+1}^{N} m_k\gamma_{\mathbf{v}_k} \\ &= \sum_{k=1}^{N} m_k\gamma_{\mathbf{v}_k} \\ &= m_0\gamma_{\mathbf{v}_0}\,. \end{aligned} \tag{3.61}$$

The relativistically invariant mass m_0 of a system of particles S, (3.38)–(3.41), leads to its associated relativistic mass $m_0\gamma_{\mathbf{v}_0}$, (3.42). We thus see that, in special relativity, while relativistically invariant mass m_0 is not additive [36], relativistic mass, $m_0\gamma_{\mathbf{v}_0}$ is additive.

3.7 The Relativistically Invariant Mass of a System in Astrophysics

The resultant relativistically invariant mass m_0, (3.39),

$$m_0 = \sqrt{\left(\sum_{k=1}^{N} m_k\right)^2 + 2\sum_{\substack{j,k=1\\ j<k}}^{N} m_j m_k(\gamma_{\ominus\mathbf{v}_j\oplus\mathbf{v}_k} - 1)} \tag{3.62}$$

of a particle system $S = S(m_k, \mathbf{v}_k, \Sigma_\mathbf{0}, N)$ comprises two distinct kinds of relativistically invariant mass that represent the Newtonian contribution and the relativistic contribution. These two distinct kinds of mass are:

1. The Newtonian mass m_{newton},

$$m_{newton} := \sum_{k=1}^{N} m_k \tag{3.63}$$

   which is the sum of the invariant, rest masses of the particles that constitute the system S, as in (3.13).
2. The dark mass m_{dark},

$$m_{dark} := \sqrt{2\sum_{\substack{j,k=1\\ j<k}}^{N} m_j m_k(\gamma_{\ominus\mathbf{v}_j\oplus\mathbf{v}_k} - 1)}\,. \tag{3.64}$$

The dark mass of a particle system S, given by (3.64), depends on the *velocity dispersion* of S, that is, on the spread of internal velocities $\mathbf{v}_{jk} = \ominus\mathbf{v}_j\oplus\mathbf{v}_k$, $1 \le j <$

$k \leq N$, of the constituent particles of S relative to each other. In other words, the dark mass in (3.64) measures the extent to which the system S deviates away from rigidity. Gravitationally, dark mass behaves just like ordinary mass, as postulated in cosmology [6, p. 37]. However, it is undetectable by all means other than gravity since it is fictitious, or virtual, in the sense that it is generated solely by relative motion between constituent objects of the system.

We thus see from (3.64) that the dark mass of a system may be viewed as a measure of mass that results solely from the velocity dispersion of the system. In astrophysics, the velocity dispersion of stars or galaxies in a cluster is estimated by measuring the radial velocities of selected constituents. Once the velocity distribution is known, the cluster's mass is calculated by using the *virial theorem* [5].

If one attributes the dark mass m_{dark} of a galaxy to the mass of a central dark object, a black hole, a correlation should result between the mass of a black hole and the velocity dispersion of its host galaxy. Indeed, Ferrarese and Merritt reported in 2000 that "The masses of supermassive black holes correlate almost perfectly with the velocity dispersion of their host bulges" [16]; and Gebhardt et al. remarked that the resulting relation is of interest "because it implies that central black hole mass is constrained and closely related to properties of the host galaxy's bulge" [19]. A recent improved version of the black hole mass (M)–velocity dispersion (σ) relation (called the M–σ relation) and black hole mass–luminosity (L) relation (called the M–L relation) was reported by Gültekin et al. [23].

Dark matter was introduced into cosmology as an ad hoc postulate, hypothesized to provide observed missing gravitational force [7]. In contrast, dark mass emerges here as a consequence of the covariance of Einstein's special theory of relativity, and it stems from relative motion between constituent objects of a system. All relative velocities between the constituent particles of a *rigid* system vanish, so that if the system S is rigid, then $\ominus\mathbf{v}_j\oplus\mathbf{v}_k = \mathbf{0}$, $j, k = 1, \ldots, N$. This, in turn, implies by (3.64) that the dark mass of a rigid system vanishes.

The mass m_{newton} and the dark mass m_{dark} of a system S are relativistically invariant, and are composed according to the Pythagorean formula

$$m_0 = \sqrt{m_{newton}^2 + m_{dark}^2}, \tag{3.65}$$

giving rise to the invariant resultant rest mass m_0 of the system S in (3.62)–(3.64).

3.8 The Relativistically Invariant Mass of a System in Particle Physics

Following the four-momentum in (3.4), p. 61, and the four-vector norm (3.10)–(3.11) we have

$$\left\| \begin{pmatrix} p_0 \\ \mathbf{p} \end{pmatrix} \right\| = \sqrt{\frac{E^2}{c^4} - \frac{\|\mathbf{p}\|^2}{c^2}} = m, \tag{3.66}$$

where, by (3.4),

$$\begin{aligned} p_0 &= m\gamma_{\mathbf{v}}, \\ E &= m\gamma_{\mathbf{v}} c^2, \\ \mathbf{p} &= m\gamma_{\mathbf{v}} \mathbf{v}. \end{aligned} \tag{3.67}$$

Assuming that both energy, E, and three-momentum, $\mathbf{p}$, are additive is equivalent to assuming that the four-momentum is additive. The latter assumption, in turn, led us to identity (3.14), p. 64, that we now write as

$$\begin{pmatrix} \frac{E}{c^2} \\ \mathbf{p} \end{pmatrix} = \sum_{k=1}^{N} \begin{pmatrix} \frac{E_k}{c^2} \\ \mathbf{p}_k \end{pmatrix} = \sum_{k=1}^{N} m_k \begin{pmatrix} \gamma_{\mathbf{v}_k} \\ \gamma_{\mathbf{v}_k} \mathbf{v}_k \end{pmatrix} = m_0 \begin{pmatrix} \gamma_{\mathbf{v}_0} \\ \gamma_{\mathbf{v}_0} \mathbf{v}_0 \end{pmatrix}. \tag{3.68}$$

Here, in (3.68),

$$\begin{aligned} E_k &= m_k \gamma_{\mathbf{v}_k} c^2, \\ \mathbf{p}_k &= m_k \gamma_{\mathbf{v}_k} \mathbf{v}_k \end{aligned} \tag{3.69}$$

are the energy and momentum of the kth particle of the system S, $k = 1, \ldots, N$, and accordingly,

$$\begin{aligned} E &= \sum_{k=1}^{N} E_k, \\ \mathbf{p} &= \sum_{k=1}^{N} \mathbf{p}_k \end{aligned} \tag{3.70}$$

are the energy and momentum of the system S.

Furthermore, as in (3.14), $\mathbf{v}_0$ is the velocity of the center of momentum frame of S relative to the rest frame $\Sigma_\mathbf{0}$, and m_0 is the resultant invariant mass of S.

Noting (3.11), p. 63, the norms of the two extreme sides of (3.68) give the equation

$$m_0 = \sqrt{\frac{E^2}{c^4} - \frac{\|\mathbf{p}\|^2}{c^2}} \tag{3.71}$$

where E and $\mathbf{p}$ are given by (3.70). Identity (3.71) demonstrates, by the relativistic four-vector formalism, that the resultant mass m_0 of a particle system S in (3.39) is relativistically invariant, being the norm of a four-vector.

Identity (3.71), written equivalently as

$$E^2 = m_0^2 c^4 + \|\mathbf{p}\|^2 c^2 \tag{3.72}$$

is known in particle physics as the *energy–momentum relation*. For a particle in its inertial rest frame, where $\mathbf{p} = \mathbf{0}$, Relation (3.72) reduces to Einstein's famous formula

$$E = m_0 c^2. \tag{3.73}$$

The energy–momentum relation (3.71)–(3.72) is used in particle physics to calculate the relativistically invariant mass m_0 of a system of particles in terms of the total energy E and momentum $\mathbf{p}$ of the system.

As an illustrative example, let us consider two particles with rest (or, Newtonian) masses m_1 and m_2, and velocities $\mathbf{v}_1$ and $\mathbf{v}_2$ relative to an inertial rest frame $\Sigma_\mathbf{0}$, respectively. If these particles were to collide and stick, the rest mass m_0 and the velocity $\mathbf{v}_0$ relative to $\Sigma_\mathbf{0}$ of the resulting composite particle would satisfy the four-momentum conservation law (3.14), p. 64, that is,

$$m_0 \begin{pmatrix} \gamma_{\mathbf{v}_0} \\ \gamma_{\mathbf{v}_0} \mathbf{v}_0 \end{pmatrix} = m_1 \begin{pmatrix} \gamma_{\mathbf{v}_1} \\ \gamma_{\mathbf{v}_1} \mathbf{v}_1 \end{pmatrix} + m_2 \begin{pmatrix} \gamma_{\mathbf{v}_2} \\ \gamma_{\mathbf{v}_2} \mathbf{v}_2 \end{pmatrix}. \tag{3.74}$$

Hence, by (3.39) and (3.65),

$$\begin{aligned} m_0 &= \sqrt{(m_1 + m_2)^2 + 2m_1 m_2 (\gamma_{\ominus \mathbf{v}_1 \oplus \mathbf{v}_2} - 1)} \\ &= \sqrt{m_{newton}^2 + m_{dark}^2}, \end{aligned} \tag{3.75}$$

where

$$\begin{aligned} m_{newton} &= m_1 + m_2, \\ m_{dark} &= 2m_1 m_2 (\gamma_{\ominus \mathbf{v}_1 \oplus \mathbf{v}_2} - 1) > 0, \end{aligned} \tag{3.76}$$

and, by (3.41),

$$\mathbf{v}_0 = \frac{m_1 \gamma_{\mathbf{v}_1} \mathbf{v}_1 + m_2 \gamma_{\mathbf{v}_2} \mathbf{v}_2}{m_1 \gamma_{\mathbf{v}_1} + m_2 \gamma_{\mathbf{v}_2}}. \tag{3.77}$$

Hence, the relativistic mass of the composite particle is $m_0 \gamma_{\mathbf{v}_0}$, where m_0 is given by (3.75), and $\mathbf{v}_0$ is given by (3.77).

It is clear from (3.75)–(3.76) that the Newtonian mass, m_{newton}, is conserved during the collision. It is only the total invariant mass, m_0, which is increased following the collision owing to the emergence of the dark mass m_{dark}.

Examples of particles that collide and stick, as described in (3.74)–(3.77), are observed in experimental searches for new particles in high-energy particle colliders.

We thus see that owing to the introduction of the relativistically invariant mass m_0 in (3.62), along with its Newtonian and dark mass components in (3.63)–(3.65), the concept of the relativistic mass fits well under the umbrella of the four-vector formalism of special relativity. Moreover, we see that the resulting dark mass emerges naturally not only in the interpretation of observations in astrophysics, demonstrated qualitatively in Sect. 3.7, but also in the interpretation of observations in particle physics, demonstrated qualitatively in this section. Naturally, we will find in

Sect. 4.2 of Chap. 4 that the relativistically invariant mass m_0 in (3.62) is what we need for the introduction of barycentric coordinates into hyperbolic geometry. The latter, in turn, is what we need for the determination of hyperbolic triangle centers.

3.9 Remarkable Analogies

In this section, we emphasize the analogies in Theorems 3.2, p. 65, and 3.3, p. 71, that the classical mass and center of momentum velocity of a particle system in (3.78a)–(3.78d) below share with their relativistic counterparts in (3.79a)–(3.79d) below.

Seeking a way to place the relativistic mass $m_0\gamma_{\mathbf{v}_0}$ of a particle system S under the umbrella of the Minkowskian four-vector formalism of special relativity, we have uncovered the novel, relativistically invariant, or rest, mass m_0 of a particle system, presented in (3.79d) below. Furthermore, following the discovery of m_0 in (3.62), we have uncovered remarkable analogies that Newtonian and Einsteinian mechanics share.

To see the analogies clearly, let us consider the following well known classical results, (3.78a)–(3.78d) below, which are involved in the determination of the Newtonian resultant mass m_0 and the classical center of momentum velocity of a Newtonian system of particles, and to which we will subsequently present our Einsteinian analogs that have been discovered in Theorem 3.2. Let

$$S = S(m_k, \mathbf{v}_k, \Sigma_\mathbf{0}, k = 1, \dots, N), \quad \mathbf{v}_k \in \mathbb{R}^n \tag{3.78a}$$

be an isolated Newtonian system of N noninteracting material particles the kth particle of which has mass m_k and Newtonian uniform velocity $\mathbf{v}_k$ relative to an inertial frame $\Sigma_\mathbf{0}$, $k = 1, \dots, N$. Furthermore, let m_0 be the resultant mass of S, considered as the mass of a virtual particle located at the center of momentum of S, and let $\mathbf{v}_0$ be the Newtonian velocity relative to $\Sigma_\mathbf{0}$ of the Newtonian center of momentum frame of S. Then we have the following well-known identities:

$$1 = \frac{1}{m_0}\sum_{k=1}^{N} m_k \tag{3.78b}$$

and

$$\begin{aligned} \mathbf{v}_0 &= \frac{1}{m_0}\sum_{k=1}^{N} m_k\mathbf{v}_k, \\ \mathbf{w} + \mathbf{v}_0 &= \frac{1}{m_0}\sum_{k=1}^{N} m_k(\mathbf{w} + \mathbf{v}_k), \end{aligned} \tag{3.78c}$$

where the binary operation $+$ is the common vector addition in $\mathbb{R}^n$, and where

$$m_0 = \sum_{k=1}^{N} m_k \tag{3.78d}$$

for $\mathbf{v}, \mathbf{w}_k \in \mathbb{R}^3$, $m_k > 0$, $k = 0, 1, \dots, N$.

In full analogy with (3.78a), let

$$S = S(m_k, \mathbf{v}_k, \Sigma_\mathbf{0}, k = 1, \dots, N), \qquad \mathbf{v}_k \in \mathbb{R}_c^n \tag{3.79a}$$

be an isolated Einsteinian system of N noninteracting material particles the kth particle of which has invariant mass m_k and Einsteinian uniform velocity $\mathbf{v}_k$ relative to an inertial frame $\Sigma_\mathbf{0}$, $k = 1, \dots, N$. Furthermore, let m_0 be the resultant mass of S, considered as the mass of a virtual particle located at the center of mass of S (calculated in (3.29)), and let $\mathbf{v}_0$ be the Einsteinian velocity relative to $\Sigma_\mathbf{0}$ of the Einsteinian center of momentum of the Einsteinian system S. Then, as shown in Theorem 3.2, the relativistic analogs of the Newtonian expressions in (3.78b)–(3.78d) are, respectively, the following Einsteinian expressions in (3.79b)–(3.79d),

$$\begin{aligned} \gamma_{\mathbf{v}_0} &= \frac{1}{m_0} \sum_{k=1}^{N} m_k \gamma_{\mathbf{v}_k}, \\ \gamma_{\mathbf{u}\oplus\mathbf{v}_0} &= \frac{1}{m_0} \sum_{k=1}^{N} m_k \gamma_{\mathbf{u}\oplus\mathbf{v}_k}, \end{aligned} \tag{3.79b}$$

and

$$\begin{aligned} \gamma_{\mathbf{v}_0} \mathbf{v}_0 &= \frac{1}{m_0} \sum_{k=1}^{N} m_k \gamma_{\mathbf{v}_k} \mathbf{v}_k, \\ \gamma_{\mathbf{w}\oplus\mathbf{v}_0} (\mathbf{w}\oplus\mathbf{v}_0) &= \frac{1}{m_0} \sum_{k=1}^{N} m_k \gamma_{\mathbf{w}\oplus\mathbf{v}_k} (\mathbf{w}\oplus\mathbf{v}_k), \end{aligned} \tag{3.79c}$$

where the binary operation $\oplus$ is the Einstein velocity addition in $\mathbb{R}_c^n$, given by (1.2), p. 4, and where

$$m_0 = \sqrt{\left(\sum_{k=1}^{N} m_k\right)^2 + 2 \sum_{\substack{j,k=1 \\ j<k}}^{N} m_j m_k (\gamma_{\ominus\mathbf{v}_j\oplus\mathbf{v}_k} - 1)} \tag{3.79d}$$

for $\mathbf{w}, \mathbf{v}_k \in \mathbb{R}_c^3$, $m_k > 0$, $k = 0, 1, \dots, N$. Here m_0 is the relativistic invariant mass of the Einsteinian system S, supposed concentrated at the relativistic center of mass of S, and $\mathbf{v}_0$ is the Einsteinian velocity relative to $\Sigma_\mathbf{0}$ of the Einsteinian center of momentum frame of the Einsteinian system S.

To conform with the Minkowskian four-vector formalism of special relativity, both m_0 and $\mathbf{v}_0$ are determined in Theorem 3.2 as the unique solution of the Minkowskian four-vector equation (3.19).

We finally wrote (3.62) as (3.65), i.e.,

$$m_0 = \sqrt{m_{newton}^2 + m_{dark}^2}, \tag{3.80}$$

viewing the relativistically invariant, or rest, mass m_0 of the system S as a Pythagorean composition of the Newtonian rest mass, m_{newton} and the dark mass, m_{dark} of S. The mass m_{dark} is *dark* in the sense that it is the mass of virtual matter that does not collide and does not emit radiation. Following observations in cosmology, one may postulate that our dark mass reveals its presence only gravitationally. We have shown qualitatively that (3.80) explains observations in both astrophysics and particle physics.

We should remark that the presence of our dark mass is predicted by theoretic special relativistic techniques. Hence, it need not account for the whole mass of dark matter observed by astrophysicists in the cosmos because there could be contributions from general relativistic considerations and, perhaps, other unknown sources.

3.10 Problems

Problem 3.1 Matrix Representation of the Lorentz Boost:
Show that the Lorentz boost $L(\mathbf{u})$, given vectorially by (3.5), p. 61, is a linear map that possesses the matrix representation (3.1), p. 60.

Part II
Mathematical Tools for Hyperbolic Geometry

In Part II of the book we employ results of Part I of the book to the study of the Cartesian–Beltrami–Klein ball model of the hyperbolic geometry of Bolyai and Lobachevsky. We pay special attention to the development of tools that will be needed in Part III of the book for the determination of hyperbolic triangle centers. The development of the tools involves the adaption of

1. Cartesian coordinates
2. Barycentric coordinates
3. Trigonometry; and
4. Vector algebra

into hyperbolic geometry.

The resulting hyperbolic barycentric coordinates, trigonometry and vectors are naturally called gyrobarycentric coordinates, gyrotrigonometry and gyrovectors, respectively.

Chapter 4
Euclidean and Hyperbolic Barycentric Coordinates

Abstract In Chap. 3, we have seen two important theorems in mechanics. These are Theorem 3.3, p. 69, about the mass and the center of momentum velocity of a particle system in classical mechanics, and Theorem 3.2, p. 64, about the mass and the center of momentum velocity of a particle system in relativistic mechanics. Theorem 3.3 naturally suggests the introduction of the concept of barycentric coordinates into Euclidean geometry. Guided by analogies, we will see in this chapter how Theorem 3.2 naturally suggests the introduction of the concept of barycentric coordinates into hyperbolic geometry, where they are called gyrobarycentric coordinates.

4.1 Euclidean Barycentric Coordinates

The notion of barycentric coordinates dates back to Möbius. The use of barycentric coordinates in Euclidean geometry is described in [74], and the historical contribution of Möbius' barycentric coordinates to vector analysis is described in [10, pp. 48–50].

In this section, we set the stage for the introduction in Sect. 4.2 of barycentric coordinates into hyperbolic geometry by illustrating the way Theorem 3.3, p. 71, suggests the introduction of barycentric coordinates into Euclidean geometry.

For any positive integer N, let $m_k \in \mathbb{R}$ be N given real numbers such that

$$\sum_{k=1}^{N} m_k \neq 0, \tag{4.1}$$

and let $A_k \in \mathbb{R}^n$ be N given points in the Euclidean n-space $\mathbb{R}^n$, $k = 1, \ldots, N$. Theorem 3.3, p. 71, states the trivial, but geometrically significant, result that the equation

$$\sum_{k=1}^{N} m_k \begin{pmatrix} 1 \\ A_k \end{pmatrix} = m_0 \begin{pmatrix} 1 \\ P \end{pmatrix} \tag{4.2}$$

A.A. Ungar, *Hyperbolic Triangle Centers*,
Fundamental Theories of Physics 166,
DOI 10.1007/978-90-481-8637-2_4, © Springer Science+Business Media B.V. 2010

for the unknowns $m_0 \in \mathbb{R}$ and $P \in \mathbb{R}^n$ possesses the unique solution given by

$$m_0 = \sum_{k=1}^{N} m_k \tag{4.3}$$

and

$$P = \frac{\sum_{k=1}^{N} m_k A_k}{\sum_{k=1}^{N} m_k}, \tag{4.4}$$

satisfying for all $X \in \mathbb{R}^n$

$$X + P = \frac{\sum_{k=1}^{N} m_k (X + A_k)}{\sum_{k=1}^{N} m_k}. \tag{4.5}$$

We view (4.4) as the representation of a point $P \in \mathbb{R}^n$ in terms of its *barycentric coordinates* m_k, $k = 1, \ldots, N$, with respect to the set of points $S = \{A_1, \ldots, A_N\}$. Identity (4.5), then, insures that the barycentric coordinate representation (4.4) of P with respect to the set S is *covariant* (or, *invariant in form*) in the following sense. The point P and the points of the set S of its barycentric coordinate representation vary together under translations. Indeed, a translation $X + A_k$ of A_k by X, $k = 1, \ldots, N$, in (4.5) results in the translation $X + P$ of P by X.

In order to insure that barycentric coordinate representations with respect to a set S are unique, we require S to be pointwise independent.

Definition 4.1 (Euclidean Pointwise Independence) A set S of N points $S = \{A_1, \ldots, A_N\}$ in $\mathbb{R}^n$, $n \geq 2$, is *pointwise independent* if the $N-1$ vectors $-A_1 + A_k$, $k = 2, \ldots, N$, are linearly independent.

We are now in the position to present the formal definition of Euclidean barycentric coordinates, as motivated by mass and center of momentum velocity of Newtonian particle systems in (3.49)–(3.50), p. 72.

Definition 4.2 (Barycentric Coordinates) Let

$$S = \{A_1, \ldots, A_N\} \tag{4.6}$$

be a pointwise independent set of N points in $\mathbb{R}^n$. The real numbers $m_1, \ldots, m_N$, satisfying

$$\sum_{k=1}^{N} m_k \neq 0 \tag{4.7}$$

are barycentric coordinates of a point $P \in \mathbb{R}^n$ with respect to the set S if

$$P = \frac{\sum_{k=1}^{N} m_k A_k}{\sum_{k=1}^{N} m_k}. \tag{4.8}$$

Barycentric coordinates are homogeneous in the sense that the barycentric coordinates $(m_1, \dots, m_N)$ of the point P in (4.8) are equivalent to the barycentric coordinates $(\lambda m_1, \dots, \lambda m_N)$ for any real nonzero number $\lambda \in \mathbb{R}$, $\lambda \neq 0$. Since in barycentric coordinates only ratios of coordinates are relevant, the barycentric coordinates $(m_1, \dots, m_N)$ are also written as $(m_1 : \dots : m_N)$.

Barycentric coordinates that are normalized by the condition

$$\sum_{k=1}^{N} m_k = 1 \tag{4.9}$$

are called *special barycentric coordinates*.

Equation (4.8) is said to be the (unique) barycentric coordinate representation of P with respect to the set S.

Theorem 4.3 (Covariance of Barycentric Coordinate Representations) *Let*

$$P = \frac{\sum_{k=1}^{N} m_k A_k}{\sum_{k=1}^{N} m_k} \tag{4.10}$$

be the barycentric coordinate representation of a point $P \in \mathbb{R}^n$ in a Euclidean n-space $\mathbb{R}^n$ with respect to a pointwise independent set $S = \{A_1, \dots, A_N\} \subset \mathbb{R}^n$. The barycentric coordinate representation (4.10) *is covariant, that is,*

$$X + P = \frac{\sum_{k=1}^{N} m_k (X + A_k)}{\sum_{k=1}^{N} m_k} \tag{4.11}$$

for all $X \in \mathbb{R}^n$, and

$$RP = \frac{\sum_{k=1}^{N} m_k R A_k}{\sum_{k=1}^{N} m_k} \tag{4.12}$$

for all $R \in SO(n)$.

Proof The proof is immediate, noting that rotations $R \in SO(n)$ of $\mathbb{R}^n$ about its origin are linear maps of $\mathbb{R}^n$. □

Following the vision of Felix Klein in his *Erlangen Program* [41], it is owing to the covariance with respect to translations and rotations that barycentric coordinate representations possess geometric significance. Indeed, translations and rotations in Euclidean geometry form the *group of motions* of the geometry, and according to Felix Klein's Erlangen Program, a geometric property is a property that remains invariant in form under the motions of the geometry.

4.2 Hyperbolic Barycentric, Gyrobarycentric, Coordinates

Guided by analogies with Sect. 4.1, in this section we introduce barycentric coordinates into hyperbolic geometry [65]. For any positive integer N, let $m_k \in \mathbb{R}$ be N given real numbers, and let $A_k \in \mathbb{R}_s^n$ be N given points in an Einstein gyrovector space $(\mathbb{R}_s^n, \oplus, \otimes)$, $k = 1, \ldots, N$, satisfying, (3.24), p. 66,

$$\sum_{k=1}^{N} m_k \gamma_{\mathbf{v}_k} > 0. \tag{4.13}$$

Theorem 3.2, p. 65, presents the result that the equation

$$\sum_{k=1}^{N} m_k \begin{pmatrix} \gamma_{A_k} \\ \gamma_{A_k} A_k \end{pmatrix} = m_0 \begin{pmatrix} \gamma_P \\ \gamma_P \, P \end{pmatrix} \tag{4.14}$$

for the unknowns $m_0 \in \mathbb{R}$ and $P \in \mathbb{R}_s^n$ possesses the unique solution given by

$$m_0 = \sqrt{\left(\sum_{k=1}^{N} m_k\right)^2 + 2 \sum_{\substack{j,k=1 \\ j<k}}^{N} m_j m_k (\gamma_{\ominus A_j \oplus A_k} - 1)} \tag{4.15}$$

$m_0 > 0$, satisfying

$$m_0 = \sqrt{\left(\sum_{k=1}^{N} m_k\right)^2 + 2 \sum_{\substack{j,k=1 \\ j<k}}^{N} m_j m_k (\gamma_{\ominus(X \oplus A_j) \oplus (X \oplus A_k)} - 1)} \tag{4.16}$$

for all $X \in \mathbb{R}_s^n$, and

$$P = \frac{\sum_{k=1}^{N} m_k \gamma_{A_k} A_k}{\sum_{k=1}^{N} m_k \gamma_{A_k}} \tag{4.17}$$

satisfying

$$X \oplus P = \frac{\sum_{k=1}^{N} m_k \gamma_{X \oplus A_k} (X \oplus A_k)}{\sum_{k=1}^{N} m_k \gamma_{X \oplus A_k}} \tag{4.18}$$

for all $X \in \mathbb{R}_s^n$.

Furthermore, Theorem 3.2, p. 65, also states that P and m_0 satisfy the two identities

$$\gamma_P = \frac{\sum_{k=1}^{N} m_k \gamma_{A_k}}{m_0} \tag{4.19}$$

and

$$\gamma_P\ P = \frac{\sum_{k=1}^{N} m_k \gamma_{A_k} A_k}{m_0} \tag{4.20}$$

and, more generally,

$$\gamma_{X\oplus P} = \frac{\sum_{k=1}^{N} m_k \gamma_{X\oplus A_k}}{m_0} \tag{4.21}$$

and

$$\gamma_{X\oplus P}(X\oplus P) = \frac{\sum_{k=1}^{N} m_k \gamma_{X\oplus A_k}(X\oplus A_k)}{m_0} \tag{4.22}$$

for all $X \in \mathbb{R}_s^n$.

We view (4.17) as the representation of a point $P \in \mathbb{R}_s^n$ in terms of its *hyperbolic barycentric coordinates* m_k, $k = 1, \ldots, N$, with respect to the set of points $S = \{A_1, \ldots, A_N\}$. Naturally in gyrolanguage, hyperbolic barycentric coordinates are called *gyrobarycentric coordinates*. Identity (4.18) insures that the *gyrobarycentric coordinate representation* (4.17) of P with respect to the set S is *gyrocovariant* in the sense of Definition 2.7, p. 57, as shown in Theorem 4.6 below. The point P and the points of the set S of its gyrobarycentric coordinate representation vary together under left gyrotranslations. Indeed, a left gyrotranslation $X\oplus A_k$ of A_k by X, $k = 1, \ldots, N$ in (4.18) results in the left gyrotranslation $X\oplus P$ of P by X.

In order to insure that gyrobarycentric coordinate representations with respect to a set S are unique, we require S to be hyperbolically pointwise independent.

Definition 4.4 (Hyperbolic Pointwise Independence) A set S of N points $S = \{A_1, \ldots, A_N\}$ in $\mathbb{R}_s^n$, $n \geq 2$, is *pointwise independent* if the $N-1$ gyrovectors in $\mathbb{R}_s^n$, $\ominus A_1\oplus A_k$, $k = 2, \ldots, N$, considered as vectors in $\mathbb{R}^n$, are linearly independent.

We are now in the position to present the formal definition of gyrobarycentric coordinates, that is, hyperbolic barycentric coordinates, as motivated by mass and center of momentum velocity of Einsteinian particle systems.

Definition 4.5 (Gyrobarycentric Coordinates) Let

$$S = \{A_1, \ldots, A_N\} \tag{4.23}$$

be a pointwise independent set of N points in $\mathbb{R}_s^n$. The real numbers $m_1, \ldots, m_N$, satisfying

$$\sum_{k=1}^{N} m_k \gamma_{A_k} > 0 \tag{4.24}$$

are gyrobarycentric coordinates of a point $P \in \mathbb{R}_s^n$ with respect to the set S if

$$P = \frac{\sum_{k=1}^{N} m_k \gamma_{A_k} A_k}{\sum_{k=1}^{N} m_k \gamma_{A_k}}. \tag{4.25}$$

Gyrobarycentric coordinates are homogeneous in the sense that the gyrobarycentric coordinates $(m_1, \ldots, m_N)$ of the point P in (4.25) are equivalent to the gyrobarycentric coordinates $(\lambda m_1, \ldots, \lambda m_N)$ for any real nonzero number $\lambda \in \mathbb{R}$, $\lambda \neq 0$. Since in gyrobarycentric coordinates only ratios of coordinates are relevant, the gyrobarycentric coordinates $(m_1, \ldots, m_N)$ are also written as $(m_1 : \cdots : m_N)$.

Gyrobarycentric coordinates that are normalized by the condition

$$\sum_{k=1}^{N} m_k = 1 \tag{4.26}$$

are called *special gyrobarycentric coordinates.*

Equation (4.25) is said to be the gyrobarycentric coordinate representation of P with respect to the set S.

Finally, the constant of the gyrobarycentric coordinate representation of P in (4.25) is $m_0 > 0$, given by

$$m_0 = \sqrt{\left(\sum_{k=1}^{N} m_k\right)^2 + 2 \sum_{\substack{j,k=1 \\ j<k}}^{N} m_j m_k (\gamma_{\ominus A_j \oplus A_k} - 1)}. \tag{4.27}$$

Theorem 4.6 (Gyrocovariance of Gyrobarycentric Coordinate Representations) *Let*

$$P = \frac{\sum_{k=1}^{N} m_k \gamma_{A_k} A_k}{\sum_{k=1}^{N} m_k \gamma_{A_k}} \tag{4.28a}$$

be a gyrobarycentric coordinate representation of a point $P \in \mathbb{R}_s^n$ in an Einstein gyrovector space $(\mathbb{R}_s^n, \oplus, \otimes)$ with respect to a pointwise independent set $S = \{A_1, \ldots, A_N\} \subset \mathbb{R}_s^n$.

Then

$$\gamma_P = \frac{\sum_{k=1}^{N} m_k \gamma_{A_k}}{m_0} \tag{4.28b}$$

and

$$\gamma_P\, P = \frac{\sum_{k=1}^{N} m_k \gamma_{A_k} A_k}{m_0} \tag{4.28c}$$

where $m_0 > 0$, *given by*

$$m_0 = \sqrt{\left(\sum_{k=1}^{N} m_k\right)^2 + 2\sum_{\substack{j,k=1\\ j<k}}^{N} m_j m_k (\gamma_{\ominus A_j \oplus A_k} - 1)} \tag{4.28d}$$

is the constant of the gyrobarycentric coordinate representation (4.28a).

Furthermore, the gyrobarycentric coordinate representation (4.28a) *and its associated identities in* (4.28b)–(4.28d) *are gyrocovariant, that is,*

$$X\oplus P = \frac{\sum_{k=1}^{N} m_k \gamma_{X\oplus A_k}(X\oplus A_k)}{\sum_{k=1}^{N} m_k \gamma_{X\oplus A_k}}, \tag{4.29a}$$

$$\gamma_{X\oplus P} = \frac{\sum_{k=1}^{N} m_k \gamma_{X\oplus A_k}}{m_0}, \tag{4.29b}$$

$$\gamma_{X\oplus P}(X\oplus P) = \frac{\sum_{k=1}^{N} m_k \gamma_{X\oplus A_k}(X\oplus A_k)}{m_0}, \tag{4.29c}$$

$$m_0 = \sqrt{\left(\sum_{k=1}^{N} m_k\right)^2 + 2\sum_{\substack{j,k=1\\ j<k}}^{N} m_j m_k (\gamma_{\ominus(X\oplus A_j)\oplus(X\oplus A_k)} - 1)} \tag{4.29d}$$

for all $X \in \mathbb{R}_s^n$, *and*

$$RP = \frac{\sum_{k=1}^{N} m_k \gamma_{RA_k} RA_k}{\sum_{k=1}^{N} m_k \gamma_{RA_k}}, \tag{4.30a}$$

$$\gamma_{RP} = \frac{\sum_{k=1}^{N} m_k \gamma_{RA_k}}{m_0}, \tag{4.30b}$$

$$\gamma_{RP}(RP) = \frac{\sum_{k=1}^{N} m_k \gamma_{RA_k}(RA_k)}{m_0}, \tag{4.30c}$$

$$m_0 = \sqrt{\left(\sum_{k=1}^{N} m_k\right)^2 + 2\sum_{\substack{j,k=1\\ j<k}}^{N} m_j m_k (\gamma_{\ominus(RA_j)\oplus(RA_k)} - 1)} \tag{4.30d}$$

for all $R \in SO(n)$.

Proof Let $P \in \mathbb{R}_s^n$ be given by (4.28a). Then, by (3.32)–(3.33), p. 68, P satisfies (4.28b). Following (4.28a)–(4.28b), P satisfies (4.28c) as well.

It follows from (4.28b)–(4.28d) that the pair (m_0, P) is a solution of the four-vector equation

$$\sum_{k=1}^{N} m_k \begin{pmatrix} \gamma_{A_k} \\ \gamma_{A_k} P \end{pmatrix} = m_0 \begin{pmatrix} \gamma_P \\ \gamma_P\, P \end{pmatrix}. \tag{4.31}$$

Hence, by Theorem 3.2, p. 65, the pair (m_0, P) is the unique solution of the four-vector equation (4.31), satisfying (4.29a)–(4.29d) for all $X \in \mathbb{R}_s^n$. Finally, noting that rotations $R \in SO(n)$ are linear maps of $\mathbb{R}^n$ that keep the norm invariant, (4.30a)–(4.30d) follow from (4.28a)–(4.28d). The proof is thus complete. □

As a way to emphasize its power and elegance, two immediate corollaries of Theorem 4.6, the first of which is a special case of the second, are presented. In these corollaries, it is sometimes convenient to use the notation $\gamma_{\mathbf{v}} = \gamma(\mathbf{v})$ for the gamma factor.

Corollary 4.7 *Let $S = \{A_1, \ldots, A_N\} \subset \mathbb{R}_s^n$ be a pointwise independent set of N points in $\mathbb{R}_s^n$, and let*

$$\frac{\sum_{k=1}^{N} m_k \gamma_{A_k} A_k}{\sum_{k=1}^{N} m_k \gamma_{A_k}} \in \mathbb{R}_s^n \tag{4.32}$$

be a gyrobarycentric coordinate representation of a point in an Einstein gyrovector space $(\mathbb{R}_s^n, \oplus, \otimes)$ with respect to the set S. Then

$$\gamma\left(\frac{\sum_{k=1}^{N} m_k \gamma_{A_k} A_k}{\sum_{k=1}^{N} m_k \gamma_{A_k}}\right) = \frac{\sum_{k=1}^{N} m_k \gamma_{A_k}}{m_0} \tag{4.33}$$

and

$$\gamma\left(\frac{\sum_{k=1}^{N} m_k \gamma_{A_k} A_k}{\sum_{k=1}^{N} m_k \gamma_{A_k}}\right) \frac{\sum_{k=1}^{N} m_k \gamma_{A_k} A_k}{\sum_{k=1}^{N} m_k \gamma_{A_k}} = \frac{\sum_{k=1}^{N} m_k \gamma_{A_k} A_k}{m_0}, \tag{4.34}$$

where m_0 is the constant of the gyrobarycentric coordinate representation (4.32), *given by* (4.28d).

More generally, Theorem 4.6 admits the following corollary:

Corollary 4.8 *Let $S = \{A_1, \ldots, A_N\} \subset \mathbb{R}_s^n$ be a pointwise independent set of N points in $\mathbb{R}_s^n$, and let*

$$\frac{\sum_{k=1}^{N} m_k \gamma_{A_k} A_k}{\sum_{k=1}^{N} m_k \gamma_{A_k}} \in \mathbb{R}_s^n \tag{4.35}$$

be a gyrobarycentric coordinate representation of a point in an Einstein gyrovector space $(\mathbb{R}_s^n, \oplus, \otimes)$ *with respect to the set* S. *Then, for all* $X \in \mathbb{R}_s^n$,

$$\gamma\left(X \oplus \frac{\sum_{k=1}^{N} m_k \gamma_{A_k} A_k}{\sum_{k=1}^{N} m_k \gamma_{A_k}}\right) = \frac{\sum_{k=1}^{N} m_k \gamma_{X \oplus A_k}}{m_0} \tag{4.36}$$

and

$$\gamma\left(X \oplus \frac{\sum_{k=1}^{N} m_k \gamma_{A_k} A_k}{\sum_{k=1}^{N} m_k \gamma_{A_k}}\right)\left(X \oplus \frac{\sum_{k=1}^{N} m_k \gamma_{A_k} A_k}{\sum_{k=1}^{N} m_k \gamma_{A_k}}\right) = \frac{\sum_{k=1}^{N} m_k \gamma_{X \oplus A_k} (X \oplus A_k)}{m_0}, \tag{4.37}$$

where m_0 *is the constant of the gyrobarycentric coordinate representation* (4.35), *given by* (4.29d) *or, equivalently, by* (4.28d).

Clearly, Corollary 4.7 is a special case of Corollary 4.8 corresponding to $X = \mathbf{0}$. Corollary 4.8 demonstrates that the identities in Corollary 4.7 are gyrocovariant with respect to left gyrotranslations. Similarly, they are also gyrocovariant with respect to rotations. Corollary 4.8 will prove crucially important for the mission of this book to generate momentum toward the hunt for gyrotriangle gyrocenters in Einstein gyrovector spaces, that is, in the Cartesian–Beltrami–Klein ball model of hyperbolic geometry.

Following the vision of Felix Klein in his *Erlangen Program* [41], it is owing to the gyrocovariance, that is, covariance with respect to left gyrotranslations and rotations, that gyrobarycentric coordinate representations are geometrically significant. Indeed, left gyrotranslations and rotations in hyperbolic geometry form the group of motions of the geometry, and according to Felix Klein's Erlangen Program, a geometric property is a property that remains invariant in form under the motions of the geometry.

Two additional corollaries of Theorem 4.6 that prove useful follow.

Corollary 4.9 *Let* $S = \{A_1, \ldots, A_N\} \subset \mathbb{R}_s^n$ *be a pointwise independent set of* N *points in* $\mathbb{R}_s^n$, *and let*

$$P = \frac{\sum_{k=1}^{N} m_k \gamma_{A_k} A_k}{\sum_{k=1}^{N} m_k \gamma_{A_k}} \tag{4.38}$$

be a gyrobarycentric coordinate representation of a point $P \in \mathbb{R}^n$ *with respect to the set* S. *Furthermore, let* m_0 *be the representation constant, given by*

$$m_0^2 = \left(\sum_{k=1}^{N} m_k\right)^2 + 2\sum_{\substack{j,k=1 \\ j<k}}^{N} m_j m_k (\gamma_{\ominus A_j \oplus A_k} - 1). \tag{4.39}$$

Then, the point P *lies in the ball* $\mathbb{R}_s^n$, $P \in \mathbb{R}_s^n$, *if and only if* $m_0^2 > 0$.

Proof Following (4.38), and (3.32)–(3.33), p. 69, with $c = s$, $\mathbf{v}_0 = P$, and $\mathbf{v}_k = A_k$, we have

$$\gamma_P = \frac{1}{\sqrt{1 - \frac{P^2}{s^2}}} = \frac{\sum_{k=1}^{N} m_k A_k}{m_0}. \tag{4.40}$$

Hence, the point P lies in the ball $\mathbb{R}_s^n$, $P \in \mathbb{R}_s^n$, if and only if γ_P is real, that is, if and only if $m_0^2 > 0$. □

Corollary 4.10 *Let $S = \{A_1, \ldots, A_N\} \subset \mathbb{R}_s^n$ be a pointwise independent set of N points in $\mathbb{R}_s^n$, and let*

$$P = \frac{\sum_{k=1}^{N} m_k \gamma_{A_k} A_k}{\sum_{k=1}^{N} m_k \gamma_{A_k}} \tag{4.41}$$

be a gyrobarycentric coordinate representation of a point $P \in \mathbb{R}^n$ with respect to the set S, with positive gyrobarycentric coordinates $m_k > 0$, $k = 1, \ldots, N$. Then, $P \in \mathbb{R}_s^n$. Moreover, P lies on the convex span of S if and only if $m_k > 0$, $k = 1, \ldots, N$.

Proof The gyrobarycentric coordinate representation (4.41) possesses the constant m_0 in (4.39). This representation constant is positive since $m_k > 0$, $k = 1, \ldots, N$, and since the gamma factors in (4.39) are greater than 1. Hence, by Corollary 4.9, $P \in \mathbb{R}_s^n$.

The gyrobarycentric combination (4.41) is positive if all the coefficients m_k, $k = 1, \ldots, N$, are positive. The set of all positive gyrobarycentric combinations of the points of the set S is the convex span of S. By convexity considerations, it is a subset of $\mathbb{R}_s^n$. Hence, P lies on the convex span of S if and only if $m_k > 0$, $k = 1, \ldots, N$. Owing to the homogeneity of gyrobarycentric coordinates, it is agreed that if all the gyrobarycentric coordinates of a point have equal signs, then the signs are selected to be positive. □

4.3 Uniqueness of Gyrobarycentric Coordinate Representations

Theorem 4.11 (Uniqueness of Gyrobarycentric Coordinate Representations) *A gyrobarycentric coordinate representation of a point in an Einstein gyrovector space $(\mathbb{R}_s^n, \oplus, \otimes)$ with respect to a pointwise independent set $S = \{A_1, \ldots, A_N\}$ is unique.*

Proof Let

$$P = \frac{\sum_{k=1}^{N} m_k \gamma_{A_k} A_k}{\sum_{k=1}^{N} m_k \gamma_{A_k}} = \frac{\sum_{k=1}^{N} m'_k \gamma_{A_k} A_k}{\sum_{k=1}^{N} m'_k \gamma_{A_k}} \in \mathbb{R}_s^n \tag{4.42}$$

be two gyrobarycentric coordinate representations of a point P with respect to a pointwise independent set $S = \{A_1, \ldots, A_N\} \subset \mathbb{R}_s^n$ in an Einstein gyrovector space $(\mathbb{R}_s^n, \oplus, \otimes)$.

Then, by the Gyrobarycentric Coordinate Representation Gyrocovariance Theorem 4.6, p. 90, with $X = \ominus A_j$ in (4.29a), along with the convenient notation

$$\begin{aligned} \mathbf{a}_{jk} &= \ominus A_j \oplus A_k, \\ \gamma_{ij} &= \gamma_{\ominus A_j \oplus A_k}, \end{aligned} \tag{4.43}$$

we have

$$\begin{aligned} \ominus A_j \oplus P &= \frac{\sum_{\substack{k=1\\k\neq j}}^N m_k \gamma_{\ominus A_j \oplus A_k} (\ominus A_j \oplus A_k)}{\sum_{k=1}^N m_k \gamma_{\ominus A_j \oplus A_k}} = \frac{\sum_{\substack{k=1\\k\neq j}}^N m_k \gamma_{jk} \mathbf{a}_{jk}}{\sum_{k=1}^N m_k \gamma_{jk}} \\ &= \frac{\sum_{\substack{k=1\\k\neq j}}^N m'_k \gamma_{jk} \mathbf{a}_{jk}}{\sum_{k=1}^N m'_k \gamma_{jk}} \end{aligned} \tag{4.44}$$

for any A_l, $1 \le j \le N$. Note that when $k = j$ in (4.44), $\mathbf{a}_{jk} = \ominus A_j \oplus A_k = \mathbf{0}$ and $\gamma_{jk} = \gamma_{\mathbf{a}_{jk}} = \gamma_{\mathbf{0}} = 1$.

The set $S = \{A_1, \ldots, A_N\} \subset \mathbb{R}_s^n \subset \mathbb{R}^n$ is pointwise independent. Hence, by Definition 4.4, p. 89, the set of gyrovectors $\mathbf{a}_{jk} = \ominus A_j \oplus A_k$, $k = 1, \ldots, N$, $k \neq j$, considered as vectors in $\mathbb{R}^n$, forms a set of $N - 1$ linearly independent vectors for each j. Owing to this linear independence,

$$m'_k = c m_k \tag{4.45}$$

for all $k = 1, \ldots, N$, where c is an unspecified nonzero constant. Since gyrobarycentric coordinates are homogeneous, a nonzero common factor of the gyrobarycentric coordinates of a representation is irrelevant. Hence, the two gyrobarycentric coordinate representations of P in (4.42) are identically the same, so that the gyrobarycentric coordinate representation (4.42) of P is unique. □

4.4 Triangle Centroid

The triangle centroid is located at the intersection of the triangle medians. In this section, we demonstrate the use of barycentric coordinates by determining the triangle centroid in $\mathbb{R}^n$.

Let $A_1A_2A_3$ be a triangle with vertices A_1, A_2 and A_3 in a Euclidean n-space $\mathbb{R}^n$, and let G be the triangle centroid, as shown in Fig. 4.1 for $n = 2$. Then, G is given by its barycentric coordinate representation (4.8), p. 86, with respect to the set $\{A_1, A_2, A_3\}$,

$$G = \frac{m_1 A_1 + m_2 A_2 + m_3 A_3}{m_1 + m_2 + m_3}, \tag{4.46}$$

where the barycentric coordinates m_1, m_2 and m_3 of G in (4.46) are to be determined in (4.52) below.

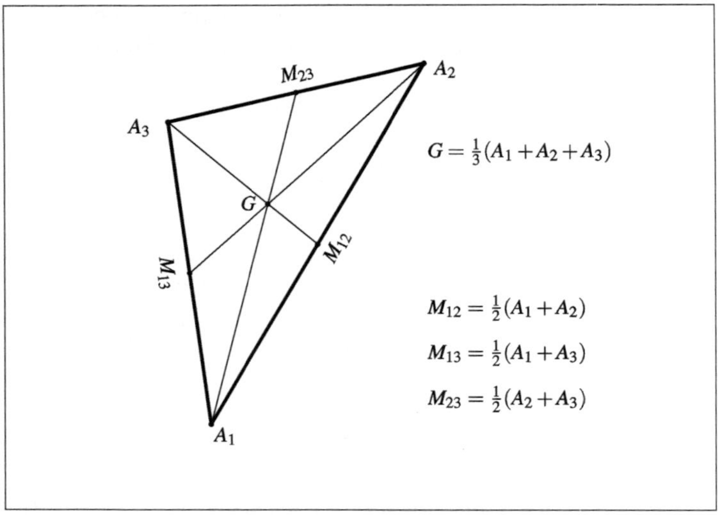

Fig. 4.1 The triangle medians and centroid in a Euclidean plane $\mathbb{R}^2$. The centroid G is the point of concurrency of the triangle medians. The midpoints M_{12}, M_{13}, and M_{23} of the three sides, A_1A_2, A_1A_3 and A_2A_3, of triangle $A_1A_2A_3$ in a Euclidean n-space $\mathbb{R}^n$ are shown here for $n = 2$, along with its medians A_1M_{23}, A_2M_{13}, A_3M_{12}, and its centroid G

The midpoint M_{12} of side A_1A_2, Fig. 4.1, is given by

$$M_{12} = \frac{A_1 + A_2}{2} \tag{4.47}$$

so that an equation of the line L_{123} through the points M_{12} and A_3 is given by

$$L_{123}(t_1) = A_3 + \left(-A_3 + \frac{A_1 + A_2}{2}\right)t_1 \tag{4.48}$$

with the parameter $t_1 \in \mathbb{R}$.

The line $L_{123}(t_1)$ contains one of the three medians of triangle $A_1A_2A_3$. Equations of the lines L_{123}, L_{231} and L_{312} that contain, respectively, the three triangle medians are therefore obtained from (4.48) by index cyclic permutations,

$$\begin{aligned} L_{123}(t_1) &= \frac{t_1}{2}A_1 + \frac{t_1}{2}A_2 + (1 - t_1)A_3, \\ L_{231}(t_2) &= \frac{t_2}{2}A_2 + \frac{t_2}{2}A_3 + (1 - t_2)A_1, \\ L_{312}(t_3) &= \frac{t_3}{2}A_3 + \frac{t_3}{2}A_1 + (1 - t_3)A_2 \end{aligned} \tag{4.49}$$

for $t_1, t_2, t_3 \in \mathbb{R}$.

The triangle centroid G is the point of concurrency of the three lines in (4.49). It is found by solving the equation $L_{123}(t_1) = L_{231}(t_2) = L_{312}(t_3)$ for the unknowns

$t_1, t_2, t_3 \in \mathbb{R}$, obtaining $t_1 = t_2 = t_3 = 2/3$. Hence, G is given by the equation

$$G = \frac{A_1 + A_2 + A_3}{3}, \tag{4.50}$$

as we see by substituting $t_1 = t_2 = t_3 = 2/3$ into (4.49).

Comparing (4.50) with (4.46) we find that the special barycentric coordinates (m_1, m_2, m_3) of G with respect to the set $\{A_1, A_2, A_3\}$ are given by

$$m_1 = m_2 = m_3 = \frac{1}{3}. \tag{4.51}$$

Accordingly, convenient barycentric coordinates $(m_1 : m_2 : m_3)$ of G are

$$(m_1 : m_2 : m_3) = (1 : 1 : 1), \tag{4.52}$$

as it is well-known in the literature; see, for instance, [29, 30].

4.5 Gyromidpoint

In this section, we demonstrate the use of gyrobarycentric coordinates by determining the gyromidpoint of gyrosegments in Einstein gyrovector spaces.

Let A_1A_2 be a gyrosegment in an Einstein gyrovector space $(\mathbb{R}^n_s, \oplus, \otimes)$, $n \geq 1$, formed by two distinct points $A_1, A_2 \in \mathbb{R}^n_s$. The gyromidpoint M_{12} of gyrosegment A_1A_2, Fig. 4.2, is the point of the gyrosegment that is equigyrodistant from A_1 and A_2, that is,

$$\|\ominus A_1 \oplus M_{12}\| = \|\ominus A_2 \oplus M_{12}\|. \tag{4.53}$$

In order to determine the gyromidpoint M_{12} of gyrosegment A_1A_2, let M_{12} be given by its gyrobarycentric coordinate representation (4.25) with respect to the set $S = \{A_1, A_2\}$,

$$M_{12} = \frac{m_1\gamma_{A_1} A_1 + m_2\gamma_{A_2} A_2}{m_1\gamma_{A_1} + m_2\gamma_{A_2}}, \tag{4.54}$$

where the gyrobarycentric coordinates m_1 and m_2 are to be determined in (4.62) below.

The constant m_0 of the gyrobarycentric coordinate representation (4.54) of M_{12} turns out to be

$$m_0 = \sqrt{(m_1 + m_2)^2 + 2m_1m_2(\gamma_{12} - 1)}, \tag{4.55}$$

according to (4.27).

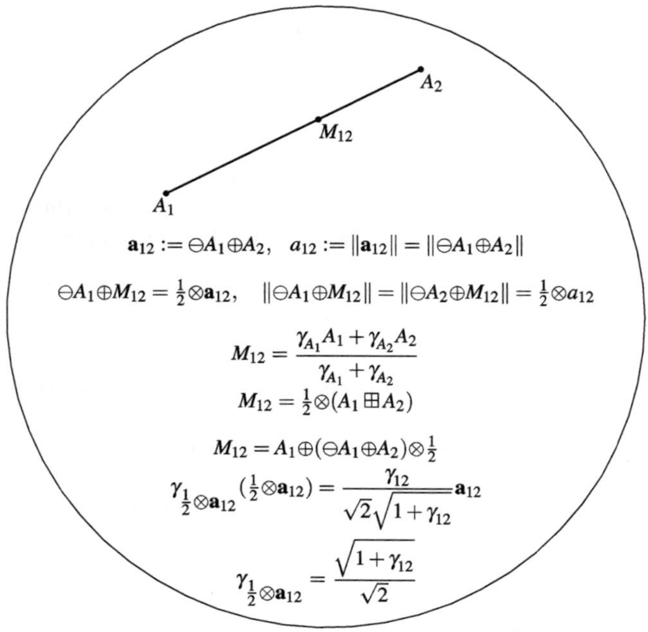

Fig. 4.2 The Einstein Gyromidpoint. The Einstein gyromidpoint M_{12} of a gyrosegment A_1A_2 in an Einstein gyrovector space $(\mathbb{R}_s^n, \oplus, \otimes)$ is shown for $n = 2$, along with several useful identities that the gyromidpoint possesses

Following the gyrocovariance of gyrobarycentric coordinate representations, Theorem 4.6, we have from (4.29a) with $X = \ominus A_1$ and $X = \ominus A_2$, respectively,

$$\ominus A_1 \oplus M_{12} = \frac{m_1 \gamma_{\ominus A_1 \oplus A_1} (\ominus A_1 \oplus A_1) + m_2 \gamma_{\ominus A_1 \oplus A_2} (\ominus A_1 \oplus A_2)}{m_1 \gamma_{\ominus A_1 \oplus A_1} + m_2 \gamma_{\ominus A_1 \oplus A_2}} = \frac{m_2 \gamma_{12} \mathbf{a}_{12}}{m_1 + m_2 \gamma_{12}},$$

$$\ominus A_2 \oplus M_{12} = \frac{m_1 \gamma_{\ominus A_2 \oplus A_1} (\ominus A_2 \oplus A_1) + m_2 \gamma_{\ominus A_2 \oplus A_2} (\ominus A_2 \oplus A_2)}{m_1 \gamma_{\ominus A_2 \oplus A_1} + m_2 \gamma_{\ominus A_2 \oplus A_2}} = \frac{m_1 \gamma_{12} \mathbf{a}_{21}}{m_1 \gamma_{21} + m_2}, \tag{4.56}$$

where, as shown in Fig. 4.2, we use the convenient notation

$$\begin{aligned} \mathbf{a}_{12} &= \ominus A_1 \oplus A_2, \qquad a_{12} = \|\mathbf{a}_{12}\|, \qquad \gamma_{12} = \gamma_{a_{12}}, \\ \mathbf{a}_{21} &= \ominus A_2 \oplus A_1, \qquad a_{21} = \|\mathbf{a}_{21}\|, \qquad \gamma_{21} = \gamma_{a_{21}}. \end{aligned} \tag{4.57}$$

We note that $a_{12} = a_{21}$, $\gamma_{12} = \gamma_{21}$, and $\gamma_{\mathbf{0}} = 0$, while, in general, $\mathbf{a}_{21} \neq \mathbf{a}_{12}$ since, by the gyrocommutative law, $\mathbf{a}_{21} = \mathrm{gyr}[\ominus A_2, A_1]\mathbf{a}_{12}$.

Taking magnitudes of the extreme sides of each of the two equations in (4.56), we have

$$\begin{aligned}\|\ominus A_1 \oplus M_{12}\| &= \frac{m_2}{m_1 + m_2\gamma_{12}}\gamma_{12}a_{12}, \\ \|\ominus A_2 \oplus M_{12}\| &= \frac{m_1}{m_1\gamma_{12} + m_2}\gamma_{12}a_{12}\end{aligned} \tag{4.58}$$

so that, by (4.58) and (4.53), we have

$$\frac{m_1}{m_1\gamma_{12} + m_2} = \frac{m_2}{m_1 + m_2\gamma_{12}}, \tag{4.59}$$

implying $m_1 = \pm m_2$.

For $m_1 = m_2 =: m$, the constant m_0 of the gyrobarycentric coordinate representation (4.54) of M_{12} is given by

$$m_0^2 = (m_1 + m_2)^2 + 2m_1m_2(\gamma_{12} - 1) = 2m^2(\gamma_{12} + 1) > 0 \tag{4.60}$$

so that, being positive, m_0^2 is acceptable.

In contrast, for $m_1 = -m_2 =: m$, the constant m_0 of the gyrobarycentric coordinate representation (4.54) of M_{12} is given by

$$m_0^2 = (m_1 + m_2)^2 + 2m_1m_2(\gamma_{12} - 1) = -2m^2(\gamma_{12} - 1) < 0 \tag{4.61}$$

so that, being negative, m_0^2 is rejected. Indeed, if $m_0^2 < 0$, then m_0 is purely imaginary so that, by (4.14), p. 88, also γ_P is purely imaginary, implying that while M_{12} lies in $\mathbb{R}^n$, it does not lie in the ball, $M_{12} \notin \mathbb{R}^n_s$. Hence, the solution $m_1 = -m_2$ of (4.59) is rejected, allowing the unique solution $m_1 = m_2$.

The unique solution for the gyrobarycentric coordinates of the midpoint M_{12} (modulo a multiplicative normalization constant) is, therefore, $(m_1 : m_2) = (m : m)$ or, equivalently,

$$(m_1 : m_2) = (1 : 1). \tag{4.62}$$

Substituting the gyrobarycentric coordinates (4.62) into (4.54), we finally express the gyromidpoint M_{12} in terms of its vertices A_1 and A_2 by the equation

$$M_{12} = \frac{\gamma_{A_1}A_1 + \gamma_{A_2}A_2}{\gamma_{A_1} + \gamma_{A_2}}. \tag{4.63}$$

Following (4.60) and (4.62)–(4.63), the constant m_0 of the gyrobarycentric coordinate representation of the gyromidpoint M_{12} in (4.63) is

$$m_0 = \sqrt{2}\sqrt{\gamma_{12} + 1}. \tag{4.64}$$

Hence, by the Gyrobarycentric Coordinate Representation Gyrocovariance Theorem 4.6, p. 90, the gyromidpoint M_{12} possesses the following three identities:

$$X\oplus M_{12}=\frac{\gamma_{X\oplus A_1}(X\oplus A_1)+\gamma_{X\oplus A_2}(X\oplus A_2)}{\gamma_{X\oplus A_1}+\gamma_{X\oplus A_2}}, \tag{4.65a}$$

$$\gamma_{X\oplus M_{12}}=\frac{\gamma_{X\oplus A_1}+\gamma_{X\oplus A_2}}{\sqrt{2}\sqrt{\gamma_{12}+1}}, \tag{4.65b}$$

$$\gamma_{X\oplus M_{12}}(X\oplus M_{12})=\frac{\gamma_{X\oplus A_1}(X\oplus A_1)+\gamma_{X\oplus A_2}(X\oplus A_2)}{\sqrt{2}\sqrt{\gamma_{12}+1}} \tag{4.65c}$$

for all $X\in\mathbb{R}^n_s$.

Following (4.65a) with $X=\ominus A_1$, by Einstein half (2.3), p. 46, we have

$$\ominus A_1\oplus M_{12}=\frac{\gamma_{\ominus A_1\oplus A_2}(\ominus A_1\oplus A_2)}{1+\gamma_{\ominus A_1\oplus A_2}}=\frac{\gamma_{12}}{1+\gamma_{12}}\mathbf{a}_{12}=\frac{1}{2}\otimes\mathbf{a}_{12} \tag{4.66}$$

so that by the scaling property (2.6), p. 46,

$$\|\ominus A_1\oplus M_{12}\|=\left\|\frac{1}{2}\otimes\mathbf{a}_{12}\right\|=\frac{1}{2}\otimes\|\mathbf{a}_{12}\|=\frac{1}{2}\otimes a_{12}. \tag{4.67}$$

Similarly, following (4.65b)–(4.65c) with $X=\ominus A_1$, we have

$$\gamma_{\ominus A_1\oplus M_{12}}=\frac{1+\gamma_{\ominus A_1\oplus A_2}}{\sqrt{2}\sqrt{1+\gamma_{12}}}=\frac{1+\gamma_{12}}{\sqrt{2}\sqrt{1+\gamma_{12}}}a=\frac{\sqrt{1+\gamma_{12}}}{\sqrt{2}} \tag{4.68}$$

and

$$\gamma_{\ominus A_1\oplus M_{12}}(\ominus A_1\oplus M_{12})=\frac{\gamma_{\ominus A_1\oplus A_2}(\ominus A_1\oplus A_2)}{\sqrt{2}\sqrt{1+\gamma_{12}}}=\frac{\gamma_{12}\mathbf{a}_{12}}{\sqrt{2}\sqrt{1+\gamma_{12}}}. \tag{4.69}$$

Hence, by (4.67) and (4.68),

$$\gamma_{\frac{1}{2}\otimes\mathbf{a}_{12}}=\frac{\sqrt{1+\gamma_{12}}}{\sqrt{2}} \tag{4.70}$$

and, by (4.66)–(4.67) and (4.69),

$$\gamma_{\frac{1}{2}\otimes\mathbf{a}_{12}}\left(\frac{1}{2}\otimes\mathbf{a}_{12}\right)=\frac{\gamma_{12}\mathbf{a}_{12}}{\sqrt{2}\sqrt{1+\gamma_{12}}}. \tag{4.71}$$

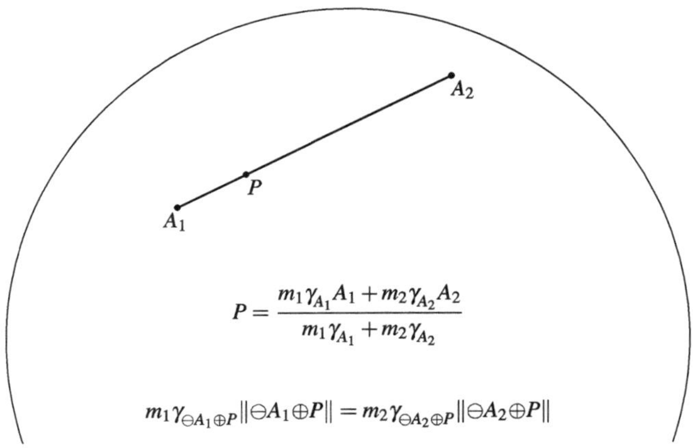

Fig. 4.3 The Hyperbolic Lever Law Relation in an Einstein gyrovector space $(\mathbb{R}^n_s, \oplus, \otimes)$. The position of a generic point P on a gyrosegment A_1A_2 is determined by its gyrobarycentric coordinates $m_k > 0$, $k = 1, 2$, with respect to the set $S = \{A_1, A_2\} \subset \mathbb{R}^n_s$. These gyrobarycentric coordinates, in turn, satisfy the hyperbolic lever law relation (4.78)

Two equivalent expressions for the gyromidpoint M_{12} in (4.63) are given by

$$\begin{aligned} M_{12} &= \frac{1}{2}\otimes(A_1 \boxplus A_2), \\ M_{12} &= A_1\oplus(\ominus A_1\oplus A_2)\otimes\frac{1}{2} \end{aligned} \tag{4.72}$$

as we see from (1.37), p. 13, and (2.15), p. 48.

The identities in (4.72) for the gyromidpoint M_{12} in (4.63) demonstrate once again that in order to capture analogies with classical results, both Einstein addition, $\oplus$, and Einstein coaddition, $\boxplus$, are required.

4.6 The Hyperbolic Lever Law Relation

Let A_1 and A_2 be any two distinct points of an Einstein gyrovector space $(\mathbb{R}^n_s.\oplus, \otimes)$, and let P be any point between them, Fig. 4.3. Then, the point P possesses a gyrobarycentric coordinate representation,

$$P = \frac{m_1\gamma_{A_1}A_1 + m_2\gamma_{A_2}A_2}{m_1\gamma_{A_1} + m_2\gamma_{A_2}} \tag{4.73}$$

with respect to the set $S = \{A_1, A_2\}$, with gyrobarycentric coordinates $m_1 > 0$ and $m_2 > 0$ that depend on the position of P between A_1 and A_2.

By the Gyrobarycentric Coordinate Representation Gyrocovariance Theorem 4.6, p. 90, with $X = \ominus A_1$ and $X = \ominus A_2$, it follows from (4.73), respectively,

$$\begin{aligned}
\gamma_{\ominus A_1 \oplus P}(\ominus A_1 \oplus P) &= \frac{m_1 \gamma_{\ominus A_1 \oplus A_1}(\ominus A_1 \oplus A_1) + m_2 \gamma_{\ominus A_1 \oplus A_2}(\ominus A_1 \oplus A_2)}{m_0} \\
&= \frac{m_2 \gamma_{\ominus A_1 \oplus A_2}(\ominus A_1 \oplus A_2)}{m_0}, \\
\gamma_{\ominus A_2 \oplus P}(\ominus A_2 \oplus P) &= \frac{m_1 \gamma_{\ominus A_2 \oplus A_1}(\ominus A_2 \oplus A_1) + m_2 \gamma_{\ominus A_2 \oplus A_2}(\ominus A_2 \oplus A_2)}{m_0} \\
&= \frac{m_1 \gamma_{\ominus A_2 \oplus A_1}(\ominus A_2 \oplus A_1)}{m_0},
\end{aligned} \tag{4.74}$$

where $m_0 > 0$ is given by

$$m_0^2 = m_1^2 + m_2^2 + 2m_1 m_2 \gamma_{\ominus A_1 \oplus A_2}. \tag{4.75}$$

Taking the norms of the extreme sides of each of the two equations in (4.74), we have

$$\begin{aligned}
\gamma_{\ominus A_1 \oplus P} \|\ominus A_1 \oplus P\| &= \frac{m_2 \gamma_{\ominus A_1 \oplus A_2}}{m_0} \|\ominus A_1 \oplus A_2\|, \\
\gamma_{\ominus A_2 \oplus P} \|\ominus A_2 \oplus P\| &= \frac{m_1 \gamma_{\ominus A_1 \oplus A_2}}{m_0} \|\ominus A_1 \oplus A_2\|,
\end{aligned} \tag{4.76}$$

noting that while, in general, $\ominus A_2 \oplus A_1 \neq \ominus A_1 \oplus A_2$, we have the equalities

$$\begin{aligned}
\|\ominus A_2 \oplus A_1\| &= \|\ominus A_1 \oplus A_2\|, \\
\gamma_{\ominus A_2 \oplus A_1} &= \gamma_{\ominus A_1 \oplus A_2}.
\end{aligned} \tag{4.77}$$

It follows from (4.76) that

$$m_1 \gamma_{\ominus A_1 \oplus P} \|\ominus A_1 \oplus P\| = m_2 \gamma_{\ominus A_2 \oplus P} \|\ominus A_2 \oplus P\| \tag{4.78}$$

or, equivalently, we have the *hyperbolic lever law relation*

$$\frac{m_1}{m_2} = \frac{\gamma_{\ominus A_2 \oplus P} \|\ominus A_2 \oplus P\|}{\gamma_{\ominus A_1 \oplus P} \|\ominus A_1 \oplus P\|}. \tag{4.79}$$

In the Euclidean limit, $s \to \infty$, gyrolengths tend to corresponding lengths and gamma factors tend to 1. Hence, in that limit, (4.78) tends to a corresponding result in Euclidean geometry,

$$m_1 \|-A_1 + P\| = m_2 \|-A_2 + P\|. \tag{4.80}$$

Equation (4.80) is the classical *lever law relation*. Let us view the two distances $\|-A_1 + P\|$ and $\|-A_2 + P\|$ in (4.80) as the lengths of the two arms of a lever supported at the point P in $\mathbb{R}^3$, and the gyrobarycentric coordinates m_1 and m_2 as masses that apply proportional forces on the lever arms in the same direction perpendicular to the arms. Then, the relation (4.80) is the *Archimedes law of the lever*, saying that the masses balance at distances inversely proportional to their magnitudes. By analogy, we call (4.78) the *hyperbolic lever law relation*.

We are now in the position to prove the following theorem about the hyperbolic lever law relation.

Theorem 4.12 (Hyperbolic Lever Law Relation) *Let P be a point between two distinct points A_1 and A_2 of an Einstein gyrovector space $(\mathbb{R}^n_s, \oplus, \otimes)$. Then, the point P possesses the gyrobarycentric coordinate representation*

$$P = \frac{m_1\gamma_{A_1}A_1 + m_2\gamma_{A_2}A_2}{m_1\gamma_{A_1} + m_2\gamma_{A_2}} \tag{4.81}$$

if and only if it obeys the hyperbolic lever law relation

$$\frac{m_1}{m_2} = \frac{\gamma_{\ominus A_2 \oplus P}\|\ominus A_2 \oplus P\|}{\gamma_{\ominus A_1 \oplus P}\|\ominus A_1 \oplus P\|}. \tag{4.82}$$

Proof The proof that (4.81) implies the hyperbolic lever law relation (4.82) is established in (4.73)–(4.79).

Conversely, let P be a point between A_1 and A_2 that satisfies the hyperbolic lever law relation (4.82). We will show that P is given by (4.81).

Being a point between A_1 and A_2, P possesses the gyrobarycentric coordinate representation

$$P = \frac{m'_1\gamma_{A_1}A_1 + m'_2\gamma_{A_2}A_2}{m'_1\gamma_{A_1} + m'_2\gamma_{A_2}} \tag{4.83}$$

with respect to the set $S = \{A_1, A_2\}$. Hence, by the first part of the proof, and since P satisfies (4.82),

$$\frac{m'_1}{m'_2} = \frac{\gamma_{\ominus A_2 \oplus P}\|\ominus A_2 \oplus P\|}{\gamma_{\ominus A_1 \oplus P}\|\ominus A_1 \oplus P\|} = \frac{m_1}{m_2}. \tag{4.84}$$

Hence, being homogeneous, the gyrobarycentric coordinates $(m_1 : m_2)$ and $(m'_1 : m'_2)$ are equal,

$$(m_1 : m_2) = (m'_1 : m'_2) \tag{4.85}$$

so that the point P in (4.83) satisfies (4.81) as desired. □

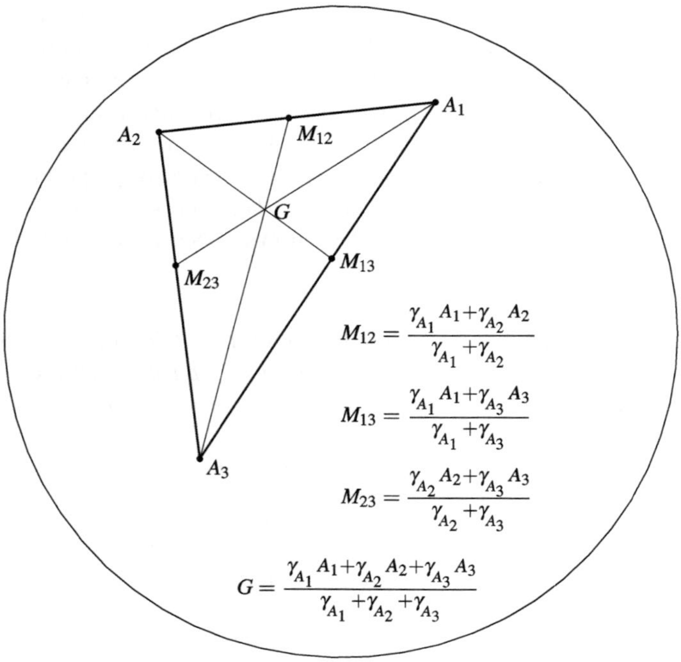

Fig. 4.4 The gyrotriangle gyromedians and gyrocentroid in an Einstein gyrovector plane $(\mathbb{R}^2_s, \oplus, \otimes)$. The gyrocentroid G is the point of concurrency of the gyrotriangle gyromedians. The gyromidpoints M_{12}, M_{13}, and M_{23} of the three sides, A_1A_2, A_1A_3 and A_2A_3, of gyrotriangle $A_1A_2A_3$ in an Einstein gyrovector space $(\mathbb{R}^n_s, \oplus, \otimes)$ are shown here for $n = 2$, along with its gyromedians A_1M_{23}, A_2M_{13}, A_3M_{12}, and its gyrocentroid G

4.7 Gyrotriangle Gyrocentroid

The hyperbolic triangle centroid is called, in gyrolanguage, the gyrotriangle gyrocentroid.

In this section, we demonstrate the use of gyrobarycentric coordinates by determining the gyrotriangle gyrocentroid in Einstein gyrovector spaces.

Definition 4.13 (Gyromedians, Gyrotriangle Gyrocentroids) A gyromedian of a gyrotriangle in an Einstein gyrovector space is the gyrosegment joining a vertex of the gyrotriangle with the gyromidpoint of the opposing side, Fig. 4.4. The gyrocentroid, G, of a gyrotriangle is the point of concurrency of the gyrotriangle gyromedians, Fig. 4.4.

Let $A_1A_2A_3$ be a gyrotriangle in an Einstein gyrovector space $(\mathbb{R}^n_s, \oplus, \otimes)$, and let the gyromidpoints of its sides be M_{12}, M_{13} and M_{23}, as shown in Fig. 4.4. Hence, by (4.63), M_{12} and, in a similar way, M_{13} and M_{23}, are given by the equations

$$\begin{aligned} M_{12} &= \frac{\gamma_{A_1} A_1 + \gamma_{A_2} A_2}{\gamma_{A_1} + \gamma_{A_2}}, \\ M_{13} &= \frac{\gamma_{A_1} A_1 + \gamma_{A_3} A_3}{\gamma_{A_1} + \gamma_{A_3}}, \\ M_{23} &= \frac{\gamma_{A_2} A_2 + \gamma_{A_3} A_3}{\gamma_{A_2} + \gamma_{A_3}}. \end{aligned} \tag{4.86}$$

The three gyromedians of gyrotriangle $A_1A_2A_3$ in Fig. 4.4 are the gyrosegments A_1M_{23}, A_2M_{13}, and A_3M_{12}. Since gyrosegments in Einstein gyrovector spaces coincide with Euclidean segments, one can employ methods of linear algebra to determine the point of concurrency, that is the gyrocentroid, of the three gyromedians of gyrotriangle $A_1A_2A_3$ in Fig. 4.4.

The details of the use of methods of linear algebra for the determination of the gyrobarycentric coordinates of the gyrotriangle gyrocentroid in Einstein gyrovector spaces are presented below.

In order to determine the gyrobarycentric coordinates of the gyrotriangle gyrocentroid in Einstein gyrovector spaces, we begin with some gyroalgebraic manipulations that reduce the task we face to a problem in linear algebra.

Let the gyrocentroid G of gyrotriangle $A_1A_2A_3$ in an Einstein gyrovector space $(\mathbb{R}^n_s, \oplus, \otimes)$, Fig. 4.4, be given by its gyrobarycentric coordinate representation, (4.25), p. 90, with respect to the set $S = \{A_1, A_2, A_3\}$ of the gyrotriangle vertices,

$$G = \frac{m_1\gamma_{A_1} A_1 + m_2\gamma_{A_2} A_2 + m_3\gamma_{A_3} A_3}{m_1\gamma_{A_1} + m_2\gamma_{A_2} + m_3\gamma_{A_3}}, \tag{4.87}$$

where the gyrobarycentric coordinates (m_1, m_2, m_3) of G in (4.87) are to be determined in (4.110) below.

Left gyrotranslating gyrotriangle $A_1A_2A_3$ by $\ominus A_1$, the gyrotriangle becomes gyrotriangle $O(\ominus A_1 \oplus A_2)(\ominus A_1 \oplus A_3)$, where $O = \ominus A_1 \oplus A_1$ is the arbitrarily selected origin of the Einstein gyrovector space $\mathbb{R}^n_s$, so that

$$O = \mathbf{0} = (0, \dots, 0) \tag{4.88}$$

with respect to the Cartesian coordinates of $\mathbb{R}^n_s$.

Following the left gyrotranslation by $\ominus A_1$, the gyrotriangle side-gyromidpoints M_{12}, M_{13} and M_{23} become, respectively, $\ominus A_1 \oplus M_{12}$, $\ominus A_1 \oplus M_{13}$ and $\ominus A_1 \oplus M_{23}$. These are calculated in (4.89) below by employing the gyrocovariance of gyrobarycentric coordinate representations, Theorem 4.6, p. 90.

Accordingly, we obtain from (4.29a), p. 91, with $X = \ominus A_1$ the following left gyrotranslation by $\ominus A_1$ of the gyromidpoints M_{12}, M_{13} and M_{23} in (4.86):

$$\begin{aligned}
\ominus A_1 \oplus M_{12} &= \ominus A_1 \oplus \frac{\gamma_{A_1} A_1 + \gamma_{A_2} A_2}{\gamma_{A_1} + \gamma_{A_2}} = \frac{\gamma_{\ominus A_1 \oplus A_2}(\ominus A_1 \oplus A_2)}{1 + \gamma_{\ominus A_1 \oplus A_2}} = \frac{\gamma_{12}\mathbf{a}_{12}}{\gamma_{12}+1}, \\
\ominus A_1 \oplus M_{13} &= \ominus A_1 \oplus \frac{\gamma_{A_1} A_1 + \gamma_{A_3} A_3}{\gamma_{A_1} + \gamma_{A_3}} = \frac{\gamma_{\ominus A_1 \oplus A_3}(\ominus A_1 \oplus A_3)}{1 + \gamma_{\ominus A_1 \oplus A_3}} = \frac{\gamma_{13}\mathbf{a}_{13}}{\gamma_{13}+1}, \\
\ominus A_1 \oplus M_{23} &= \ominus A_1 \oplus \frac{\gamma_{A_2} A_2 + \gamma_{A_3} A_3}{\gamma_{A_2} + \gamma_{A_3}} \\
&= \frac{\gamma_{\ominus A_1 \oplus A_2}(\ominus A_1 \oplus A_2) + \gamma_{\ominus A_1 \oplus A_3}(\ominus A_1 \oplus A_3)}{\gamma_{\ominus A_1 \oplus A_2} + \gamma_{\ominus A_1 \oplus A_3}} \\
&= \frac{\gamma_{12}\mathbf{a}_{12} + \gamma_{13}\mathbf{a}_{13}}{\gamma_{12} + \gamma_{13}}.
\end{aligned} \tag{4.89}$$

As in (4.57), in (4.89) we use the convenient notation

$$\mathbf{a}_{ij} = \ominus A_i \oplus A_j, \qquad a_{ij} = \|\mathbf{a}_{ij}\|, \qquad \gamma_{ij} = \gamma_{a_{ij}} \tag{4.90}$$

for $i, j = 1, 2, 3$, noting that $a_{ij} = a_{ji}$, $\gamma_{ij} = \gamma_{ji}$.

Note that, by Definition 4.5, p. 89, the set of points $S = \{A_1, A_2, A_3\}$ is pointwise independent in an Einstein gyrovector space $(\mathbb{R}^n_s, \oplus, \otimes)$. Hence, the two gyrovectors $\mathbf{a}_{12} = \ominus A_1 \oplus A_2$ and $\mathbf{a}_{13} = \ominus A_1 \oplus A_3$ in $\mathbb{R}^n_s \subset \mathbb{R}^n$ in (4.89), considered as vectors in $\mathbb{R}^n$, are linearly independent in $\mathbb{R}^n$.

Similarly to the gyroalgebra in (4.89), under a left gyrotranslation by $\ominus A_1$ the gyrocentroid G in (4.87) becomes

$$\begin{aligned}
\ominus A_1 \oplus G &= \frac{m_2\gamma_{\ominus A_1 \oplus A_2}(\ominus A_1 \oplus A_2) + m_3\gamma_{\ominus A_1 \oplus A_3}(\ominus A_1 \oplus A_3)}{m_1 + m_2\gamma_{\ominus A_1 \oplus A_2} + m_3\gamma_{\ominus A_1 \oplus A_3}} \\
&= \frac{m_2\gamma_{12}\mathbf{a}_{12} + m_3\gamma_{13}\mathbf{a}_{13}}{m_1 + m_2\gamma_{12} + m_3\gamma_{13}}.
\end{aligned} \tag{4.91}$$

The gyromedian of the left gyrotranslated gyrotriangle $O(\ominus A_1 \oplus A_2)(\ominus A_1 \oplus A_3)$ that joins the vertex

$$\ominus A_1 \oplus A_1 = O = \mathbf{0} \tag{4.92}$$

with the gyromidpoint of its opposing side, as calculated in (4.89),

$$\ominus A_1 \oplus M_{23} = \frac{\gamma_{12}\mathbf{a}_{12} + \gamma_{13}\mathbf{a}_{13}}{\gamma_{12} + \gamma_{13}} \tag{4.93}$$

is contained in the Euclidean line

$$L_1 = O + \big(-O + \{\ominus A_1 \oplus M_{23}\}\big)t_1 = \frac{\gamma_{12}\mathbf{a}_{12} + \gamma_{13}\mathbf{a}_{13}}{\gamma_{12} + \gamma_{13}} t_1 \tag{4.94}$$

where $t_1 \in \mathbb{R}$ is the line parameter. This line passes through the point $O = \mathbf{0} \in \mathbb{R}_s^n \subset \mathbb{R}^n$ when $t_1 = 0$, and it passes through the point

$$\ominus A_1 \oplus M_{23} = \frac{\gamma_{12}\mathbf{a}_{12} + \gamma_{13}\mathbf{a}_{13}}{\gamma_{12} + \gamma_{13}} \in \mathbb{R}_s^n \subset \mathbb{R}^n \tag{4.95}$$

when $t_1 = 1$.

Similarly to (4.92)–(4.94), the gyromedian of the left gyrotranslated gyrotriangle $O(\ominus A_1 \oplus A_2)(\ominus A_1 \oplus A_3)$ that joins the vertex

$$\ominus A_1 \oplus A_2 = \mathbf{a}_{12} \tag{4.96}$$

with the gyromidpoint of its opposing side, as calculated in (4.89),

$$\ominus A_1 \oplus M_{13} = \frac{\gamma_{13}\mathbf{a}_{13}}{\gamma_{13} + 1} \tag{4.97}$$

is contained in the Euclidean line

$$L_2 = \mathbf{a}_{12} + \big(-\mathbf{a}_{12} + \{\ominus A_1 \oplus M_{13}\}\big)t_2 = \mathbf{a}_{12} + \left(-\mathbf{a}_{12} + \frac{\gamma_{13}\mathbf{a}_{13}}{\gamma_{13} + 1}\right)t_2 \tag{4.98}$$

where $t_2 \in \mathbb{R}$ is the line parameter. This line passes through the point $\mathbf{a}_{12} \in \mathbb{R}_s^n \subset \mathbb{R}^n$ when $t_2 = 0$, and it passes through the point $\ominus A_1 \oplus M_{13} = \gamma_{13}\mathbf{a}_{13}/(\gamma_{13} + 1) \in \mathbb{R}_s^n \subset \mathbb{R}^n$ when $t_2 = 1$.

Similarly to (4.92)–(4.94), and similarly to (4.96)–(4.98), the gyromedian of the left gyrotranslated gyrotriangle $O(\ominus A_1 \oplus A_2)(\ominus A_1 \oplus A_3)$ that joins the vertex

$$\ominus A_1 \oplus A_3 = \mathbf{a}_{13} \tag{4.99}$$

with the gyromidpoint of its opposing side, as calculated in (4.89),

$$\ominus A_1 \oplus M_{12} = \frac{\gamma_{12}\mathbf{a}_{12}}{\gamma_{12} + 1} \tag{4.100}$$

is contained in the Euclidean line

$$L_3 = \mathbf{a}_{13} + \big(-\mathbf{a}_{13} + \{\ominus A_1 \oplus M_{12}\}\big)t_3 = \mathbf{a}_{13} + \left(-\mathbf{a}_{13} + \frac{\gamma_{12}\mathbf{a}_{12}}{\gamma_{12} + 1}\right)t_3 \tag{4.101}$$

where $t_3 \in \mathbb{R}$ is the line parameter. This line passes through the point $\mathbf{a}_{13} \in \mathbb{R}_s^n \subset \mathbb{R}^n$ when $t_3 = 0$, and it passes through the point $\ominus A_1 \oplus M_{12} = \gamma_{12}\mathbf{a}_{12}/(\gamma_{12}+1) \in \mathbb{R}_s^n \subset \mathbb{R}^n$ when $t_3 = 1$.

Hence, if the gyrocentroid G exists, its left gyrotranslated gyrocentroid, $-\ominus A_1 \oplus G$, given by (4.91), is contained in each of the three Euclidean lines L_k, $k = 1, 2, 3$, in (4.94), (4.98) and (4.101). Formalizing, if G exists then the point P in (4.91), i.e.,

$$P = \ominus A_1 \oplus G = \frac{m_2\gamma_{12}\mathbf{a}_{12} + m_3\gamma_{13}\mathbf{a}_{13}}{m_1 + m_2\gamma_{12} + m_3\gamma_{13}}, \tag{4.102}$$

lies on each of the lines L_k, $k = 1, 2, 3$. Imposing the normalization condition $m_1 + m_2 + m_3 = 1$ of special gyrobarycentric coordinates, (4.102) can be simplified by means of the resulting equation $m_1 = 1 - m_2 - m_3$, obtaining

$$P = \ominus A_1 \oplus G = \frac{m_2\gamma_{12}\mathbf{a}_{12} + m_3\gamma_{13}\mathbf{a}_{13}}{1 + m_2(\gamma_{12} - 1) + m_3(\gamma_{13} - 1)}. \tag{4.103}$$

Since the point P lies on each of the three lines L_k, $k = 1, 2, 3$, there exist values $t_{k,0}$ of the line parameters t_k, $k = 1, 2, 3$, respectively, such that

$$\begin{aligned}
&P - \frac{\gamma_{12}\mathbf{a}_{12} + \gamma_{13}\mathbf{a}_{13}}{\gamma_{12} + \gamma_{13}} t_{1,0} = 0, \\
&P - \mathbf{a}_{12} - \left(-\mathbf{a}_{12} + \frac{\gamma_{13}\mathbf{a}_{13}}{\gamma_{13} + 1}\right) t_{2,0} = 0, \\
&P - \mathbf{a}_{13} - \left(-\mathbf{a}_{13} + \frac{\gamma_{12}\mathbf{a}_{12}}{\gamma_{12} + 1}\right) t_{3,0} = 0.
\end{aligned} \tag{4.104}$$

The kth equation in (4.104), $k = 1, 2, 3$, is equivalent to the condition that point P lies on line L_k.

The system of equations (4.104) was obtained by methods of gyroalgebra, and will be solved below by a common method of linear algebra.

Substituting P from (4.103) into (4.104), and rewriting each equation in (4.104) as a linear combination of $\mathbf{a}_{12}$ and $\mathbf{a}_{13}$ equals zero, one obtains the following linear homogeneous system of three gyrovector equations

$$\begin{aligned}
c_{11}\mathbf{a}_{12} + c_{12}\mathbf{a}_{13} &= \mathbf{0}, \\
c_{21}\mathbf{a}_{12} + c_{22}\mathbf{a}_{13} &= \mathbf{0}, \\
c_{31}\mathbf{a}_{12} + c_{32}\mathbf{a}_{13} &= \mathbf{0},
\end{aligned} \tag{4.105}$$

where each coefficient c_{ij}, $i = 1, 2, 3$, $j = 1, 2$, is a function of γ_{12}, γ_{13}, γ_{23}, and the five unknowns m_2, m_3, and $t_{k,0}$, $k = 1, 2, 3$.

Since the set $S = \{A_1, A_2, A_3\}$ is pointwise independent, the two gyrovectors $\mathbf{a}_{12} = \ominus A_1 \oplus A_2$ and $\mathbf{a}_{13} = \ominus A_1 \oplus A_3$ in $\mathbb{R}^n_s$, considered as vectors in $\mathbb{R}^n$, are linearly independent. Hence, each coefficient c_{ij} in (4.105) equals zero. Accordingly, the three gyrovector equations in (4.105) are equivalent to the following six scalar equations,

$$c_{11} = c_{12} = c_{21} = c_{22} = c_{31} = c_{32} = 0 \tag{4.106}$$

for the five unknowns m_2, m_3 and $t_{k,0}$, $k = 1, 2, 3$.

Explicitly, the six scalar equations in (4.106) are equivalent to the following six equations:

$$\begin{aligned}
\big[1 + m_2(\gamma_{12} - 1) + m_3(\gamma_{13} - 1)\big]t_{1,0} - m_2(\gamma_{12} + \gamma_{13}) &= 0, \\
\big[1 + m_2(\gamma_{12} - 1) + m_3(\gamma_{13} - 1)\big]t_{1,0} - m_3(\gamma_{12} + \gamma_{13}) &= 0, \\
\big[1 + m_2(\gamma_{12} - 1) + m_3(\gamma_{13} - 1)\big]t_{2,0} - m_3(\gamma_{13} + 1) &= 0, \\
\big[1 + m_2(\gamma_{12} - 1) + m_3(\gamma_{13} - 1)\big]t_{2,0} - m_3(\gamma_{13} - 1) + m_2 - 1 &= 0, \\
\big[1 + m_2(\gamma_{12} - 1) + m_3(\gamma_{13} - 1)\big]t_{3,0} - m_2(\gamma_{12} + 1) &= 0, \\
\big[1 + m_2(\gamma_{12} - 1) + m_3(\gamma_{13} - 1)\big]t_{3,0} - m_2(\gamma_{12} - 1) + m_3 - 1 &= 0.
\end{aligned} \tag{4.107}$$

The unique solution of (4.107) is given by

$$\begin{aligned}
t_{1,0} &= \frac{\gamma_{12} + \gamma_{13}}{\gamma_{12} + \gamma_{13} + 1}, \\
t_{2,0} &= \frac{\gamma_{13} + 1}{\gamma_{12} + \gamma_{13} + 1}, \\
t_{3,0} &= \frac{\gamma_{12} + 1}{\gamma_{12} + \gamma_{13} + 1},
\end{aligned} \tag{4.108}$$

and

$$m_2 = m_3 = \frac{1}{3} \tag{4.109}$$

so that by the normalization condition $m_1 + m_2 + m_3 = 1$, also $m_1 = 1/3$.

Hence, the special gyrobarycentric coordinates of the gyrocentroid of a gyrotriangle $A_1A_2A_3$ with respect to the pointwise independent set $\{A_1, A_2, A_3\}$ in an Einstein gyrovector space $(\mathbb{R}^n_s, \oplus, \otimes)$, Fig. 4.4, are given by $(m_1, m_2, m_3) = (\frac{1}{3}, \frac{1}{3}, \frac{1}{3})$,

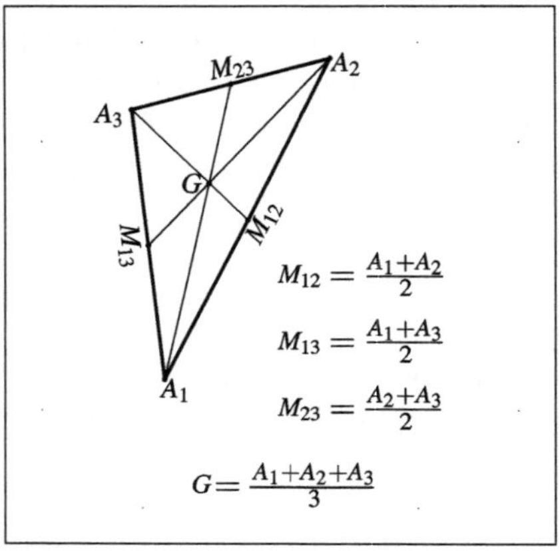

Fig. 4.5 The triangle medians M and centroid G in a Euclidean plane $\mathbb{R}^2$. Analogies with their counterpart in an Einstein gyrovector plane, Fig. 4.6, are obvious

so that convenient gyrobarycentric coordinates of the gyrotriangle gyrocentroid are

$$(m_1 : m_2 : m_3) = (1, 1, 1). \tag{4.110}$$

Finally, following (4.110) and (4.87), the gyrocentroid of a gyrotriangle $A_1A_2A_3$ in an Einstein gyrovector space $(\mathbb{R}_s^n, \oplus, \otimes)$ is given by the equation, Fig. 4.4,

$$G = \frac{\gamma_{A_1} A_1 + \gamma_{A_2} A_2 + \gamma_{A_3} A_3}{\gamma_{A_1} + \gamma_{A_2} + \gamma_{A_3}}. \tag{4.111}$$

The similarity between the gyrotriangle gyrocentroid G of gyrotriangle $A_1A_2A_3$ in (4.111) and the gyromidpoint M_{12} of gyrosegment A_1A_2 in (4.86) is remarkable. The extension to higher dimensions is now obvious. Indeed, the gyrocentroid G_t of a gyrotetrahedron $A_1A_2A_3A_4$ in Einstein gyrovector spaces $\mathbb{R}_s^n$, $n \geq 3$, is given by [63, (6.338)]

$$G_t = \frac{\gamma_{A_1} A_1 + \gamma_{A_2} A_2 + \gamma_{A_3} A_3 + \gamma_{A_4} A_4}{\gamma_{A_1} + \gamma_{A_2} + \gamma_{A_3} + \gamma_{A_4}}. \tag{4.112}$$

4.8 Analogies Between Centroids and Gyrocentroids

Analogies like those shown between Figs. 4.5 and 4.6 form the right tool in the study of hyperbolic triangle centers. Guided by the analogies, we extend the old and familiar study of triangle centers into the new study of gyrotriangle gyrocenters. The analogies shown between Figs. 4.5 and 4.6 are enhanced by the result that gyrotriangle gyrocenters are gyrocovariant just as triangle centers are covariant.

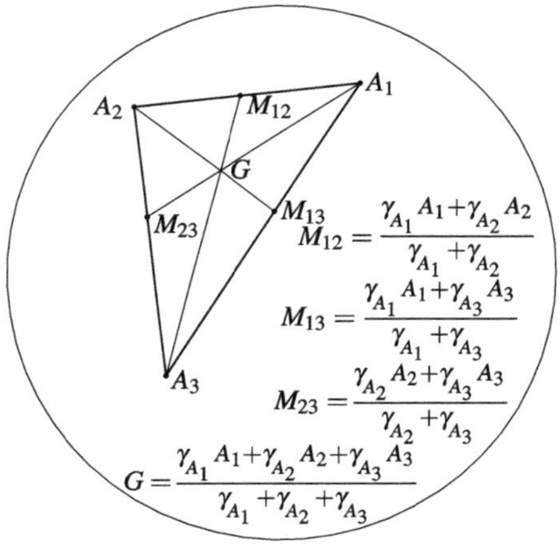

Fig. 4.6 The gyrotriangle gyromedians M and gyrocentroid G in an Einstein gyrovector plane $(\mathbb{R}_s^2, \oplus, \otimes)$. A reduction to the Euclidean plane is obtained when gamma factors are 1

In particular, triangle centroids are covariant. They are covariant under the action of translations,

$$X + G = \frac{(X + A_1) + (X + A_2) + (X + A_3)}{3} \tag{4.113}$$

for all $A_1, A_2, A_3, X \in \mathbb{R}^n$, and under the action of rotations,

$$RG = \frac{RA_1 + RA_2 + RA_3}{3} \tag{4.114}$$

for all $R \in SO(n)$.

In full analogy, gyrotriangle gyrocentroids are gyrocovariant. They are covariant under the action of left gyrotranslations,

$$X \oplus G = \frac{\gamma_{X\oplus A_1}(X\oplus A_1) + \gamma_{X\oplus A_2}(X\oplus A_2) + \gamma_{X\oplus A_3}(X\oplus A_3)}{\gamma_{X\oplus A_1} + \gamma_{X\oplus A_2} + \gamma_{X\oplus A_3}} \tag{4.115}$$

for all $A_1, A_2, A_3, X \in \mathbb{R}_s^n$, and under the action of rotations,

$$RG = \frac{\gamma_{RA_1} RA_1 + \gamma_{RA_2} RA_2 + \gamma_{RA_3} RA_3}{\gamma_{RA_1} + \gamma_{RA_2} + \gamma_{RA_3}} \tag{4.116}$$

for all $R \in SO(n)$. The harmonious interplay between Einstein addition, $\oplus$, in $\mathbb{R}_s^n$ and vector addition, +, in $\mathbb{R}^n$ as shown, for instance, in (4.115) is crucially important for our special relativistic approach to the study of hyperbolic triangle centers.

4.9 Gyrodistance in Gyrobarycentric Coordinates

Let P and P' be two pints in an Einstein gyrovector space $(\mathbb{R}_s^n, \oplus, \otimes)$ with gyrobarycentric coordinate representations

$$
\begin{aligned}
P &= \frac{\sum_{k=1}^{N} m_k \gamma_{A_k} A_k}{\sum_{k=1}^{N} m_k \gamma_{A_k}}, \\
P' &= \frac{\sum_{k=1}^{N} m'_k \gamma_{A_k} A_k}{\sum_{k=1}^{N} m'_k \gamma_{A_k}}
\end{aligned}
\tag{4.117}
$$

with respect to the set of points $S = \{A_1, \ldots, A_N\}$.

By the gyrocovariance in (4.29b), p. 91, of γ_P, with $X = \ominus P'$, and following (4.117), we obtain the two gamma factors in (4.118a), (4.118b), (4.119a), (4.119b) below:

$$
\gamma_{\ominus P' \oplus P} = \frac{\sum_{j=1}^{N} m_j \gamma_{\ominus P' \oplus A_j}}{m_0} = \frac{\sum_{j=1}^{N} m_j \gamma_{\ominus A_j \oplus P'}}{m_0}, \tag{4.118a}
$$

where, following (4.15), p. 88, $m_0 > 0$ is given by

$$
m_0^2 = \left(\sum_{k=1}^{N} m_k \right)^2 + 2 \sum_{\substack{j,k=1 \\ j<k}}^{N} m_j m_k (\gamma_{\ominus A_j \oplus A_k} - 1) \tag{4.118b}
$$

and

$$
\gamma_{\ominus A_j \oplus P'} = \frac{\sum_{k=1}^{N} m'_k \gamma_{\ominus A_j \oplus A_k}}{m'_0}, \tag{4.119a}
$$

where, as in (4.118b), $m'_0 > 0$ is given by

$$
(m'_0)^2 = \left(\sum_{k=1}^{N} m'_k \right)^2 + 2 \sum_{\substack{j,k=1 \\ j<k}}^{N} m_j m'_k (\gamma_{\ominus A_j \oplus A_k} - 1). \tag{4.119b}
$$

Note that while, in general, $\ominus P' \oplus A_j \neq \ominus A_j \oplus P'$, their norms are equal, $\|\ominus P' \oplus A_j\| = \|\ominus A_j \oplus P'\|$, implying the equality of their gamma factors, thus justifying the second equation in (4.118a).

Substituting (4.119a) into the extreme right-hand side of (4.118a), we obtain the gamma factor of $\ominus P' \oplus P$ in terms of the parameters of the gyrobarycentric coordinate representations (4.117), according to the following theorem:

Theorem 4.14 (Gyrodistance in Gyrobarycentric Coordinates) *Let P and P' be two pints in an Einstein gyrovector space* $(\mathbb{R}_s^n, \oplus, \otimes)$ *with gyrobarycentric coordinate representations* (4.117), (4.118a), (4.118b) *with respect to the set of points* $S = \{A_1, \ldots, A_N\}$.

Then, the squared gyrodistance between the points P and P' is given by the equation

$$\|\ominus P' \oplus P\|^2 = s^2 \frac{\gamma^2_{\ominus P' \oplus P} - 1}{\gamma^2_{\ominus P' \oplus P}}, \tag{4.120}$$

where

$$\begin{aligned}
\gamma_{\ominus P' \oplus P} &= \frac{1}{m_0 m'_0} \sum_{j=1}^{N} \sum_{k=1}^{N} m_j m'_k \gamma_{\ominus A_j \oplus A_k} \\
&= \frac{1}{m_0 m'_0} \Bigg\{ \sum_{\substack{j,k=1 \\ j<k}}^{N} m_j m'_k \gamma_{\ominus A_j \oplus A_k} + \sum_{\substack{j,k=1 \\ j>k}}^{N} m_j m'_k \gamma_{\ominus A_j \oplus A_k} \\
&\quad + \sum_{\substack{j,k=1 \\ j=k}}^{N} m_j m'_k \gamma_{\ominus A_j \oplus A_k} \Bigg\} \\
&= \frac{1}{m_0 m'_0} \Bigg\{ \sum_{\substack{j,k=1 \\ j<k}}^{N} (m_j m'_k + m'_j m_k) \gamma_{\ominus A_j \oplus A_k} + \sum_{i=1}^{N} m_i m'_i \Bigg\}.
\end{aligned} \tag{4.121}$$

Proof The gamma factor $\gamma_{\ominus P' \oplus P}$ is obtained by substituting (4.119a) into the extreme right-hand side of (4.118a). The resulting gyrodistance $\|\ominus P' \oplus P\|$ in (4.120) between the points P and P' is obtained from the gamma factor $\gamma_{\ominus P' \oplus P}$ in (4.121) by means of Identity (1.9), p. 5. □

4.10 Gyrolines in Gyrobarycentric Coordinates

Let

$$P_k = \frac{m_{k1}\gamma_{A_1} A_1 + m_{k2}\gamma_{A_2} A_2 + m_{k3}\gamma_{A_3} A_3}{m_{k1}\gamma_{A_1} + m_{k2}\gamma_{A_2} + m_{k3}\gamma_{A_3}} \in \mathbb{R}_s^n, \tag{4.122}$$

$k = 1, 2$, be two points represented by their gyrobarycentric coordinates with respect to the set $S = \{A_1, A_2, A_3\}$ of the vertices of a reference gyrotriangle $A_1 A_2 A_3$ in an Einstein gyrovector space $(\mathbb{R}_s^n, \oplus, \otimes)$, and let

$$P = \frac{m_1\gamma_{A_1} A_1 + m_2\gamma_{A_2} A_2 + m_3\gamma_{A_3} A_3}{m_1\gamma_{A_1} + m_2\gamma_{A_2} + m_3\gamma_{A_3}} \in \mathbb{R}^n_s \tag{4.123}$$

be a generic point spanned by S in $\mathbb{R}^n_s$.

Gyrolines in an Einstein gyrovector space $\mathbb{R}^n_s$ are Euclidean straight segments in the Euclidean space $\mathbb{R}^n$. Hence, the point P lies on the gyroline P_1P_2 that passes through the points P_1 and P_2 only if there exists $t \in \mathbb{R}$ such that

$$P = P_1 + (-P_1 + P_2)t \in \mathbb{R}^n_s. \tag{4.124}$$

Substituting (4.122)–(4.123) into (4.124), we obtain an equation of the form

$$c_1A_1 + c_2A_2 + c_3A_3 = \mathbf{0}. \tag{4.125}$$

By the definition of gyrobarycentric coordinates, Definition 4.5, p. 89, the points A_1, A_2 and A_3 of $\mathbb{R}^n_s$ are pointwise linearly independent in $\mathbb{R}^n$, so that (4.125) is equivalent to the three equations $c_1 = c_2 = c_3 = 0$. These three equations, in turn, determine the gyrobarycentric coordinates of P in (4.123) as functions of parameters of the reference gyrotriangle $A_1A_2A_3$ and the gyroline parameter, t, up to an unspecified nonzero common factor that we can arbitrarily select, obtaining

$$\begin{aligned}
m_1 &= m_{11}m_{21}\gamma_{A_1} + \{m_{11}m_{22} + (-m_{11}m_{22} + m_{12}m_{21})t\}\gamma_{A_2} \\
&\quad + \{m_{11}m_{23} + (-m_{11}m_{23} + m_{13}m_{21})t\}\gamma_{A_3}, \\
m_2 &= m_{12}m_{22}\gamma_{A_2} + \{m_{12}m_{23} + (-m_{12}m_{23} + m_{13}m_{22})t\}\gamma_{A_3} \\
&\quad + \{m_{12}m_{21} + (-m_{12}m_{21} + m_{11}m_{22})t\}\gamma_{A_1}, \\
m_3 &= m_{13}m_{23}\gamma_{A_3} + \{m_{13}m_{21} + (-m_{13}m_{21} + m_{11}m_{23})t\}\gamma_{A_1} \\
&\quad + \{m_{13}m_{22} + (-m_{13}m_{22} + m_{12}m_{23})t\}\gamma_{A_2}.
\end{aligned} \tag{4.126}$$

Note that each equation in (4.126) can be obtained from the others by an index permutation.

The line parameter $t \in \mathbb{R}$ in (4.126), which results from (4.124), is restricted to permissible values, that is, values for which $P \in \mathbb{R}^n_s \subset \mathbb{R}^n$. Thus, in particular, any $0 \le t \le 1$ is permissible, giving rise to a point on the gyrosegment P_1P_2.

The constant m_0 of the gyrobarycentric coordinate representation (4.123) of P is given by

$$m_0^2 = m_1^2 + m_2^2 + m_3^2 + 2(m_1m_2\gamma_{12} + m_1m_3\gamma_{13} + m_2m_3\gamma_{23}) \tag{4.127}$$

and, hence, it is a function of the parameter $t \in \mathbb{R}$. By Theorem 4.9, p. 93, the point $P \in \mathbb{R}^n$ lies in $\mathbb{R}^n_s$ if and only if $m_0^2 > 0$, that is, if and only if m_0 is real and nonzero.

4.11 Problems

Problem 4.1 A System of Equations:
Show that the six scalar equations in (4.106), p. 109, are equivalent to the six equations (4.107), p. 109.

Problem 4.2 Gyrolines in Gyrobarycentric Coordinates:
Derive the gyrobarycentric coordinates m_1, m_2, m_3 in (4.126), p. 114, by the way outlined in Sect. 4.10.

Chapter 5
Gyrovectors

Abstract In this chapter vectors are introduced into hyperbolic geometry, where they are called *gyrovectors*. Gyrovectors are equivalence classes that add according to the gyroparallelogram law in full analogy with vectors, which are equivalence classes that add according to the parallelogram law. A gyroparallelogram, in turn, is a gyroquadrangle the two gyrodiagonals of which intersect at their gyromidpoints in full analogy with a parallelogram, which is a quadrangle the two diagonals of which intersect at their midpoints.

5.1 Points and Vectors in Euclidean Geometry

The relationship between points and vectors in the Euclidean geometry of $\mathbb{R}^n$ involves equivalence relations and equivalence classes. Their definitions are, therefore, presented below.

Definition 5.1 (Equivalence Relations and Classes) A relation on a nonempty set S is a subset R of $S \times S$, $R \subset S \times S$, written as $a \sim b$ if $(a, b) \in R$. A relation $\sim$ on a set S is

1. Reflexive if $a \sim a$ for all $a \in S$.
2. Symmetric if $a \sim b$ implies $b \sim a$ for all $a, b \in S$.
3. Transitive if $a \sim b$ and $b \sim c$ imply $a \sim c$ for all $a, b, c \in S$.

A relation is an equivalence relation if it is reflexive, symmetric and transitive.

An equivalence relation $\sim$ on a set S gives rise to equivalence classes. The equivalence class of $a \in S$ is the subset $\{x \in S : x \sim a\}$ of S of all the elements $x \in S$ that are related to a by the relation $\sim$.

Two equivalence classes in a set S with an equivalence relation $\sim$ are either equal or disjoint, and the union of all the equivalence classes in S equals S. Accordingly, we say that the equivalence classes of a set S with an equivalence relation form a *partition* of S.

A.A. Ungar, *Hyperbolic Triangle Centers*,
Fundamental Theories of Physics 166,
DOI 10.1007/978-90-481-8637-2_5,

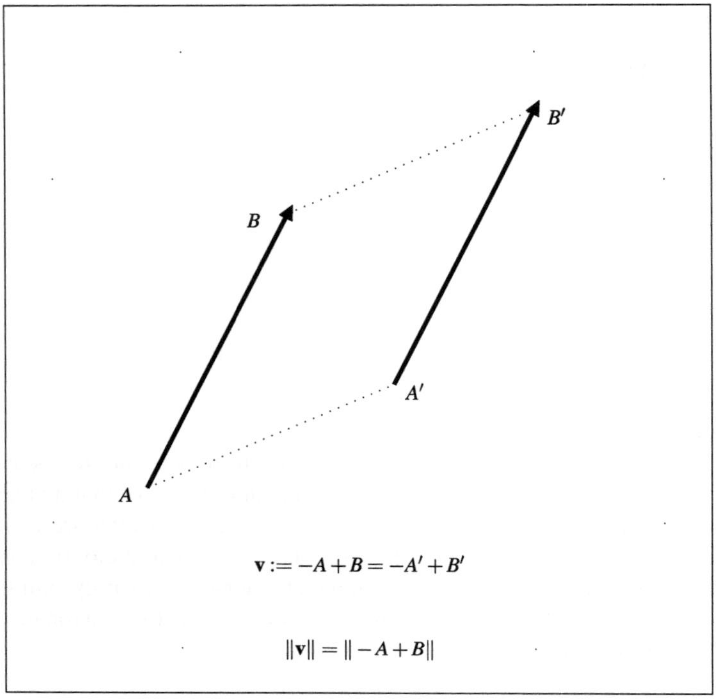

Fig. 5.1 The vectors $-A + B$ and $-A' + B'$ have equal values, that is, $-A + B = -A' + B'$, in a Euclidean space $\mathbb{R}^n$. As such, these two vectors are equivalent and, hence, indistinguishable in their vector space and its underlying Euclidean geometry. Two equivalent nonzero vectors in Euclidean geometry are parallel, and possess equal lengths, as shown here for $n = 2$. It turns out that two equivalent vectors $-A + B$ and $-A' + B'$ are parallel and have equal lengths, forming the parallelogram $AA'B'B$

Points of $\mathbb{R}^n$, denoted by capital italic letters A, B, etc., give rise to vectors in $\mathbb{R}^n$, denoted by bold roman lowercase letters $\mathbf{u}$, $\mathbf{v}$, etc. Any two ordered points $A, B \in \mathbb{R}^n$ give rise to a unique rooted vector $\mathbf{v} \in \mathbb{R}^n$, rooted at the point A. It has a *tail* at the point A and a *head* at the point B, Fig. 5.1, and it has the *value* $-A + B$,

$$\mathbf{v} = -A + B. \tag{5.1}$$

The length $\|\mathbf{v}\|$ of the rooted vector $\mathbf{v} = -A + B$ is the distance between its tail, A, and its head, B, that is,

$$\|\mathbf{v}\| = \| -A + B\|. \tag{5.2}$$

Two rooted vectors $-A + B$ and $-C + D$ are equivalent if they have the same value, $-A + B = -C + D$, that is,

$$-A + B \sim -C + D \quad \text{if and only if} \quad -A + B = -C + D. \tag{5.3}$$

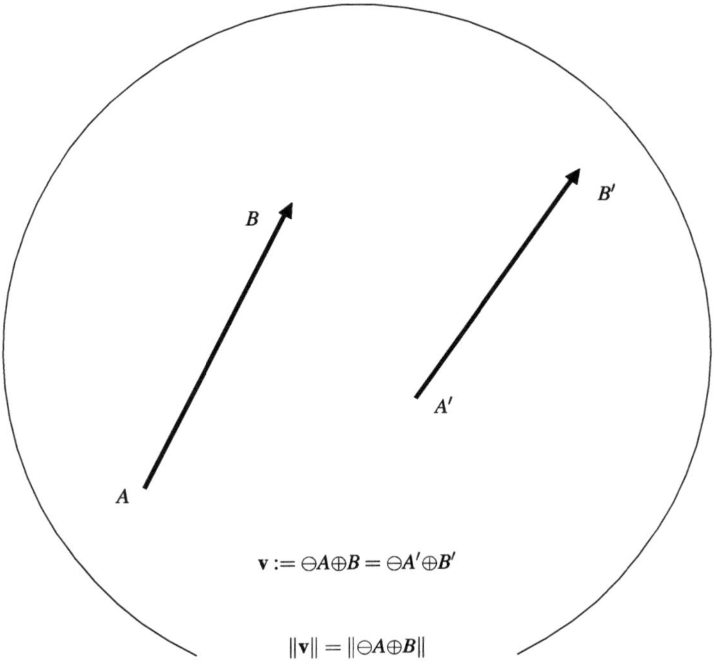

Fig. 5.2 The rooted gyrovectors $\ominus A \oplus B$ and $\ominus A' \oplus B'$ in an Einstein gyrovector space $(\mathbb{R}^n_s, \oplus, \otimes)$ have equal values, $\ominus A \oplus B = \ominus A' \oplus B'$. Hence, they are equivalent so that they represent the same gyrovector. Accordingly, as gyrovectors, $\ominus A \oplus B$ and $\ominus A' \oplus B'$ are indistinguishable in their gyrovector space and its underlying hyperbolic geometry. As an important disanalogy with Euclidean geometry, the gyroquadrangle $AA'B'B$ is not a hyperbolic parallelogram; see Fig. 5.3. For the generation of equivalent rooted gyrovectors in a given gyrovector space, see [63] or [64, Chap. 3]

The relation $\sim$ in (5.3) between rooted vectors is reflexive, symmetric and transitive. Hence, it is an equivalence relation that gives rise to equivalence classes of rooted vectors. To liberate rooted vectors from their roots, we define a *vector* to be an equivalence class of rooted vectors. The vector $-A + B$ is thus a representative of all rooted vectors with value $-A + B$. Thus, for instance, the two distinct rooted vectors $-A + B$ and $-A' + B'$ in Fig. 5.1 possess the same value so that, as vectors, they are indistinguishable. As it is well-known, vectors add according to the parallelogram addition law [64] so that vectors in Euclidean geometry are, in fact, equivalence classes of ordered pairs of points that add according to the parallelogram law.

A point $P \in \mathbb{R}^n$ is identified with the vector $-O + P$, O being the arbitrarily selected origin of the space $\mathbb{R}^n$. Hence, the algebra of vectors can be applied to points as well.

In full analogy with Fig. 5.1, Fig. 5.2 presents two equivalent rooted gyrovectors, studied in Sect. 5.2.

5.2 Points and Gyrovectors in Hyperbolic Geometry

Points of an Einstein gyrovector space $(\mathbb{R}_s^n, \oplus, \otimes)$, denoted by capital italic letters A, B, etc., give rise to gyrovectors in $\mathbb{R}_s^n$, denoted by bold roman lowercase letters $\mathbf{u}$, $\mathbf{v}$, etc. Any two ordered points $A, B \in \mathbb{R}_s^n$ give rise to a unique rooted gyrovector $\mathbf{v} \in \mathbb{R}_s^n$, rooted at the point A. It has a *tail* at the point A and a *head* at the point B, and it has the value $\ominus A \oplus B$,

$$\mathbf{v} = \ominus A \oplus B. \tag{5.4}$$

The gyrolength $\|\mathbf{v}\|$ of the rooted gyrovector $\mathbf{v} = \ominus A \oplus B$ is the gyrodistance between its tail, A, and its head, B, that is,

$$\|\mathbf{v}\| = \|\ominus A \oplus B\|. \tag{5.5}$$

Two rooted gyrovectors $\ominus A \oplus B$ and $\ominus A' \oplus B'$, Fig. 5.2, are equivalent if they have the same value, $\ominus A \oplus B = \ominus A' \oplus B'$, that is,

$$\ominus A \oplus B \sim \ominus A' \oplus B' \quad \text{if and only if} \quad \ominus A \oplus B = \ominus A' \oplus B'. \tag{5.6}$$

The relation $\sim$ in (5.6) between rooted gyrovectors in Einstein gyrovector spaces is reflexive, symmetric and transitive. As such, it is an equivalence relation that gives rise to equivalence classes of rooted gyrovectors. To liberate rooted gyrovectors from their roots, we define a *gyrovector* to be an equivalence class of rooted gyrovectors. The gyrovector $\ominus A \oplus B$ is thus a representative of all rooted gyrovectors with value $\ominus A \oplus B$. Thus, for instance, the two distinct rooted gyrovectors $\ominus A \oplus B$ and $\ominus A' \oplus B'$ in Fig. 5.2 possess the same value so that, as gyrovectors in an Einstein gyrovector space, they are indistinguishable.

Vectors add according to the parallelogram addition law. Accordingly, vectors in Euclidean geometry are equivalence classes of ordered pairs of points that add according to the parallelogram law. In full analogy, we wish to establish gyrovectors in Einstein gyrovector spaces as equivalence classes of ordered pairs of points that add according to the "gyroparallelogram" law.

Hence, in order to introduce vectors into hyperbolic geometry we have to define the hyperbolic parallelogram that we, naturally, call a *gyroparallelogram.* At first glance, the term *hyperbolic parallelogram* sounds as a contradiction in terms, since parallelism is denied in hyperbolic geometry. However, in full analogy with parallelograms, gyroparallelograms are defined in Sect. 5.3.

5.3 Einstein Gyroparallelogram

Definition 5.2 (Gyroparallelograms) Let A, B and C be any three points in a gyrovector space $(\mathbb{R}_s^n, \oplus, \otimes)$. Then, the four points A, B, C, D in $\mathbb{R}_s^n$ are the vertices of the gyroparallelogram $ABDC$, ordered either clockwise or counterclockwise, Fig. 5.3, if D satisfies the gyroparallelogram condition

$$D = (B \boxplus C) \ominus A. \tag{5.7}$$

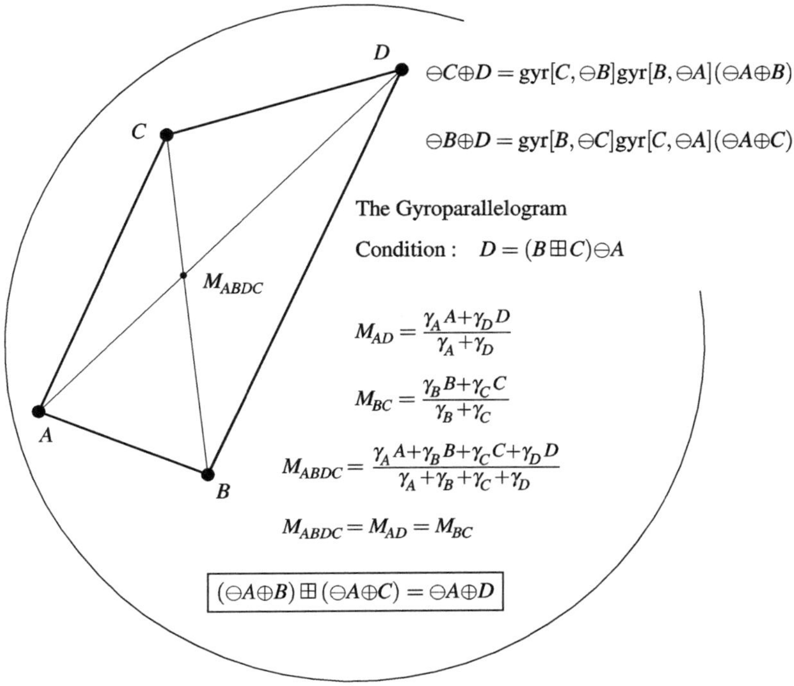

Fig. 5.3 The Einstein gyroparallelogram and its addition law in an Einstein gyrovector space $(\mathbb{R}^n_s, \oplus, \otimes)$. This figure is identical to Fig. 5.4 with one exception. Here points are denoted by capital italic letters, A, B, C, D, paving the way to Fig. 5.4 where gyrodifferences of points are recognized as gyrovectors, denoted by bold roman lowercase letters **u**, **v**, **w**

The gyroparallelogram is degenerate if the three points A, B and C are gyrocollinear.

If the gyroparallelogram $ABDC$ is non-degenerate, then the two vertices in each of the pairs (A, D) and (B, C) are said to be opposite to one another. The gyrosegments of adjacent vertices, AB, BD, DC and CA, are the sides of the gyroparallelogram. The gyrosegments AD and BC that link opposite vertices of the non-degenerate gyroparallelogram $ABDC$ are the gyrodiagonals of the gyroparallelogram.

The gyrocenter $\mathbf{m}_{ABDC}$ of the gyroparallelogram $ABDC$, Fig. 5.3, is the gyromidpoint of each of its two gyrodiagonals, so that $\mathbf{m}_{ABDC} = \mathbf{m}_{AD} = \mathbf{m}_{BC}$ (see Theorem 5.3 below).

In what seemingly sounds like a contradiction in terms, we have extended in Definition 5.2 the Euclidean parallelogram into hyperbolic geometry where the parallel postulate is denied. The resulting gyroparallelogram shares remarkable analogies with its Euclidean counterpart, giving rise to the gyroparallelogram law of gyrovector addition, which is commutative and fully analogous to the common parallelogram law of vector addition in Euclidean geometry. A gyroparallelogram in an

Einstein gyrovector plane, that is, in the Cartesian–Beltrami–Klein disc model of hyperbolic geometry, is presented in Figs. 5.3–5.4.

Theorem 5.3 (Gyroparallelogram Symmetries) *Every vertex of the gyroparallelogram $ABDC$, Fig. 5.3, satisfies the gyroparallelogram condition* (5.7), *that is,*

$$\begin{aligned} A &= (B \boxplus C) \ominus D, \\ B &= (A \boxplus D) \ominus C, \\ C &= (A \boxplus D) \ominus B, \\ D &= (B \boxplus C) \ominus A. \end{aligned} \tag{5.8}$$

Furthermore, the two gyrodiagonals of a non-degenerate gyroparallelogram are concurrent, the concurrency point being the gyromidpoint of each of the two gyrodiagonals.

Proof The last equation in (5.8) follows from the gyroparallelogram condition (5.7) in Definition 5.2 of the gyroparallelogram. By the right cancellation law (1.79), p. 21, this equation is equivalent to the equation

$$A \boxplus D = B \boxplus C. \tag{5.9}$$

By (1.37), p. 13, the coaddition $\boxplus$ in Einstein gyrovector spaces is commutative. Hence, by the right cancellation law (1.78), p. 21, the equation in (5.9) is equivalent to each of the equations in (5.8), thus verifying (5.8).

Equation (5.9) implies the *gyroparallelogram gyromidpoint condition*

$$\frac{1}{2} \otimes (A \boxplus D) = \frac{1}{2} \otimes (B \boxplus C). \tag{5.10}$$

By (4.72), p. 101, the left- and the right-hand side of (5.10) are, respectively, the gyromidpoint of the gyrodiagonal AD and the gyrodiagonal BC. Hence, (5.10) implies that the gyromidpoints of the two gyrodiagonals of the gyroparallelogram coincide. □

A gyroquadrangle is a not self-intersecting gyropolygon with four sides and four vertices in an Einstein gyrovector space. It is non-degenerate if any three of its vertices are non-gyrocollinear. The following theorem characterizes non-degenerate gyroquadrangles which are non-degenerate gyroparallelograms.

Theorem 5.4 *A non-degenerate gyroquadrangle $ABDC$ in an Einstein gyrovector space is a non-degenerate gyroparallelogram if and only if its gyrodiagonals AD and BC intersect at their gyromidpoints.*

Proof The gyrodiagonals AD and BC of a non-degenerate gyroquadrangle $ABDC$, Fig. 5.3, in an Einstein gyrovector space intersect at their gyromidpoints if and only

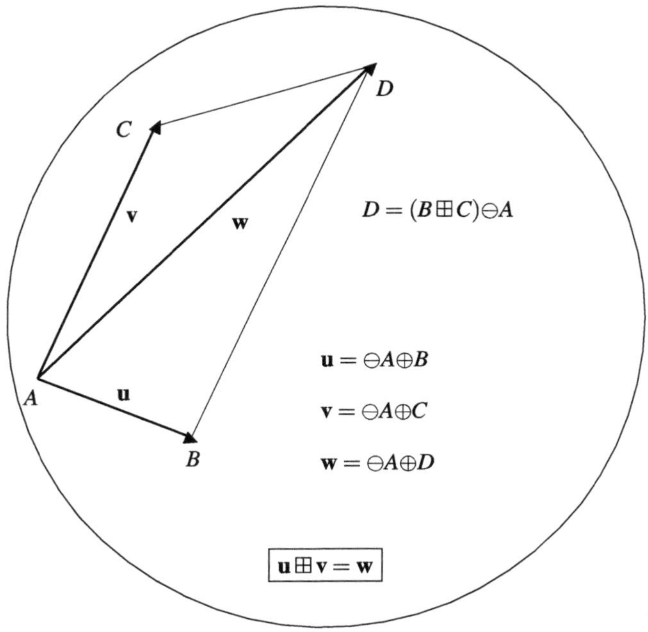

Fig. 5.4 The Einstein gyroparallelogram law of gyrovector addition. Let $A, B, C \in \mathbb{R}_s^n$ be any three points of an Einstein gyrovector space $(\mathbb{R}_s^n, \oplus, \otimes)$, giving rise to the two gyrovectors $\mathbf{u} = \ominus A \oplus B$ and $\mathbf{v} = \ominus A \oplus C$. Furthermore, let D be a point of the gyrovector space such that $ABDC$ is a gyroparallelogram, that is, $D = (B \boxplus C) \ominus A$ by Definition 5.2 of the gyroparallelogram. Then, Einstein coaddition of gyrovectors $\mathbf{u}$ and $\mathbf{v}$, $\mathbf{u} \boxplus \mathbf{v} = \mathbf{w}$, expresses the gyroparallelogram law, where $\mathbf{w} = \ominus A \oplus D$. Einstein coaddition, $\boxplus$, thus gives rise to the gyroparallelogram addition law of Einsteinian velocities, which is commutative and fully analogous to the parallelogram addition law of Newtonian velocities. Einsteinian velocities are, thus, gyrovectors that add according to the gyroparallelogram law just as Newtonian velocities are vectors that add according to the parallelogram law. Like vectors, a gyrovector $\ominus A \oplus B$ in an Einstein gyrovector space $(\mathbb{R}_s^n, \oplus, \otimes)$, $n = 2, 3$, is described graphically as a straight arrow from the tail A to the head B

if (5.10) is satisfied. The latter, in turn, is satisfied if and only if the gyroparallelogram condition (5.7) is satisfied, as explained in the proof of Theorem 5.3. Finally, by Definition 5.2, the gyroparallelogram condition is satisfied if and only if the gyroquadrangle $ABDC$ is a gyroparallelogram. $\square$

5.4 The Gyroparallelogram Law

The gyroparallelogram law of gyrovector addition gives a commutative binary operation between gyrovectors in Einstein gyrovector spaces, Fig. 5.4. As demonstrated in Theorem 5.5 below, it is fully analogous to the common parallelogram law of vector addition in Euclidean geometry.

Theorem 5.5 (The Gyroparallelogram (Addition) Law) *Let $ABDC$ be a gyroparallelogram in an Einstein gyrovector space $(\mathbb{R}^n_s, \oplus, \otimes)$, Fig. 5.4. Then*

$$(\ominus A \oplus B) \boxplus (\ominus A \oplus C) = \ominus A \oplus D. \tag{5.11}$$

Proof By Corollary 1.45, p. 41, and by the gyroparallelogram condition (5.7) we have

$$(\ominus A \oplus B) \boxplus (\ominus A \oplus C) = \ominus A \oplus \{(B \boxplus C) \ominus A\} = \ominus A \oplus D. \tag{5.12}$$

□

In the next theorem, we will uncover the relationship between opposite sides of the gyroparallelogram.

Theorem 5.6 *Opposite sides of a gyroparallelogram $ABDC$ in an Einstein gyrovector space $(\mathbb{R}^n_s, \oplus, \otimes)$, Figs. 5.3–5.4, are equal modulo gyrations,*

$$\begin{aligned} \ominus C \oplus D &= \mathrm{gyr}[C, \ominus B]\,\mathrm{gyr}[B, \ominus A](\ominus A \oplus B) = \mathrm{gyr}[C, \ominus B](B \ominus A), \\ \ominus B \oplus D &= \mathrm{gyr}[B, \ominus C]\,\mathrm{gyr}[C, \ominus A](\ominus A \oplus C) = \mathrm{gyr}[B, \ominus C](C \ominus A) \end{aligned} \tag{5.13}$$

and, equivalently,

$$\begin{aligned} \ominus C \oplus D &= \ominus\, \mathrm{gyr}[C, \ominus B](\ominus B \oplus A), \\ \ominus C \oplus A &= \ominus\, \mathrm{gyr}[C, \ominus B](\ominus B \oplus D). \end{aligned} \tag{5.14}$$

In particular, two opposite sides of a gyroparallelogram are congruent, having equal gyrolengths,

$$\begin{aligned} \|\ominus A \oplus B\| &= \|\ominus C \oplus D\|, \\ \|\ominus A \oplus C\| &= \|\ominus B \oplus D\|. \end{aligned} \tag{5.15}$$

Proof By Theorem 1.13, p. 18, we have

$$\ominus A \oplus D = (\ominus A \oplus C) \oplus \mathrm{gyr}[\ominus A, C](\ominus C \oplus D), \tag{5.16}$$

and by Theorem 5.5, p. 124, noting the definition of the gyrogroup cooperation, we have

$$\ominus A \oplus D = (\ominus A \oplus C) \boxplus (\ominus A \oplus B) = (\ominus A \oplus C) \oplus \mathrm{gyr}[\ominus A \oplus C, A \ominus B](\ominus A \oplus B). \tag{5.17}$$

Comparing the right-hand side of (5.16) with the extreme right-hand side of (5.17), and employing a left cancellation, we have

$$\mathrm{gyr}[\ominus A \oplus C, A \ominus B](\ominus A \oplus B) = \mathrm{gyr}[\ominus A, C](\ominus C \oplus D). \tag{5.18}$$

Identity (5.18) can be written in terms of Identity (1.180), p. 39, as

$$\mathrm{gyr}[A, \ominus C]\,\mathrm{gyr}[C, \ominus B]\,\mathrm{gyr}[B, \ominus A](\ominus A \oplus B) = \mathrm{gyr}[\ominus A, C](\ominus C \oplus D) \qquad (5.19)$$

which is reducible to the first identity in (5.13) by eliminating $\mathrm{gyr}[A, \ominus C]$ on both sides of (5.19). Similarly, interchanging B and C, one can verify the second identity in (5.13).

The equivalence between (5.13) and (5.14) follows from the gyroautomorphic inverse property and a gyration inversion.

Finally, (5.15) follows from (5.13) since gyrations preserve the gyrolength. □

The Newtonian counterpart of the gyroparallelogram addition law (5.11) gives the parallelogram addition law

$$(-A + B) + (-A + C) = -A + D \quad \text{(Euclidean Geometry)} \qquad (5.20)$$

with A, B, D, C being the vertices of a Euclidean parallelogram, where $-A + B$, $-A + C$, and $-A + D$ are three Euclidean vectors that emanate from the point A, and where the point D satisfies the parallelogram condition analogous to the gyroparallelogram condition (5.7),

$$D = B + C - A \quad \text{(Euclidean Geometry)} \qquad (5.21)$$

in $\mathbb{R}^n$.

The parallelogram condition (5.21) is equivalent to the *parallelogram midpoint condition*

$$\frac{1}{2}(A + D) = \frac{1}{2}(B + C) \qquad (5.22)$$

according to which the diagonals AD and BC of the parallelogram $ABDC$ in $\mathbb{R}^n$ intersect each other at their midpoints. This equivalence is in full analogy with the equivalence between the gyroparallelogram condition (5.7) and its associated gyroparallelogram gyromidpoint condition (5.10).

In classical mechanics, Newtonian velocity vectors and the parallelogram addition law of vectors are determined by the same binary operation, $+$, as we see from (5.20). In contrast, in relativistic mechanics, Einsteinian velocity gyrovectors and the gyroparallelogram addition law of gyrovectors are determined by two different binary operations, $\oplus$ and $\boxplus$, that interplay harmoniously, as we see from (5.11). Indeed, following (5.11), as shown graphically in Fig. 5.4, gyrovectors are determined by Einstein addition, $\oplus$, which is noncommutative, while gyrovector addition is given by Einstein coaddition, $\boxplus$, (1.37), p. 13, which is commutative. The introduction of Einstein coaddition $\boxplus$ into special relativity thus extends significantly the unfinished symphony of Einstein addition $\oplus$.

Experimental evidence that supports the physical significance of Einstein gyroparallelogram addition law (5.11) is provided by the relativistic interpretation of the cosmological stellar aberration phenomenon, as indicated in the Epilogue of this book, Chap. 10, and as explained in detail in [63, Chap. 13].

Chapter 6
Gyrotrigonometry

Abstract Trigonometry is the study of relations between the triangle three sides and three angles. In full analogy, gyrotrigonometry is the study of relations between the gyrotriangle three sides and three gyroangles. In this chapter, gyrotrigonometry is studied in Einstein gyrovector spaces or, equivalently, in the Cartesian–Beltrami–Klein ball model of hyperbolic geometry. Remarkably, the elementary gyrotrigonometric functions coincide with the common elementary trigonometric functions. They, however, stem from a system of two distinct Pythagorean identities that each right-gyroangled gyrotriangle possesses. In the transition from hyperbolic to Euclidean geometry, these two distinct Pythagorean identities degenerate into the single Pythagorean identity that each right-angled triangle possesses in Euclidean geometry.

6.1 Gyroangles

The standard notation that we use with a gyrotriangle $A_1A_2A_3$ in $\mathbb{R}^n_s$ is presented in Fig. 6.1 for $n = 2$. In this notation, an Einstein gyrotriangle $A_1A_2A_3$ has (i) three vertices, A_1, A_2 and A_3; (ii) three gyroangles, α_1, α_2 and α_3; and (iii) three sides, which form the three gyrovectors $\mathbf{a}_{12}$, $\mathbf{a}_{23}$ and $\mathbf{a}_{31}$; with respective (iv) three side-gyrolengths a_{12}, a_{23} and a_{31}, as defined in (6.1) below and shown in Fig. 6.1.

The standard notation that we use with a gyrotriangle $A_1A_2A_3$, presented in Fig. 6.1, involves the following equations:

$$
\begin{aligned}
\mathbf{a}_{12} &= \ominus A_1 \oplus A_2, & a_{12} &= \|\mathbf{a}_{12}\|, & \gamma_{21} &= \gamma_{12} = \gamma_{a_{12}} = \gamma_{\mathbf{a}_{12}}, \\
\mathbf{a}_{13} &= \ominus A_1 \oplus A_3, & a_{13} &= \|\mathbf{a}_{13}\|, & \gamma_{31} &= \gamma_{13} = \gamma_{a_{13}} = \gamma_{\mathbf{a}_{13}}, \\
\mathbf{a}_{23} &= \ominus A_2 \oplus A_3, & a_{23} &= \|\mathbf{a}_{23}\|, & \gamma_{32} &= \gamma_{23} = \gamma_{a_{23}} = \gamma_{\mathbf{a}_{23}}.
\end{aligned}
\tag{6.1}
$$

Let $\ominus A_1 \oplus A_2$ and $\ominus A_1 \oplus A_3$ be two rooted gyrovectors with a common tail A_1, see Fig. 6.1. They include a gyroangle $\alpha_1 = \angle A_2A_1A_3 = \angle A_3A_1A_2$, the radian

A.A. Ungar, *Hyperbolic Triangle Centers*,
Fundamental Theories of Physics 166,
DOI 10.1007/978-90-481-8637-2_6,

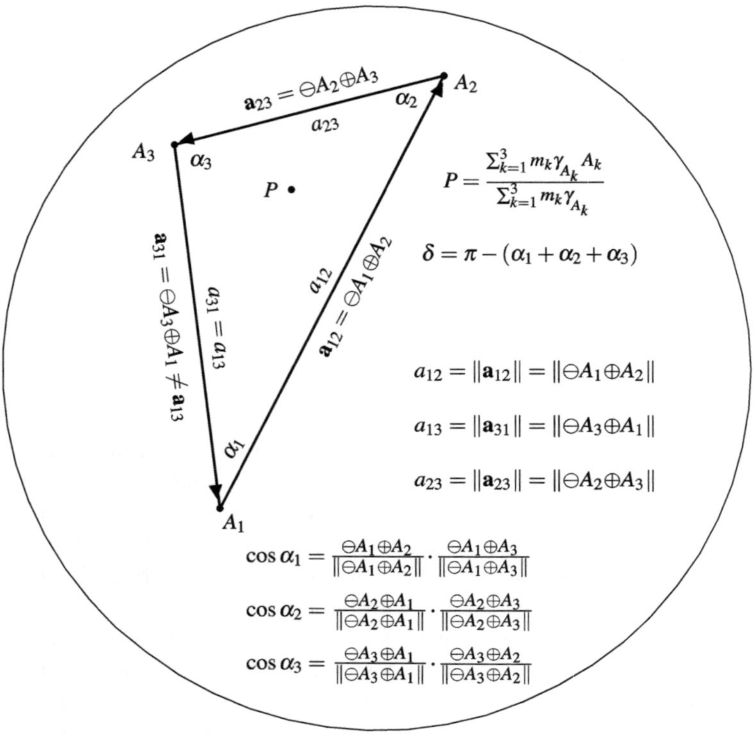

Fig. 6.1 The gyrotriangle $A_1A_2A_3$ in an Einstein gyrovector space $(\mathbb{R}_s^n, \oplus, \otimes)$ is shown for $n = 2$, along with its associated standard *index notation*. The gyrotriangle vertices, A_1, A_2 and A_3, are any non-gyrocollinear points of $\mathbb{R}_s^n$. Its sides are presented graphically as gyrosegments that join the vertices. They form gyrovectors, $\mathbf{a}_{ij}$, side-gyrolengths, $a_{ij} = \|\mathbf{a}_{ij}\|$, $1 \le i, j \le 3$, $i \ne j$, and gyroangles, α_k, $k = 1, 2, 3$. The gyrotriangle gyroangle sum is less than π, the difference, $\delta = \pi - (\alpha_1 + \alpha_2 + \alpha_3)$, being the gyrotriangular defect. The gyrocosine function of the gyrotriangle gyroangles is presented. Remarkably, it assumes a form that is fully analogous to the cosine function of Euclidean trigonometry. The point P is a generic point on the interior of the gyrotriangle, with positive gyrobarycentric coordinates $(m_1 : m_1 : m_3)$ with respect to the gyrotriangle vertices, (4.17), p. 88

measure of which is given by the equation

$$\cos\alpha_1 = \frac{\ominus A_1 \oplus A_2}{\|\ominus A_1 \oplus A_2\|} \cdot \frac{\ominus A_1 \oplus A_3}{\|\ominus A_1 \oplus A_3\|} \tag{6.2}$$

shown in Fig. 6.1. Accordingly,

$$\alpha_1 = \cos^{-1} \frac{\ominus A_1 \oplus A_2}{\|\ominus A_1 \oplus A_2\|} \cdot \frac{\ominus A_1 \oplus A_3}{\|\ominus A_1 \oplus A_3\|}, \tag{6.3}$$

where cos and $\cos^{-1} = \arccos$ are the standard cosine and arccosine functions of trigonometry. In the context of gyroangles, as in (6.2)–(6.3), we refer these functions of trigonometry to as the functions *gyrocosine* and *gyroarccosine* of gyrotrigonometry.

Gyrotriangle gyroangle sum in hyperbolic geometry is less than π. The difference, δ,

$$\delta = \pi - (\alpha_1 + \alpha_2 + \alpha_3) \tag{6.4}$$

being the triangular *defect*.

Theorem 6.1 (Gyroangle Gyroinvariance) *Gyroangles are gyroinvariant.*

Proof Following Definition 2.7, p. 57, we have to show that for any gyroangle α in an Einstein gyrovector space $(\mathbb{R}^n_s, \oplus, \otimes)$, $\cos\alpha$ remains invariant under both left gyrotranslations and rotations of $\mathbb{R}^n_s$.

The gyroangle α_1 is invariant under the action of left gyrotranslations. Indeed, by (1.65), p. 18, and (2.9), p. 47, we have

$$\begin{aligned}
\cos\alpha'_1 &= \frac{\ominus(X\oplus A_1)\oplus(X\oplus A_2)}{\|\ominus(X\oplus A_1)\oplus(X\oplus A_2)\|}\cdot\frac{\ominus(X\oplus A_1)\oplus(X\oplus A_3)}{\|\ominus(X\oplus A_1)\oplus(X\oplus A_3)\|} \\
&= \frac{\mathrm{gyr}[X, A_1](\ominus A_1\oplus A_2)}{\|\mathrm{gyr}[X, A_1](\ominus A_1\oplus A_2)\|}\cdot\frac{\mathrm{gyr}[X, A_1](\ominus A_1\oplus A_3)}{\|\mathrm{gyr}[X, A_1](\ominus A_1\oplus A_3)\|} \\
&= \frac{\ominus A_1\oplus A_2}{\|\ominus A_1\oplus A_2\|}\cdot\frac{\ominus A_1\oplus A_3}{\|\ominus A_1\oplus A_3\|} \\
&= \cos\alpha_1
\end{aligned} \tag{6.5}$$

for all $A_1, A_2, A_3, X \in \mathbb{R}^n_s$.

Similarly, the gyroangle α_1 is invariant under the action of rotations of $\mathbb{R}^n_s$ about its origin. Indeed, by (2.30b), p. 56, we have

$$\begin{aligned}
\cos\alpha''_1 &= \frac{\ominus RA_1\oplus RA_2}{\|\ominus RA_1\oplus RA_2\|}\cdot\frac{\ominus RA_1\oplus RA_3}{\|\ominus RA_1\oplus RA_3\|} \\
&= \frac{R(\ominus A_1\oplus A_2)}{\|R(\ominus A_1\oplus A_2)\|}\cdot\frac{R(\ominus A_1\oplus A_3)}{\|R(\ominus A_1\oplus A_3)\|} \\
&= \frac{\ominus A_1\oplus A_2}{\|\ominus A_1\oplus A_2\|}\cdot\frac{\ominus A_1\oplus A_3}{\|\ominus A_1\oplus A_3\|} \\
&= \cos\alpha_1
\end{aligned} \tag{6.6}$$

for all $A_1, A_2, A_3 \in \mathbb{R}^n$ and $R \in SO(n)$, since rotations $R \in SO(n)$ preserve the inner product and the norm in $\mathbb{R}^n_s$. □

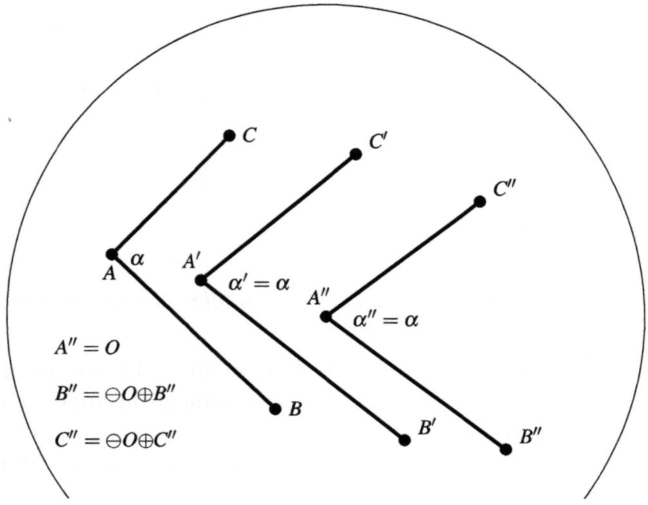

$$\cos\alpha = \frac{\ominus A \oplus B}{\|\ominus A \oplus B\|} \cdot \frac{\ominus A \oplus C}{\|\ominus A \oplus C\|} = \frac{\ominus A' \oplus B'}{\|\ominus A' \oplus B'\|} \cdot \frac{\ominus A' \oplus C'}{\|\ominus A' \oplus C'\|} = \frac{B''}{\|B''\|} \cdot \frac{C''}{\|C''\|}$$

Fig. 6.2 The Origin of any Einstein Gyrovector Space is Conformal. A gyroangle α is formed by two intersecting gyrorays in an Einstein gyrovector space $(\mathbb{R}_s^n, \oplus, \otimes)$. The Einstein gyrovector space is equipped with Cartesian coordinates that it inherits from its Euclidean space $\mathbb{R}^n$, with origin $O = \mathbf{0} = (0, \dots, 0)$, as shown in Figs. 2.3–2.4, p. 52. Gyroangles are invariant under left gyrotranslations, see (6.5). Shown are two successive left gyrotranslations of the gyroangle $\alpha = \angle BAC$ into $\alpha = \angle B'A'C'$, and the latter into $\alpha = \angle B''A''C'' = \angle B''OC''$, so that $\cos\alpha = (B''/\|B''\|)\cdot(C''/\|C''\|)$ is identical with its Euclidean counterpart. Hence, the origin of an Einstein gyrovector space is "conformal" in the sense that the measure of any (Einsteinian) gyroangle $\angle B''OC''$ in $\mathbb{R}_s^n$ with vertex at the origin O of $\mathbb{R}_s^n$ coincides with the measure of its corresponding (Euclidean) angle $\angle B''OC''$ in $\mathbb{R}^n$ with vertex at the origin O of $\mathbb{R}^n$

Being invariant under the motions of $\mathbb{R}_s^n$, which are left gyrotranslations and rotations about the origin, gyroangles are geometric objects of the hyperbolic geometry of Einstein gyrovector spaces.

6.2 The Gyroangle–Angle Relationship

By Theorem 6.1, gyroangles in Einstein gyrovector spaces are invariant under left gyrotranslations. Hence, any gyroangle can be left gyrotranslated without distortion into a special position where its vertex coincides with the origin, O, of the Einstein gyrovector space. Interestingly, once located in a special position, the gyroangle coincides with its Euclidean counterpart, as illustrated in Fig. 6.2.

Following Fig. 6.2, let $\alpha = \angle BAC$ be an arbitrary gyroangle in an Einstein gyrovector space $(\mathbb{R}_s^n, \oplus, \otimes)$, with vertex A different from the origin O of $\mathbb{R}_s^n$, and let $X \in \mathbb{R}_s^n$ be any point. A left gyrotranslation of A, B, C by X results in the left gyrotranslated gyroangle $\alpha' = \angle B'A'C'$ where

$$A' = X \oplus A, \qquad B' = X \oplus B, \qquad C' = X \oplus C \tag{6.7}$$

and where, by Theorem 6.1, $\alpha' = \alpha$.

It should be noted that it is the hyperbolic measures of gyroangles α' and α that are equal. Their Euclidean measures, in general, are not equal as one can see in Fig. 6.2.

Let us now left gyrotranslate the points A', B', $C' \in \mathbb{R}_s^n$ by $\ominus B'$, so that gyroangle α' becomes $\alpha'' = \angle B''A''C''$ where

$$A'' = \ominus B' \oplus A', \qquad B'' = \ominus B' \oplus B' = O = \mathbf{0}, \qquad C'' = \ominus B' \oplus C' \tag{6.8}$$

and where, by Theorem 6.1, $\alpha'' = \alpha'$.

Accordingly, by Theorem 6.1, $\cos\alpha$ is given by each of the inner products between unit gyrovectors in (6.9),

$$\cos\alpha = \frac{\ominus A \oplus B}{\|\ominus A \oplus B\|} \cdot \frac{\ominus A \oplus C}{\|\ominus A \oplus C\|} = \frac{\ominus A' \oplus B'}{\|\ominus A' \oplus B'\|} \cdot \frac{\ominus A' \oplus C'}{\|\ominus A' \oplus C'\|} = \frac{B''}{\|B''\|} \cdot \frac{C''}{\|C''\|}. \tag{6.9}$$

Remarkably, the inner product between unit gyrovectors in $\mathbb{R}_s^n$ on the extreme right-hand side of (6.9) can be viewed as an inner product between unit vectors in $\mathbb{R}^n$. Hence,

(i) On the one hand, any gyroangle α in an Einstein gyrovector space can be left gyrotranslated without distorting its measure into a special position where its vertex coincides with the origin, O, of the space; and
(ii) On the other hand, the measure of a gyroangle α in $\mathbb{R}_s^n$ with vertex O equals its measure when viewed as a Euclidean angle in $\mathbb{R}^n$

The special positions of a gyroangle are positions where the gyroangle vertex coincides with the origin O of its gyrovector space. The measure of a gyroangle in a special position in $\mathbb{R}_s^n$ equals its measure when viewed as a corresponding angle in the Euclidean geometry of $\mathbb{R}^n$. Hence, the origin of an Einstein gyrovector space $\mathbb{R}_s^n$ is said to be *conformal*.

The result that every gyroangle can be moved without distortion into special positions where it can be viewed, without distorting its measure, as a Euclidean angle is crucially important in the gyrotrigonometry of Einstein gyrovector spaces. It implies that every trigonometric identity of trigonometric functions remains valid in gyrotrigonometry, giving rise to a corresponding gyrotrigonometric identity of gyrotrigonometric functions. Accordingly, we refer these identities to as trigonometric/gyrotrigonometric identities. Thus, for instance, the familiar trigonometric identities

$$\begin{aligned} \cos^2\alpha + \sin^2\alpha &= 1, \\ \cot\frac{\alpha}{2} &= \frac{1+\cos\alpha}{\sin\alpha}, \end{aligned} \tag{6.10}$$

$\alpha \in \mathbb{R}$, remain valid in gyrotrigonometry as well. To see this, one may left gyrotranslate the gyroangle α into a special position, where it can be viewed as an angle satisfying (6.10). This angle α can be left gyrotranslated back to its original position, where it can no longer be viewed as an angle. But, since gyroangles are invariant under left gyrotranslations, gyroangle α still obeys (6.10).

As a remarkable consequence of this intimate relationship between trigonometric and gyrotrigonometric functions, we will find in this book the following result:

Gyrobarycentric coordinates of gyrotriangle gyrocenters that are determined in terms of gyrotriangle gyroangles survive unimpaired, in form, in the transition from hyperbolic to Euclidean geometry. The converse is, however, not true for the following reason:

Barycentric coordinates of triangle centers that are determined in terms of triangle angles need not survive a transition from Euclidean to hyperbolic geometry since, typically, trigonometric expressions for triangle centers embody the assumption that triangle angle sum is π. Clearly, a triangle trigonometric expression that embodies this assumption need not survive in hyperbolic geometry.

6.3 The Law of Gyrocosines

Let $\ominus A \oplus B$ and $\ominus A \oplus C$ be two gyrovectors that form two sides of gyrotriangle ABC and include the gyrotriangle gyroangle α in an Einstein gyrovector space $(\mathbb{R}^n_s, \oplus, \otimes)$, as shown in Fig. 6.3 for $n=2$.

By the Gyrotranslation Theorem 1.14, p. 18,

$$\ominus(\ominus A \oplus B) \oplus (\ominus A \oplus C) = \mathrm{gyr}[\ominus A, B](\ominus B \oplus C). \tag{6.11}$$

Since gyrations preserve the norm, (1.33), p. 10,

$$\big\| \ominus(\ominus A \oplus B) \oplus (\ominus A \oplus C) \big\| = \big\| \mathrm{gyr}[\ominus A, B](\ominus B \oplus C) \big\| = \| \ominus B \oplus C \|. \tag{6.12}$$

In the notation of Fig. 6.3 for gyrotriangle ABC, (6.12) is written as

$$\| \ominus \mathbf{c} \oplus \mathbf{b} \| = \| \mathbf{a} \|, \tag{6.13}$$

implying

$$\gamma_{\ominus \mathbf{c} \oplus \mathbf{b}} = \gamma_{\mathbf{a}} = \gamma_a. \tag{6.14}$$

By (1.8), p. 5, we have

$$\gamma_{\ominus \mathbf{c} \oplus \mathbf{b}} = \gamma_b \gamma_c \left(1 - \frac{\mathbf{b} \cdot \mathbf{c}}{s^2} \right). \tag{6.15}$$

The gyrocosine of gyroangle $\alpha = \angle BAC$ of gyrotriangle ABC in Fig. 6.3 is given by

$$\cos \alpha = \frac{\ominus A \oplus B}{\| \ominus A \oplus B \|} \cdot \frac{\ominus A \oplus C}{\| \ominus A \oplus C \|} = \frac{\mathbf{c}}{c} \cdot \frac{\mathbf{b}}{b} \tag{6.16}$$

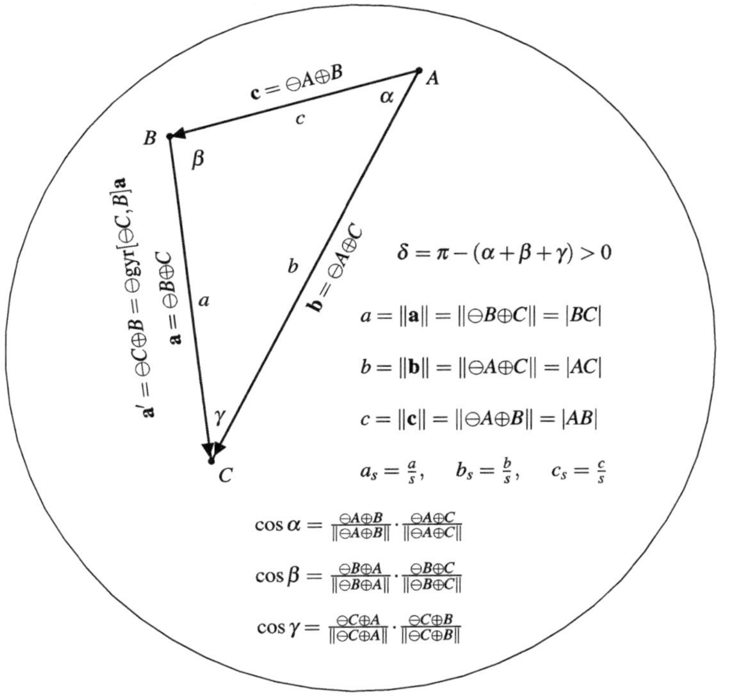

Fig. 6.3 Gyrotriangle ABC, along with its standard notation, in an Einstein gyrovector space. The notation that we use with a gyrotriangle ABC, its gyrovector sides, and its gyroangles in an Einstein gyrovector space $(\mathbb{R}^n_s, \oplus, \otimes)$ is shown here for the Einstein gyrovector plane $(\mathbb{R}^2_s, \oplus, \otimes)$

so that

$$\frac{\mathbf{b}\cdot\mathbf{c}}{s^2} = \frac{bc}{s^2}\cos\alpha = b_s c_s \cos\alpha, \tag{6.17}$$

where $b_s = b/s$, etc.

Following (6.14)–(6.17), we have

$$\gamma_a = \gamma_b \gamma_c (1 - b_s c_s \cos\alpha), \tag{6.18}$$

where a, b, c are the side-gyrolengths of gyrotriangle ABC in an Einstein gyrovector space $(\mathbb{R}^n_s, \oplus, \otimes)$, as shown in Fig. 6.3 for $n = 2$.

Identity (6.18) is the *law of gyrocosines* in the gyrotrigonometry of Einstein gyrovector spaces. As in trigonometry, it is useful for calculating one side, a, of a gyrotriangle ABC, Fig. 6.3, when the gyroangle α opposite to side a and the other two sides (that is, their gyrolengths), b and c, are known.

Remarkably, in the Euclidean limit of large s, $s \to \infty$, gamma factors tend to 1 and the law of gyrocosines (6.18) reduces to the trivial identity $1 = 1$. Hence, (6.18) has no immediate Euclidean counterpart, thus presenting a disanalogy between hyperbolic and Euclidean geometry. As a result, each of Theorems 6.2 and 6.5 below has no Euclidean counterpart as well.

6.4 The *SSS* to *AAA* Conversion Law

Let ABC be a gyrotriangle in an Einstein gyrovector space $(\mathbb{R}_s^n, \otimes, \oplus)$ with its standard notation in Fig. 6.3. According to (6.18), the gyrotriangle ABC possesses the following three identities, each of which represents its law of gyrocosines,

$$\begin{aligned}\gamma_a &= \gamma_b\gamma_c(1 - b_s c_s \cos\alpha),\\ \gamma_b &= \gamma_a\gamma_c(1 - a_s c_s \cos\beta),\\ \gamma_c &= \gamma_a\gamma_b(1 - a_s b_s \cos\gamma).\end{aligned} \tag{6.19}$$

Like Euclidean triangles, the gyroangles of a gyrotriangle are uniquely determined by its sides. Solving the system (6.19) of three identities for the three unknowns $\cos\alpha$, $\cos\beta$ and $\cos\gamma$, and employing (1.9), p. 5, we obtain the following theorem:

Theorem 6.2 (The Law of Gyrocosines; The *SSS* to *AAA* Conversion Law) *Let ABC be a gyrotriangle in an Einstein gyrovector space $(\mathbb{R}_s^n, \oplus, \otimes)$. Then, in the gyrotriangle notation in Fig.* 6.3,

$$\begin{aligned}\cos\alpha &= \frac{-\gamma_a + \gamma_b\gamma_c}{\gamma_b\gamma_c b_s c_s} = \frac{-\gamma_a + \gamma_b\gamma_c}{\sqrt{\gamma_b^2 - 1}\sqrt{\gamma_c^2 - 1}},\\ \cos\beta &= \frac{-\gamma_b + \gamma_a\gamma_c}{\gamma_a\gamma_c a_s c_s} = \frac{-\gamma_b + \gamma_a\gamma_c}{\sqrt{\gamma_a^2 - 1}\sqrt{\gamma_c^2 - 1}},\\ \cos\gamma &= \frac{-\gamma_c + \gamma_a\gamma_b}{\gamma_a\gamma_b a_s b_s} = \frac{-\gamma_c + \gamma_a\gamma_b}{\sqrt{\gamma_a^2 - 1}\sqrt{\gamma_b^2 - 1}}.\end{aligned} \tag{6.20}$$

The identities in (6.20) form the *SSS* (Side–Side–Side) to *AAA* (gyroAngle–gyroAngle–gyroAngle) conversion law in Einstein gyrovector spaces. This law is useful for calculating the gyroangles of a gyrotriangle in an Einstein gyrovector space when its sides (that is, its side-gyrolengths) are known.

In full analogy with the trigonometry of triangles, the *gyrosine* of a gyrotriangle gyroangle α is nonnegative, given by the equation

$$\sin\alpha = \sqrt{1 - \cos^2\alpha} \ge 0. \tag{6.21}$$

Hence, it follows from Theorem 6.2 that the gyrosines of the gyrotriangle gyroangles in that Theorem are given by

$$\begin{aligned}\sin\alpha &= \frac{\sqrt{1+2\gamma_a\gamma_b\gamma_c-\gamma_a^2-\gamma_b^2-\gamma_c^2}}{\sqrt{\gamma_b^2-1}\sqrt{\gamma_c^2-1}},\\ \sin\beta &= \frac{\sqrt{1+2\gamma_a\gamma_b\gamma_c-\gamma_a^2-\gamma_b^2-\gamma_c^2}}{\sqrt{\gamma_a^2-1}\sqrt{\gamma_c^2-1}},\\ \sin\gamma &= \frac{\sqrt{1+2\gamma_a\gamma_b\gamma_c-\gamma_a^2-\gamma_b^2-\gamma_c^2}}{\sqrt{\gamma_a^2-1}\sqrt{\gamma_b^2-1}}.\end{aligned} \tag{6.22}$$

Any gyrotriangle gyroangle α satisfies $0 < \alpha < \pi$, so that $\sin\alpha > 0$ for any gyrotriangle gyroangle. Following (6.22), we have the inequality

$$1+2\gamma_a\gamma_b\gamma_c-\gamma_a^2-\gamma_b^2-\gamma_c^2 > 0 \tag{6.23}$$

for any gyrotriangle in an Einstein gyrovector space, in the notation of Theorem 6.2 and Fig. 6.3.

Identities (6.22) immediately give rise to the identities

$$\frac{\sin\alpha}{\sqrt{\gamma_a^2-1}} = \frac{\sin\beta}{\sqrt{\gamma_b^2-1}} = \frac{\sin\gamma}{\sqrt{\gamma_c^2-1}} \tag{6.24}$$

that form the law of gyrosines that we will study in Theorem 6.9, p. 140.

6.5 Inequalities for Gyrotriangles

Elegant inequalities for gyrotriangles in Einstein gyrovector spaces result immediately from (6.22), as we see in the following theorem:

Theorem 6.3 *Let ABC be a gyrotriangle in an Einstein gyrovector space with the notation in Fig.* 6.3, *p.* 133, *for its side-gyrolengths a, b, c. Then*

$$\gamma_a^2+\gamma_b^2+\gamma_c^2-1 < 2\gamma_a\gamma_b\gamma_c \le \gamma_a^2+\gamma_b^2\gamma_c^2. \tag{6.25}$$

Proof The left inequality in (6.25) follows immediately from (6.23).

The inequality $\sin^2\alpha \le 1$ implies, by means of (6.22), the inequality

$$\frac{1+2\gamma_a\gamma_b\gamma_c-\gamma_a^2-\gamma_b^2-\gamma_c^2}{(\gamma_b^2-1)(\gamma_c^2-1)} \le 1 \tag{6.26}$$

which, in turn, implies

$$2\gamma_a\gamma_b\gamma_c \le \gamma_a^2 + \gamma_b^2\gamma_c^2. \tag{6.27}$$

□

An interesting chain of inequalities that results from (6.25), for instance, is

$$0 < \gamma_c^2 - 1 < 2\gamma_a\gamma_b\gamma_c - \gamma_a^2 - \gamma_b^2 < \gamma_a^2 + \gamma_b^2\gamma_c^2 - \gamma_a^2 - \gamma_b^2 = \gamma_b^2(\gamma_c^2 - 1) \tag{6.28}$$

and, hence,

$$1 < \frac{2\gamma_a\gamma_b\gamma_c - \gamma_a^2 - \gamma_b^2}{\gamma_c^2 - 1} < \gamma_b^2. \tag{6.29}$$

Equality is attained in Inequality (6.27) when, and only when, $\alpha = \pi/2$. In this case, gyrotriangle ABC is a right gyroangled gyrotriangle, satisfying the Einstein–Pythagoras identity $\gamma_a = \gamma_b\gamma_c$, (6.57), p. 144, that will be studied in Sect. 6.10.

Gyrotriangle gyroangles vary over the open interval $(0, \pi)$. Accordingly, we present the following definition about gyrotriangles and their gyroangles.

Definition 6.4 (Acute, Right, Obtuse Gyroangles and Gyrotriangles)

1. An acute gyroangle θ is a gyroangle measuring between 0 and $\pi/2$ radians, $0 < \theta < \pi/2$.
2. A right gyroangle θ is a gyroangle measuring $\pi/2$ radians, $\theta = \pi/2$.
3. An obtuse gyroangle θ is a gyroangle measuring between $\pi/2$ and π radians, $\pi/2 < \theta < \pi$.
4. A gyrotriangle in which all three gyroangles are acute, $0 < \alpha, \beta, \gamma < \pi/2$, is acute.
5. A gyrotriangle in which one gyroangle is a right gyroangle, $\theta = \pi/2$, is right.
6. A gyrotriangle which has an obtuse gyroangle, $\pi/2 < \theta < \pi$, is obtuse.

It follows from (6.20), and from the result that the trigonometric cosine function, $\cos\alpha$, and the gyrotrigonometric gyrocosine function, ambiguously also denoted $\cos\alpha$, have the same behavior, that we have important equalities and inequalities for gyrotriangles in Einstein gyrovector spaces. These are:

$$\begin{cases} \gamma_{\mathbf{b}}\ \gamma_{\mathbf{c}} > \gamma_{\mathbf{a}} & \text{if and only if } \alpha < \frac{\pi}{2} \text{ (Acute);} \\ \gamma_{\mathbf{b}}\ \gamma_{\mathbf{c}} = \gamma_{\mathbf{a}} & \text{if and only if } \alpha = \frac{\pi}{2} \text{ (Right);} \\ \gamma_{\mathbf{b}}\ \gamma_{\mathbf{c}} < \gamma_{\mathbf{a}} & \text{if and only if } \alpha > \frac{\pi}{2} \text{ (Obtuse)} \end{cases} \tag{6.30}$$

and, similarly,

$$\begin{cases} \gamma_{\mathbf{a}}\ \gamma_{\mathbf{c}} > \gamma_{\mathbf{b}} & \text{if and only if } \beta < \frac{\pi}{2} \text{ (Acute);} \\ \gamma_{\mathbf{a}}\ \gamma_{\mathbf{c}} = \gamma_{\mathbf{b}} & \text{if and only if } \beta = \frac{\pi}{2} \text{ (Right);} \\ \gamma_{\mathbf{a}}\ \gamma_{\mathbf{c}} < \gamma_{\mathbf{b}} & \text{if and only if } \beta > \frac{\pi}{2} \text{ (Obtuse)} \end{cases} \tag{6.31}$$

and

$$\begin{cases} \gamma_{\mathbf{a}}\ \gamma_{\mathbf{b}} > \gamma_{\mathbf{c}} & \text{if and only if } \gamma < \frac{\pi}{2} \text{ (Acute);} \\ \gamma_{\mathbf{a}}\ \gamma_{\mathbf{b}} = \gamma_{\mathbf{c}} & \text{if and only if } \gamma = \frac{\pi}{2} \text{ (Right);} \\ \gamma_{\mathbf{a}}\ \gamma_{\mathbf{b}} < \gamma_{\mathbf{c}} & \text{if and only if } \gamma > \frac{\pi}{2} \text{ (Obtuse)} \end{cases} \tag{6.32}$$

where **a**, **b**, **c** are the sides of any given gyrotriangle in an Einstein gyrovector space and α, β, γ are their respective opposing gyroangles, as shown in Fig. 6.3, p. 133. The equalities in (6.30)–(6.32) correspond to right gyroangles, giving rise to the Einstein–Pythagoras identity (6.57), p. 144, that will be studied in Sect. 6.10.

6.6 The *AAA* to *SSS* Conversion Law

Unlike Euclidean triangles, the side-gyrolengths of a gyrotriangle are uniquely determined by its gyroangles, as the following theorem demonstrates.

Theorem 6.5 (The *AAA* to *SSS* Conversion Law I) *Let ABC be a gyrotriangle in an Einstein gyrovector space $(\mathbb{R}^n_s, \oplus, \otimes)$. Then, in the gyrotriangle notation in Fig.* 6.3,

$$\begin{aligned} \gamma_a &= \frac{\cos\alpha + \cos\beta\cos\gamma}{\sin\beta\sin\gamma}, \\ \gamma_b &= \frac{\cos\beta + \cos\alpha\cos\gamma}{\sin\alpha\sin\gamma}, \\ \gamma_c &= \frac{\cos\gamma + \cos\alpha\cos\beta}{\sin\alpha\sin\beta}, \end{aligned} \tag{6.33}$$

where, following (6.21), *the gyrosine of the gyrotriangle gyroangle* α, $\sin\alpha$, *is given by the nonnegative value of* $\sqrt{1-\cos^2\alpha}$, *etc.*

Proof Let ABC be a gyrotriangle in an Einstein gyrovector space $(\mathbb{R}^n_s, \otimes, \oplus)$ with its standard notation in Fig. 6.3. It follows straightforwardly from the *SSS* to *AAA* conversion law (6.20) that

$$\left(\frac{\cos\alpha + \cos\beta\cos\gamma}{\sin\beta\sin\gamma}\right)^2 = \frac{(\cos\alpha + \cos\beta\cos\gamma)^2}{(1-\cos^2\beta)(1-\cos^2\gamma)} = \gamma_a^2, \tag{6.34}$$

implying the first identity in (6.33). The remaining two identities in (6.33) are obtained from (6.20) in a similar way by vertex cyclic permutations. □

The identities in (6.33) form the *AAA* to *SSS* conversion law. This law is useful for calculating the sides (that is, the side-gyrolengths) of a gyrotriangle in an Einstein gyrovector space when its gyroangles are known. Thus, for instance, γ_a is obtained from the first identity in (6.33), and a is obtained from γ_a by Identity (1.9), p. 5.

Solving the third identity in (6.33) for $\cos\gamma$, we have

$$\begin{aligned}\cos\gamma &= -\cos\alpha\cos\beta + \gamma_c \sin\alpha\sin\beta \\ &= -\cos(\alpha+\beta) + (\gamma_c - 1)\sin\alpha\sin\beta,\end{aligned} \tag{6.35}$$

implying

$$\cos\gamma = \cos(\pi - \alpha - \beta) + (\gamma_c - 1)\sin\alpha\sin\beta. \tag{6.36}$$

In the Euclidean limit of large s, $s \to \infty$, γ_c tends to 1, so that the gyrotrigonometric identity (6.36) in hyperbolic geometry reduces to the trigonometric identity

$$\cos\gamma = \cos(\pi - \alpha - \beta) \quad \text{(Euclidean Geometry)} \tag{6.37}$$

in Euclidean geometry. The latter, in turn, is equivalent to the familiar result,

$$\alpha + \beta + \gamma = \pi \quad \text{(Euclidean Geometry)} \tag{6.38}$$

of Euclidean geometry, according to which the triangle angle sum is π.

As an immediate application of the *SSS* to *AAA* conversion law (6.20)–(6.22) and the *AAA* to *SSS* conversion law (6.33) we present the following two theorems:

Theorem 6.6 (The Isosceles Gyrotriangle Theorem) *A gyrotriangle is isosceles (that is, it has two sides congruent) if and only if the two gyroangles opposing its two congruent sides are congruent.*

Proof Using the gyrotriangle notation in Fig. 6.3, p. 133, if gyrotriangle ABC has two sides, say a and b, congruent, $a = b$, then $\gamma_a = \gamma_b$ so that, by (6.20)–(6.22), $\cos\alpha = \cos\beta$ and $\sin\alpha = \sin\beta$ implying $\alpha = \beta$.

Conversely, if gyrotriangle ABC has two gyroangles, say α and β, congruent, then by (6.33), $\gamma_a = \gamma_b$, implying $a = b$. □

Theorem 6.7 (The Equilateral Gyrotriangle Theorem) *A gyrotriangle is equilateral (that is, it has all three sides congruent) if and only if the gyrotriangle is equigyroangular (that is, it has all three gyroangles congruent).*

Proof Using the gyrotriangle notation in Fig. 6.3, p. 133, if gyrotriangle ABC has all three sides congruent, $a = b = c$, then $\gamma_a = \gamma_b = \gamma_c$ so that, by (6.20)–(6.22), $\cos\alpha = \cos\beta = \cos\gamma$ and $\sin\alpha = \sin\beta = \sin\gamma$, implying $\alpha = \beta = \gamma$.

Conversely, if gyrotriangle ABC has all three gyroangles congruent, then $\alpha = \beta = \gamma$ implying, by (6.33), $\gamma_a = \gamma_b = \gamma_c$. The latter implies $a = b = c$ by means of (1.9), p. 5. □

Trigonometry deals with relationships between sides (that is, side-lengths) and angles of triangles in Euclidean vector spaces $\mathbb{R}^n$, and with the trigonometric functions, $\sin\alpha$, $\cos\alpha$, etc., that describe these relationships.

In full analogy, Gyrotrigonometry deals with relationships between sides (that is, side-gyrolengths) and gyroangles of gyrotriangles in Einstein gyrovector spaces, and with the gyrotrigonometric functions, also denoted $\sin\alpha$, $\cos\alpha$, etc., that describe these relationships.

A most important point of gyrotrigonometry is that its gyrotrigonometric functions $\sin\alpha$, $\cos\alpha$, etc., possess the same properties that their Euclidean counterparts possess. Hence, it is justified to use in gyrotrigonometry the same notation commonly used in trigonometry. As a result, one can use computer algebra for symbolic manipulation, like Mathematica and Maple, to manipulate gyrotrigonometric expressions in gyrotrigonometry. Surely, a computer software like Mathematica or Maple is designed to deal with trigonometry, but it can be used to deal with gyrotrigonometry as well owing to the above mentioned important point.

Interesting gyrotriangle identities that follow from (6.33) by gyrotrigonometric identities, and which can easily be obtained by the use of Mathematica or Maple in trigonometry, are in the notation of Theorem 6.5,

$$\begin{aligned} \gamma_a^2 - 1 &= 4\frac{F(\alpha,\beta,\gamma)}{\sin^2\beta\sin^2\gamma}, \\ \gamma_b^2 - 1 &= 4\frac{F(\alpha,\beta,\gamma)}{\sin^2\alpha\sin^2\gamma}, \\ \gamma_c^2 - 1 &= 4\frac{F(\alpha,\beta,\gamma)}{\sin^2\alpha\sin^2\beta}, \\ \gamma_a\gamma_b\gamma_c - 1 &= 4\frac{F(\alpha,\beta,\gamma)(1+\cos\alpha\cos\beta\cos\gamma)}{\sin^2\alpha\sin^2\beta\sin^2\gamma}, \end{aligned} \tag{6.39}$$

where $F(\alpha,\beta,\gamma)$ is a symmetric functions of α, β and γ, given by

$$\begin{aligned} F(\alpha,\beta,\gamma) &:= \cos\frac{\alpha+\beta+\gamma}{2}\cos\frac{\alpha-\beta-\gamma}{2}\cos\frac{-\alpha+\beta-\gamma}{2}\cos\frac{-\alpha-\beta+\gamma}{2} \\ &= \frac{1}{4}\big(2\cos\alpha\cos\beta\cos\gamma + \cos^2\alpha + \cos^2\beta + \cos^2\gamma - 1\big) \\ &= \frac{1}{4}\frac{(1+2\gamma_a\gamma_b\gamma_c - \gamma_a^2 - \gamma_b^2 - \gamma_c^2)^2}{(\gamma_a^2-1)(\gamma_b^2-1)(\gamma_c^2-1)}. \end{aligned} \tag{6.40}$$

By (1.9), p. 5, (6.39) and (6.33),

$$\begin{aligned} a_s &= \frac{\sqrt{\gamma_a^2-1}}{\gamma_a} \\ &= 2\frac{\sqrt{F(\alpha,\beta,\gamma)}}{\cos\alpha+\cos\beta\cos\gamma}, \end{aligned} \tag{6.41}$$

thus obtaining the *AAA* to *SSS* conversion law in the following theorem:

Theorem 6.8 (The *AAA* to *SSS* Conversion Law II) *Let* ABC *be a gyrotriangle in an Einstein gyrovector space* $(\mathbb{R}^n_s, \oplus, \otimes)$. *Then, in the gyrotriangle notation in Fig.* 6.3,

$$a_s = 2\frac{\sqrt{F(\alpha, \beta, \gamma)}}{\cos\alpha + \cos\beta\cos\gamma},$$
$$b_s = 2\frac{\sqrt{F(\alpha, \beta, \gamma)}}{\cos\beta + \cos\alpha\cos\gamma}, \tag{6.42}$$
$$c_s = 2\frac{\sqrt{F(\alpha, \beta, \gamma)}}{\cos\gamma + \cos\alpha\cos\beta}.$$

Useful identities that follow from (6.33) and (6.42) are:

$$\gamma_a a_s = 2\frac{\sqrt{F(\alpha, \beta, \gamma)}}{\sin\beta\sin\gamma},$$
$$\gamma_b b_s = 2\frac{\sqrt{F(\alpha, \beta, \gamma)}}{\sin\alpha\sin\gamma}, \tag{6.43}$$
$$\gamma_c c_s = 2\frac{\sqrt{F(\alpha, \beta, \gamma)}}{\sin\alpha\sin\beta}.$$

6.7 The Law of Gyrosines

Theorem 6.9 (The Law of Gyrosines) *Let* ABC *be a gyrotriangle in an Einstein gyrovector space* $(\mathbb{R}^n_s, \oplus, \otimes)$. *Then, in the gyrotriangle notation in Fig.* 6.3, *p.* 133,

$$\frac{\sin\alpha}{\gamma_a a} = \frac{\sin\beta}{\gamma_b b} = \frac{\sin\gamma}{\gamma_c c} = \frac{1}{s}\sqrt{\frac{1 + 2\gamma_a\,\gamma_b\,\gamma_c\, - \gamma_a^2 - \gamma_b^2 - \gamma_c^2}{(\gamma_a^2 - 1)(\gamma_b^2 - 1)(\gamma_c^2 - 1)}}. \tag{6.44}$$

Proof It follows from (6.20)–(6.21) and Identity (1.9), p. 5, that

$$\left(\frac{\sin\alpha}{\gamma_a\ a}\right)^2 = \frac{1}{s^2}\frac{1-\cos^2\alpha}{\gamma_a^2 - 1} = \frac{1}{s^2}\frac{1 + 2\gamma_a\,\gamma_b\,\gamma_c\, - \gamma_a^2 - \gamma_b^2 - \gamma_c^2}{(\gamma_a^2 - 1)(\gamma_b^2 - 1)(\gamma_c^2 - 1)},$$
$$\left(\frac{\sin\beta}{\gamma_b\ b}\right)^2 = \frac{1}{s^2}\frac{1-\cos^2\beta}{\gamma_b^2 - 1} = \frac{1}{s^2}\frac{1 + 2\gamma_a\,\gamma_b\,\gamma_c\, - \gamma_a^2 - \gamma_b^2 - \gamma_c^2}{(\gamma_a^2 - 1)(\gamma_b^2 - 1)(\gamma_c^2 - 1)}, \tag{6.45}$$
$$\left(\frac{\sin\gamma}{\gamma_c\ c}\right)^2 = \frac{1}{s^2}\frac{1-\cos^2\gamma}{\gamma_c^2 - 1} = \frac{1}{s^2}\frac{1 + 2\gamma_a\,\gamma_b\,\gamma_c\, - \gamma_a^2 - \gamma_b^2 - \gamma_c^2}{(\gamma_a^2 - 1)(\gamma_b^2 - 1)(\gamma_c^2 - 1)}.$$

The result (6.44) of the theorem follows immediately from (6.45). □

One should note that the extreme right-hand side of each equation in (6.45) is symmetric in a, b, c.

The law of gyrosines (6.44), excluding the extreme right-hand side in (6.44), is fully analogous to the law of sines, to which it reduces in the Euclidean limit, $s \to \infty$, of large s, when gamma factors tend to 1, side-gyrolengths tend to side-lengths, and gyroangles tend to angles.

6.8 The *ASA* to *SAS* Conversion Law

Theorem 6.10 (The *ASA* to *SAS* Conversion Law) *Let ABC be a gyrotriangle in an Einstein gyrovector space $(\mathbb{R}^n_s, \oplus, \otimes)$. Then, in the gyrotriangle notation in Fig.* 6.3, *p.* 133,

$$\begin{aligned} a_s &= \frac{\gamma_c c_s \sin\alpha}{\cos\alpha \sin\beta + \gamma_c \sin\alpha \cos\beta}, \\ b_s &= \frac{\gamma_c c_s \sin\beta}{\cos\beta \sin\alpha + \gamma_c \sin\beta \cos\alpha}, \\ \cos\gamma &= \cos(\pi - \alpha - \beta) + (\gamma_c - 1)\sin\alpha \sin\beta. \end{aligned} \tag{6.46}$$

Proof Noting (1.9), p. 5, it follows from the third identity in (6.33) and from (6.20) that

$$\begin{aligned} &\left(\frac{\gamma_c c_s \sin\alpha}{\cos\alpha \sin\beta + \gamma_c \sin\alpha \cos\beta}\right)^2 \\ &= \left(\frac{\gamma_c c_s \sin\alpha \sin\beta}{\cos\alpha \sin^2\beta + \gamma_c \sin\alpha \sin\beta \cos\beta}\right)^2 \\ &= \frac{(\gamma_c^2 - 1)(1 - \cos^2\alpha)(1 - \cos^2\beta)}{(\cos\alpha(1 - \cos^2\beta) + (\cos\gamma + \cos\alpha\cos\beta)\cos\beta)^2} \\ &= \frac{(\gamma_c^2 - 1)(1 - \cos^2\alpha)(1 - \cos^2\beta)}{(\cos\alpha + \cos\beta\cos\gamma)^2} \\ &= a_s^2, \end{aligned} \tag{6.47}$$

thus obtaining the first identity in (6.46). The second identity in (6.46), for b_s, is obtained from the first identity, for a_s, by interchanging α and β. Finally, the third identity in (6.46) has been verified in (6.36). □

Employing Identity (1.9), p. 5, for c_s^2, the first two identities in (6.46) imply

$$a_s b_s = \frac{(\gamma_c^2 - 1)\sin\alpha \sin\beta}{(\cos\alpha \sin\beta + \gamma_c \sin\alpha \cos\beta)(\cos\beta \sin\alpha + \gamma_c \sin\beta \cos\alpha)}, \tag{6.48}$$

thus obtaining an identity that proves useful in calculating the gyrotriangle defect in (6.49).

The system of identities (6.46) of Theorem 6.10 gives the *ASA* (gyroAngle–Side–gyroAngle) to *SAS* (Side–gyroAngle–Side) conversion law. It is useful for calculating two sides and their included gyroangle of a gyrotriangle in an Einstein gyrovector space when the remaining two gyroangles and the side included between them are known.

6.9 Gyrotriangle Defect

Some algebraic manipulations are too difficult to be performed by hand, but can easily be accomplished by computer algebra, that is, a computer software for symbolic manipulation, like Mathematica or Maple. It follows by straightforward substitution from (6.33) and (6.48), using computer algebra, that

$$\begin{aligned}\frac{\gamma_a\gamma_b a_s b_s \sin\gamma}{(1+\gamma_a)(1+\gamma_b)-\gamma_a\gamma_b a_s b_s\cos\gamma} &= \cot\frac{\alpha+\beta+\gamma}{2}\\ &= \tan\frac{\pi-(\alpha+\beta+\gamma)}{2}\\ &= \tan\frac{\delta}{2},\end{aligned} \tag{6.49}$$

where

$$\delta = \pi - (\alpha+\beta+\gamma) \tag{6.50}$$

is the gyrotriangular *defect* of the gyrotriangle ABC, presented in Fig. 6.3, p. 133.

Identity (6.49) is useful for calculating the defect of a gyrotriangle in an Einstein gyrovector space when two side-gyrolengths and their included gyroangle of the gyrotriangle are known.

Let us now substitute $\cos\gamma$ from (6.20) and a_s^2 and b_s^2 from Identities like (1.9), p. 5, into $\tan^2(\delta/2)$ of (6.49), obtaining the identity

$$\begin{aligned}\tan^2\frac{\delta}{2} &= \frac{\gamma_a^2\gamma_b^2 a_s^2 b_s^2(1-\cos^2\gamma)}{((1+\gamma_a)(1+\gamma_b)-\gamma_a\gamma_b a_s b_s\cos\gamma)^2}\\ &= \frac{1+2\gamma_a\gamma_b\gamma_c-\gamma_a^2-\gamma_b^2-\gamma_c^2}{(1+\gamma_a+\gamma_b+\gamma_c)^2}\end{aligned} \tag{6.51}$$

which leads to the following theorem:

Theorem 6.11 (The Gyrotriangular Defect I) *Let ABC be a gyrotriangle in an Einstein gyrovector space $(\mathbb{R}^n_s, \oplus, \otimes)$ with the standard gyrotriangle notation in*

Fig. 6.3, *p*. 133. *Then the gyrotriangular defect* δ *of* ABC *is given by the equation*

$$\tan\frac{\delta}{2} = \frac{\sqrt{1+2\gamma_a\gamma_b\gamma_c - \gamma_a^2 - \gamma_b^2 - \gamma_c^2}}{1+\gamma_a+\gamma_b+\gamma_c}. \tag{6.52}$$

In the Newtonian–Euclidean limit of large s, $s \to \infty$, gamma factors tend to 1 so that (6.52) tends to $\tan(\delta/2) = 0$ and, hence, δ tends to 0. Hence, the triangular defects of Euclidean triangles vanish, as expected.

Theorem 6.12 (The Gyrotriangular Defect II) *Let* ABC *be a gyrotriangle in an Einstein gyrovector space* $(\mathbb{R}_s^n, \oplus, \otimes)$, *with the standard gyrotriangle notation in Fig*. 6.3, *p*. 133. *Then the gyrotriangular defect* δ *of gyrotriangle* ABC *is given by the equation*

$$\tan\frac{\delta}{2} = \frac{p\sin\gamma}{1-p\cos\gamma}, \tag{6.53}$$

where $\gamma, \delta, p > 0$, *and*

$$p^2 = \frac{\gamma_a - 1}{\gamma_a + 1}\frac{\gamma_b - 1}{\gamma_b + 1}. \tag{6.54}$$

Proof Employing (1.9), p. 5, the first equation in (6.51) can be written as (6.53) where $\gamma, \delta, p > 0$, and where p is given by (6.54). □

The gyrotriangular defect formula (6.52) of Theorem 6.11 is useful for calculating the defect of a gyrotriangle in an Einstein gyrovector space when the three sides of the gyrotriangle are known.

The gyrotriangular defect formula (6.53) of Theorem 6.12 is useful for calculating the defect of a gyrotriangle in an Einstein gyrovector space when two sides and their included gyroangle of the gyrotriangle are known.

6.10 Right Gyrotriangles

Theorem 6.13 (The Einstein–Pythagoras Theorem) *Let* ABC *be a gyrotriangle in an Einstein gyrovector space* $(\mathbb{R}_s^n, \oplus, \otimes)$. *It is right gyroangled with legs* **a** *and* **b** *and hypotenuse* **c**, *Fig*. 6.4, *if and only if*

$$\gamma_a\gamma_b = \gamma_c. \tag{6.55}$$

Proof Let ABC be a right gyrotriangle in an Einstein gyrovector space $(\mathbb{R}_s^n, \oplus, \otimes)$ with the right gyroangle $\gamma = \pi/2$, as shown in Fig. 6.4 for $n = 2$. It follows from (6.33) with $\gamma = \pi/2$ that the sides a, b and c of gyrotriangle ABC in Fig. 6.4 are

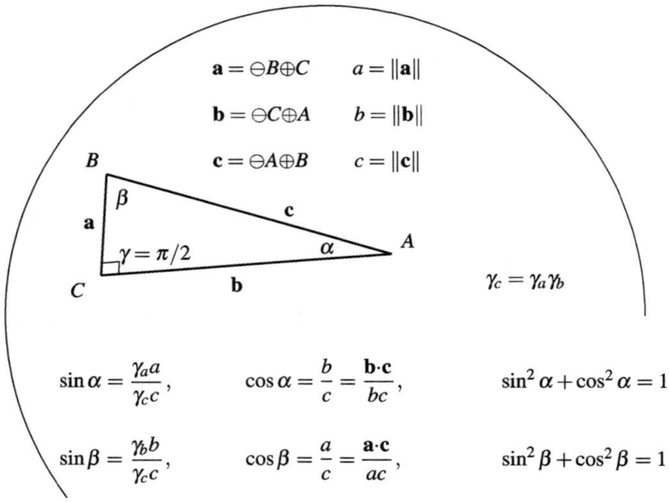

Fig. 6.4 Gyrotrigonometry in an Einstein gyrovector plane $(\mathbb{R}_s^2, \oplus, \otimes)$

related to the acute gyroangles α and β of the gyrotriangle by the equations

$$\begin{aligned}\gamma_a &= \frac{\cos\alpha}{\sin\beta},\\ \gamma_b &= \frac{\cos\beta}{\sin\alpha},\\ \gamma_c &= \frac{\cos\alpha\cos\beta}{\sin\alpha\sin\beta}.\end{aligned} \tag{6.56}$$

The identities in (6.56) imply the Einstein–Pythagoras Identity

$$\gamma_a\ \gamma_b = \gamma_c \tag{6.57}$$

for a right gyrotriangle ABC with hypotenuse c and legs a and b in an Einstein gyrovector space, Fig. 6.4. The converse statement is obvious owing to the one-to-one correspondence between gyrotriangle gyroangles and side-gyrolengths. □

It follows from (6.57) that $\gamma_a^2\gamma_b^2 = \gamma_c^2 = (1 - c_s^2)^{-1}$, implying that the gyrolength of the hypotenuse is given by the equation

$$c_s = \frac{\sqrt{\gamma_a^2\gamma_b^2 - 1}}{\gamma_a\gamma_b}. \tag{6.58}$$

In terms of *rapidities*, (6.57) takes the familiar standard form,

$$\cosh\phi_a \cosh\phi_b = \cosh\phi_c \tag{6.59}$$

of the hyperbolic Pythagorean theorem [22, p. 334], where $\cosh\phi_a := \gamma_a$, etc.

In the special case of a right gyrotriangle in an Einstein gyrovector space, as in Fig. 6.4 where $\gamma = \pi/2$, it follows from Einstein–Pythagoras Identity (6.57) that the gyrotriangle defect identity (6.51) specializes to the right gyroangled gyrotriangle defect identity

$$\tan^2\frac{\delta}{2} = \frac{\gamma_a - 1}{\gamma_a + 1}\frac{\gamma_b - 1}{\gamma_b + 1}. \tag{6.60}$$

6.11 Gyrotrigonometry

Right angled triangles in Euclidean geometry along with the Pythagorean Identity that each right triangle obeys are useful for the presentation of the elementary trigonometric functions. In full analogy, right gyroangled gyrotriangles in the hyperbolic geometry of Einstein gyrovector spaces along with the two Einsteinian–Pythagorean Identities that each right gyrotriangle obeys are useful for the presentation of the elementary gyrotrigonometric functions, as shown below.

Let a, b and c be the respective gyrolengths of the two legs $\mathbf{a}$, $\mathbf{b}$ and the hypotenuse $\mathbf{c}$ of a right gyrotriangle ABC in an Einstein gyrovector space $(\mathbb{R}^n_s, \oplus, \otimes)$, Fig. 6.4. By (1.9), p. 5, and (6.56) we have

$$\begin{aligned}\left(\frac{a}{c}\right)^2 &= \frac{(\gamma_a^2 - 1)/\gamma_a^2}{(\gamma_c^2 - 1)/\gamma_c^2} = \cos^2\beta,\\ \left(\frac{b}{c}\right)^2 &= \frac{(\gamma_b^2 - 1)/\gamma_b^2}{(\gamma_c^2 - 1)/\gamma_c^2} = \cos^2\alpha,\end{aligned} \tag{6.61}$$

where γ_a , γ_b and γ_c are related by (6.57).

Similarly, by (1.9), p. 5, and (6.56) we also have

$$\begin{aligned}\left(\frac{\gamma_a a}{\gamma_c c}\right)^2 &= \frac{\gamma_a^2 - 1}{\gamma_c^2 - 1} = \sin^2\alpha,\\ \left(\frac{\gamma_b b}{\gamma_c c}\right)^2 &= \frac{\gamma_b^2 - 1}{\gamma_c^2 - 1} = \sin^2\beta.\end{aligned} \tag{6.62}$$

Identities (6.61) and (6.62) imply

$$\begin{aligned}\left(\frac{a}{c}\right)^2 + \left(\frac{\gamma_b b}{\gamma_c c}\right)^2 &= 1,\\ \left(\frac{\gamma_a a}{\gamma_c c}\right)^2 + \left(\frac{b}{c}\right)^2 &= 1\end{aligned} \tag{6.63}$$

so that, as shown in Fig. 6.4,

$$\begin{aligned}\cos\alpha &= \frac{b}{c},\\ \cos\beta &= \frac{a}{c},\end{aligned} \tag{6.64}$$

and

$$\begin{aligned}\sin\alpha &= \frac{\gamma_a a}{\gamma_c c},\\ \sin\beta &= \frac{\gamma_b b}{\gamma_c c}.\end{aligned} \tag{6.65}$$

Interestingly, we see from (6.64)–(6.65) that the gyrocosine function of an acute gyroangle of a right gyrotriangle in an Einstein gyrovector space has the same form as its Euclidean counterpart, the cosine function. In contrast, it is only modulo gamma factors that the gyrosine function has the same form as its Euclidean counterpart.

Identities (6.63) give rise to the following two distinct Einsteinian–Pythagorean identities:

$$\begin{aligned}a^2 + \left(\frac{\gamma_b}{\gamma_c}\right)^2 b^2 &= c^2,\\ \left(\frac{\gamma_a}{\gamma_c}\right)^2 a^2 + b^2 &= c^2\end{aligned} \tag{6.66}$$

for a right gyrotriangle with hypotenuse c and legs a and b in an Einstein gyrovector space. The two distinct Einsteinian–Pythagorean identities in (6.66) that each Einsteinian right gyrotriangle obeys converge in the Newtonian–Euclidean limit of large s, $s \to \infty$, to the single Pythagorean identity

$$a^2 + b^2 = c^2 \quad \text{(Euclidean Geometry)} \tag{6.67}$$

that each Euclidean right-angled triangle obeys.

Some explorers believe that "in the hyperbolic model the Pythagorean theorem is not valid" [69, p. 363], so that "the Pythagorean theorem is strictly Euclidean." It is therefore interesting to realize that while Euclidean geometry possesses a single Pythagorean identity, (6.67), hyperbolic geometry possesses two Pythagorean identities, (6.66), that capture the missing analogy.

As an application of the gyrotrigonometry formed by (6.64)–(6.65) in Einstein gyrovector spaces, we verify the following theorem:

Theorem 6.14 (The Base–Gyroaltitude Gyrotriangle Theorem) *Let ABC be a gyrotriangle with sides a, b, c in an Einstein gyrovector space $(\mathbb{R}^n_s, \oplus, \otimes)$, Fig. 6.3, p. 133, and let h_a, h_b, h_c be the three gyroaltitudes of ABC drawn, respectively,*

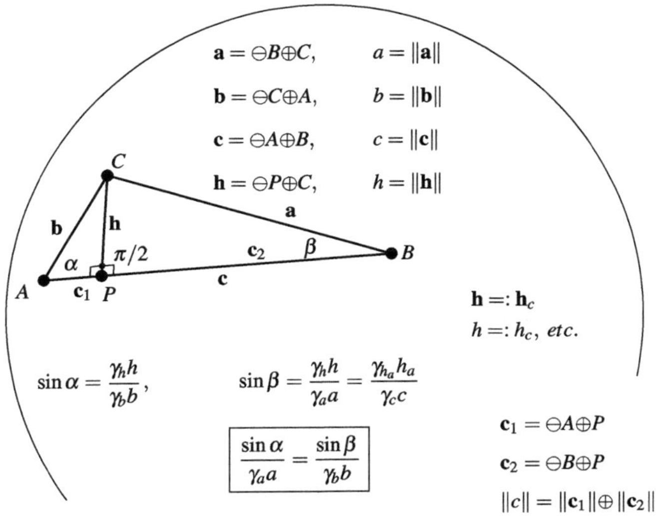

Fig. 6.5 The Law of Gyrosines. Drawing gyrotriangle gyroaltitudes in Einstein Gyrovector Spaces $(\mathbb{R}_s^n, \oplus, \otimes)$, and employing a basic gyrotrigonometric formula, shown in Fig. 6.4, we obtain the Law of Gyrosines, (6.44), in full analogy with the derivation of its Euclidean counterpart, the Law of Sines. The orthogonal projection of $\mathbf{a}$ on $\mathbf{c}' = \ominus B \oplus A$ is $\mathbf{c}_2$ and, similarly, the orthogonal projection of $\mathbf{b}' = \ominus A \oplus C$ on $\mathbf{c}$ is $\mathbf{c}_1$. Clearly, $\|\mathbf{b}'\| = \|\mathbf{b}\|$, and $\|\mathbf{c}'\| = \|\mathbf{c}\|$. It follows from the gyrotriangle equality that $\|\mathbf{c}\| = \|\mathbf{c}_1\| \oplus \|\mathbf{c}_2\|$

from vertices A, B, C perpendicular to their opposite sides a, b, c or their extension (For instance, $h = h_c$ in Fig. 6.5). Then

$$\gamma_a a \gamma_{h_a} h_a = \gamma_b b \gamma_{h_b} h_b = \gamma_c c \gamma_{h_c} h_c. \tag{6.68}$$

Proof By (6.65) with the notation of Fig. 6.5, we have $\gamma_{h_a} h_a = \gamma_c \; c \sin\beta$. Hence, by (1.9), p. 5, (6.20), p. 134, and Identity (6.51), p. 142, for $\tan(\delta/2)$, we have the following chain of equations that culminates in a most unexpected, elegant result

$$\begin{aligned}
\gamma_a^2 a^2 \gamma_{h_a}^2 h_a^2 &= \gamma_a^2 a^2 \gamma_c^2 c^2 \sin^2\beta \\
&= s^4\big(\gamma_a^2 - 1\big)\big(\gamma_c^2 - 1\big)\big(1 - \cos^2\beta\big) \\
&= s^4\big(\gamma_a^2 - 1\big)\big(\gamma_c^2 - 1\big)\left(1 - \frac{(\gamma_a\gamma_c - \gamma_b)^2}{(\gamma_a^2 - 1)(\gamma_c^2 - 1)}\right) \\
&= s^4\big(1 + 2\gamma_a\gamma_b\gamma_c - \gamma_a^2 - \gamma_b^2 - \gamma_c^2\big) \\
&= s^4(1 + \gamma_a + \gamma_b + \gamma_c)^2 \tan^2\frac{\delta}{2},
\end{aligned} \tag{6.69}$$

thus verifying the first identity in (6.70) below:

$$\begin{aligned}\gamma_a a \gamma_{h_a} h_a &= s^2(1+\gamma_a+\gamma_b+\gamma_c)\tan(\delta/2),\\ \gamma_b b \gamma_{h_b} h_b &= s^2(1+\gamma_a+\gamma_b+\gamma_c)\tan(\delta/2),\\ \gamma_c c \gamma_{h_c} h_c &= s^2(1+\gamma_a+\gamma_b+\gamma_c)\tan(\delta/2).\end{aligned} \tag{6.70}$$

The remaining two identities in (6.70) follow from the first by interchanging a and b, and by interchanging a and c, respectively, noting that $\tan(\delta/2)$ is symmetric in a, b, c by Theorem 6.11, p. 142.

Finally, the *gyrotriangle base–gyroaltitude identity* (6.68) of the theorem follows from the three identities in (6.70). □

Guided by analogies with the triangle area, the identities in (6.70) suggest the following definition of the gyrotriangle gyroarea:

Definition 6.15 (The Gyrotriangle Gyroarea) Let ABC be a gyrotriangle in an Einstein gyrovector space $(\mathbb{R}^n_s, \oplus, \otimes)$. The gyroarea $|ABC|$ of gyrotriangle ABC is given by the equation

$$|ABC| = 2s^2 \tan\frac{\delta}{2} \tag{6.71}$$

where δ is the defect of gyrotriangle ABC.

Definition 6.15, along with the identities in (6.70), gives rise to the following gyrotriangle gyroarea theorem:

Theorem 6.16 (The Gyrotriangle Gyroarea) *Let ABC be a gyrotriangle in an Einstein gyrovector space $(\mathbb{V}_s, \oplus, \otimes)$. Then, in the notation of Fig. 6.5, the gyroarea $|ABC|$ of gyrotriangle ABC is given by each of the equations*

$$\begin{aligned}|ABC| &= \frac{2}{1+\gamma_a+\gamma_b+\gamma_c}\gamma_a a \gamma_{h_a} h_a\\ &= \frac{2}{1+\gamma_a+\gamma_b+\gamma_c}\gamma_b b \gamma_{h_b} h_b\\ &= \frac{2}{1+\gamma_a+\gamma_b+\gamma_c}\gamma_c c \gamma_{h_c} h_c,\end{aligned} \tag{6.72}$$

where $h_c = h$ is the gyroaltitude of gyrotriangle ABC drawn from side **c**, *as shown in Fig. 6.5, and where h_a and h_b are the gyroaltitudes of gyrotriangle ABC drawn from sides* **a** *and* **b**, *respectively.*

In the Euclidean limit of large c, $c \to \infty$, gamma factors tend to 1, and, accordingly, the gyrotriangle gyroarea (6.72) tends to its Euclidean counterpart, the triangle area, given by the following equations:

$$|ABC| = \frac{1}{2}ah_a = \frac{1}{2}bh_b = \frac{1}{2}ch_c. \quad \text{(Euclidean Geometry)} \tag{6.73}$$

Along with analogies that the gyrotriangle gyroarea (6.72) shares with its Euclidean counterpart (6.73), there is a remarkable disanalogy. While two triangles with equal bases and equal altitudes have equal areas, two gyrotriangles with equal bases and equal gyroaltitudes need not have equal gyroareas. It is this disanalogy between triangle area and gyrotriangle gyroarea that led some explorers to believe that [39, p. 349] "it is impossible to define an area function (in hyperbolic geometry) which has even a minimal resemblance to the Euclidean area function." The title "*Why Euclidean area measure fails in the noneuclidean plane*" [48] of a paper is, thus, unjustified.

6.12 Problems

Problem 6.1 Solving a System of Three Equations:
Solve the system (6.19), p. 134, of three identities for the three unknowns $\cos\alpha$, $\cos\beta$ and $\cos\gamma$, and employ (1.9), p. 5, to derive the *SSS* to *AAA* Conversion Law (6.20).

Problem 6.2 Translating cosines into sines:
Derive the three equations in (6.22), p. 135, from the three equations in (6.20).

Problem 6.3 The *AAA* to *SSS* Conversion Law I:
Derive (6.34), p. 137, from the *SSS* to *AAA* conversion law (6.20), p. 134.

Problem 6.4 Derivation of Identities by Using Mathematica:
Derive (6.39), p. 139, p. 139, from (6.33), p. 137 by employing gyrotrigonometric/trigonometric identities. The use of the computer system Mathematica is recommended.

Problem 6.5 The *ASA* to *SAS* Conversion Law:
Show that (6.47), p. 141, follows from the third identity in (6.33), p. 137, and from (6.20), p. 134, noting (1.9), p. 5.

Problem 6.6 Substitutions:
Substitute $\cos\gamma$ from (6.20), p. 134, and a_s^2 and b_s^2 from Identities like (1.9), p. 5, into $\tan^2(\delta/2)$ of (6.49), p. 142, to obtain (6.51), p. 142.

Problem 6.7 The Gyrotriangular Defect II:
Show that the first equation in (6.51), p. 142, can be written as (6.53), p. 143.

Problem 6.8 The Gyrotriangular Defect II:
Show that Identity (6.60), p. 145, follows from (6.53), p. 143, with $\gamma = \pi/2$.

Problem 6.9 The Base-Gyroaltitude Gyrotriangle Theorem:
Derive (6.69), p. 147, by (1.9), p. 5, (6.20), p. 134, and Identity (6.51), p. 142, for $\tan(\delta/2)$.

Part III
Hyperbolic Triangle Centers

In Part III of the book we employ the tools developed in Part II of the book to determine various hyperbolic triangle centers and some relationships between them.

In particular, the following gyrotriangle gyrocenters and related points are determined in terms of their gyrobarycentric coordinate representation with respect to their vertices:

1. The gyrotriangle circumgyrocenter and circumgyroradius
2. The gyrotriangle ingyrocenter and ingyroradius
3. The ingyrocircle points of tangency
4. The gyrotriangle Gergonne gyropoint
5. The gyrotriangle orthogyrocenter
6. The gyrotriangle exgyrocenters and exgyroradii
7. The gyrotriangle Nagel gyropoint
8. The gyrotriangle exgyrocenter-point-of-tangency Gyrocenter; and
9. The gyrotriangle Brocard gyropoints

Chapter 7
Gyrotriangle Gyrocenters

Abstract Interest in triangle centers has long history, the classical ones being the triangle centroid, orthocenter, incenter and circumcenter. A list of more than 3000 triangle centers is found in Kimberling (Clark Kimberling's Encyclopedia of Triangle Centers—ETC, 2010). Hyperbolic triangles and their centers are of interest as well (Bottema in Can. J. Math. 10:502–506, 1958; Vermeer in Topol. Appl. 152(3):226–242, 2005; Demirel and Soyturk in Novi Sad J. Math. 38(2):33–39, 2008; Sonmez in Algebras Groups Geom. 26(1):75–79, 2009). The special relativistic approach of this book enables hyperbolic triangle centers to be determined along with relationships between them.

The hyperbolic triangle circumcenter, incenter and orthocenter are called, in gyrolanguage, the gyrotriangle circumgyrocenter, ingyrocenter and orthogyrocenter, respectively. These gyrocenters are determined in this chapter in terms of their gyrobarycentric coordinate representations with respect to the vertices of their reference gyrotriangles.

7.1 Gyrotriangle Circumgyrocenter

Definition 7.1 The circumgyrocenter, O, of a gyrotriangle is the point in the gyrotriangle gyroplane equigyrodistant from the three gyrotriangle vertices.

Let O be the circumgyrocenter of gyrotriangle $A_1A_2A_3$ in an Einstein gyrovector space $(\mathbb{R}_s^n, \oplus, \otimes)$, Fig. 7.1, and let $(m_1 : m_2 : m_3)$ be its gyrobarycentric coordinates with respect to the set $S = \{A_1, A_2, A_3\}$, (4.25), p. 90, so that

$$O = \frac{m_1\gamma_{A_1} A_1 + m_2\gamma_{A_2} A_2 + m_3\gamma_{A_3} A_3}{m_1\gamma_{A_1} + m_2\gamma_{A_2} + m_3\gamma_{A_3}}. \tag{7.1}$$

The gyrobarycentric coordinates m_1, m_2 and m_3 are to be determined in (7.8) below, in terms of gamma factors of the gyrotriangle sides and, alternatively in (7.16), in terms of the gyrotriangle gyroangles.

A.A. Ungar, *Hyperbolic Triangle Centers*,
Fundamental Theories of Physics 166,
DOI 10.1007/978-90-481-8637-2_7, © Springer Science+Business Media B.V. 2010

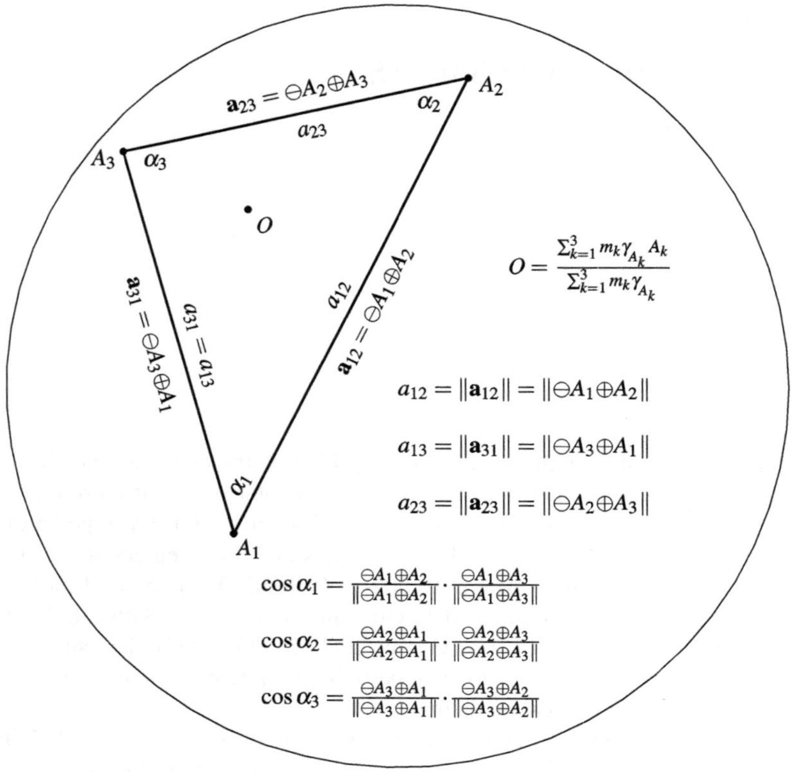

Fig. 7.1 The circumgyrocenter O of gyrotriangle $A_1A_2A_3$ in an Einstein gyrovector space $(\mathbb{R}^n_s, \oplus, \otimes)$, $n = 2$, is shown along with its standard notation. Here $\|\ominus A_1 \oplus O\| = \|\ominus A_2 \oplus O\| = \|\ominus A_3 \oplus O\|$, where O is the gyrotriangle circumgyrocenter, given by its gyrobarycentric coordinate representation (7.18), with respect to the set $S = \{A_1, A_2, A_3\}$

Following the gyrocovariance of gyrobarycentric coordinate representations, Theorem 4.6, we have from Identity (4.29b), p. 91, with $X = \ominus A_1$, using the standard gyrotriangle index notation, shown in Fig. 7.1, in Fig. 6.1, p. 128, and in (6.1), p. 127,

$$\begin{aligned}
\gamma_{\ominus A_1 \oplus O} &= \frac{m_1 \gamma_{\ominus A_1 \oplus A_1} + m_2 \gamma_{\ominus A_1 \oplus A_2} + m_3 \gamma_{\ominus A_1 \oplus A_3}}{m_0} \\
&= \frac{m_1 + m_2 \gamma_{12} + m_3 \gamma_{13}}{m_0},
\end{aligned} \tag{7.2}$$

where by (4.15), p. 88, the circumgyrocenter constant $m_0 > 0$ with respect to the set of the gyrotriangle vertices is given by the equation

$$m_0^2 = m_1^2 + m_2^2 + m_3^2 + 2(m_1 m_2 \gamma_{12} + m_1 m_3 \gamma_{13} + m_2 m_3 \gamma_{23}). \tag{7.3}$$

Hence, similarly, by the gyrocovariance of gyrobarycentric coordinate representations, Identity (4.29b), p. 91, of Theorem 4.6 with $X=\ominus A_1$, with $X=\ominus A_2$, and with $X=\ominus A_3$, we have, respectively,

$$\begin{aligned}\gamma_{\ominus A_1\oplus O} &= \frac{m_1+m_2\gamma_{12}+m_3\gamma_{13}}{m_0},\\ \gamma_{\ominus A_2\oplus O} &= \frac{m_1\gamma_{12}+m_2+m_3\gamma_{23}}{m_0},\\ \gamma_{\ominus A_3\oplus O} &= \frac{m_1\gamma_{13}+m_2\gamma_{23}+m_3}{m_0}.\end{aligned}\tag{7.4}$$

The condition that the circumgyrocenter O is equigyrodistant from its gyrotriangle vertices A_1, A_2, and A_3 implies

$$\gamma_{\ominus A_1\oplus O}=\gamma_{\ominus A_2\oplus O}=\gamma_{\ominus A_3\oplus O}.\tag{7.5}$$

Equations (7.4) and (7.5), along with the normalization condition $m_1+m_2+m_3=1$, yield the following system of three equations for the three unknowns m_1, m_2, and m_3,

$$\begin{aligned}m_1+m_2+m_3&=1,\\ m_1+m_2\gamma_{12}+m_3\gamma_{13}&=m_1\gamma_{13}+m_2\gamma_{23}+m_3,\\ m_1\gamma_{12}+m_2+m_3\gamma_{23}&=m_1\gamma_{13}+m_2\gamma_{23}+m_3\end{aligned}\tag{7.6}$$

which can be written as the matrix equation,

$$\begin{pmatrix}1 & 1 & 1\\ 1-\gamma_{13} & \gamma_{12}-\gamma_{23} & \gamma_{13}-1\\ \gamma_{12}-\gamma_{13} & 1-\gamma_{23} & \gamma_{23}-1\end{pmatrix}\begin{pmatrix}m_1\\ m_2\\ m_3\end{pmatrix}=\begin{pmatrix}1\\ 0\\ 0\end{pmatrix}.\tag{7.7}$$

Solving (7.7) for the unknowns m_1, m_2, and m_3, we have

$$\begin{aligned}m_1&=\frac{1}{D}(\gamma_{12}+\gamma_{13}-\gamma_{23}-1)(\gamma_{23}-1),\\ m_2&=\frac{1}{D}(\gamma_{12}-\gamma_{13}+\gamma_{23}-1)(\gamma_{13}-1),\\ m_3&=\frac{1}{D}(-\gamma_{12}+\gamma_{13}+\gamma_{23}-1)(\gamma_{12}-1),\end{aligned}\tag{7.8}$$

where D is the determinant of the 3×3 matrix in (7.7),

$$\begin{aligned}D&=2(\gamma_{12}\gamma_{13}+\gamma_{12}\gamma_{23}+\gamma_{13}\gamma_{23})-(\gamma_{12}^2-1)-(\gamma_{13}^2-1)-(\gamma_{23}^2-1)\\ &\quad-2(\gamma_{12}+\gamma_{13}+\gamma_{23}).\end{aligned}\tag{7.9}$$

Hence, the circumgyrocenter O of gyrotriangle $A_1A_2A_3$ is given by (7.1) where gyrobarycentric coordinates m_1, m_2, and m_3 are given by (7.8). Since in gyrobarycentric coordinates only ratios of coordinates are relevant, the gyrobarycentric coordinates, m_1, m_2, and m_3 in (7.8) can be simplified by removing their common factor $1/D$.

Gyrobarycentric coordinates, m_1, m_2, and m_3, of the circumgyrocenter O of gyrotriangle $A_1A_2A_3$ are thus given by the equations

$$\begin{aligned}
m_1' &= (\gamma_{12} + \gamma_{13} - \gamma_{23} - 1)(\gamma_{23} - 1), \\
m_2' &= (\gamma_{12} - \gamma_{13} + \gamma_{23} - 1)(\gamma_{13} - 1), \\
m_3' &= (-\gamma_{12} + \gamma_{13} + \gamma_{23} - 1)(\gamma_{12} - 1).
\end{aligned} \tag{7.10}$$

Hence, by (7.3) along with the gyrobarycentric coordinates in (7.10), we have

$$\begin{aligned}
m_0^2 = v\big\{(\gamma_{12} + \gamma_{13} + \gamma_{23} - 1)^2 - 2\big(\gamma_{12}^2 + \gamma_{13}^2 + \gamma_{23}^2 - 1\big)\big\} \\
\times \big(1 + 2\gamma_{12}\gamma_{13}\gamma_{23} - \gamma_{12}^2 - \gamma_{13}^2 - \gamma_{23}^2\big).
\end{aligned} \tag{7.11}$$

According to Corollary 4.9, p. 93, the gyrotriangle $A_1A_2A_3$ in Fig. 7.1 possesses a circumgyrocenter if and only if $m_0^2 > 0$.

The second factor of m_0^2 in (7.11) is positive for any gyrotriangle $A_1A_2A_3$ in an Einstein gyrovector space, by Inequality (6.23), p. 135. Hence, as we see from (7.11), $m_0^2 > 0$ if and only if the points A_1, A_2, and A_3 obey the *circumgyrocircle condition*

$$(\gamma_{12} + \gamma_{13} + \gamma_{23} - 1)^2 > 2\big(\gamma_{12}^2 + \gamma_{13}^2 + \gamma_{23}^2 - 1\big). \tag{7.12}$$

Gamma factors of gyrotriangle side gyrolengths are related to its gyroangles by the equations, (6.33), p. 137,

$$\begin{aligned}
\gamma_{23} &= \frac{\cos\alpha_1 + \cos\alpha_2\cos\alpha_3}{\sin\alpha_2\sin\alpha_3}, \\
\gamma_{13} &= \frac{\cos\alpha_2 + \cos\alpha_1\cos\alpha_3}{\sin\alpha_1\sin\alpha_3}, \\
\gamma_{12} &= \frac{\cos\alpha_3 + \cos\alpha_1\cos\alpha_2}{\sin\alpha_1\sin\alpha_2}.
\end{aligned} \tag{7.13}$$

Substituting these from (7.13) into (7.10), we obtain

$$\begin{aligned}
m_1' &= F\sin\left(\frac{-\alpha_1 + \alpha_2 + \alpha_3}{2}\right)\sin\alpha_1, \\
m_2' &= F\sin\left(\frac{\alpha_1 - \alpha_2 + \alpha_3}{2}\right)\sin\alpha_2, \\
m_3' &= F\sin\left(\frac{\alpha_1 + \alpha_2 - \alpha_3}{2}\right)\sin\alpha_3,
\end{aligned} \tag{7.14}$$

where the common factor F in (7.14) is given by the equation

$$F = 2^3 \frac{\cos^2(\frac{\alpha_1+\alpha_2+\alpha_3}{2})\cos(\frac{-\alpha_1+\alpha_2+\alpha_3}{2})\cos(\frac{\alpha_1-\alpha_2+\alpha_3}{2})\cos(\frac{\alpha_1+\alpha_2-\alpha_3}{2})}{\sin\alpha_1 \sin\alpha_2 \sin\alpha_3}. \tag{7.15}$$

Since in gyrobarycentric coordinates only ratios of coordinates are relevant, the gyrobarycentric coordinates, m_1', m_2', and m_3' in (7.14) can be simplified by removing their common factor F. Hence, gyrobarycentric coordinates, m_1'', m_2'', and m_3'', of the circumgyrocenter O of gyrotriangle $A_1A_2A_3$, expressed in terms of the gyrotriangle gyroangles are given by the equations

$$\begin{aligned} m_1'' &= \sin\left(\frac{-\alpha_1+\alpha_2+\alpha_3}{2}\right)\sin\alpha_1, \\ m_2'' &= \sin\left(\frac{\alpha_1-\alpha_2+\alpha_3}{2}\right)\sin\alpha_2, \\ m_3'' &= \sin\left(\frac{\alpha_1+\alpha_2-\alpha_3}{2}\right)\sin\alpha_3. \end{aligned} \tag{7.16}$$

By Corollary 4.10, p. 94, the circumgyrocenter O, (7.1), lies on the interior of its gyrotriangle $A_1A_2A_3$ if and only if all its gyrobarycentric coordinates are positive. Hence, we see from the gyrobarycentric coordinates (7.16) of O that the circumgyrocenter O lies on the interior of its gyrotriangle $A_1A_2A_3$ if and only if the largest gyroangle of the gyrotriangle has measure less than the sum of the measures of the other two gyroangles. This result is known in hyperbolic geometry; see, for instance, [28, p. 132], where the result is proved synthetically.

Expressing Inequality (7.12) gyrotrigonometrically, by means of (7.13), it can be shown that $m_0^2 > 0$ if and only if

$$\begin{aligned} &\cos\frac{3\alpha_1-\alpha_2-\alpha_3}{2} + \cos\frac{-\alpha_1+3\alpha_2-\alpha_3}{2} + \cos\frac{-\alpha_1-\alpha_2+3\alpha_3}{2} \\ &\quad > \cos\frac{\alpha_1+\alpha_2+\alpha_3}{2}. \end{aligned} \tag{7.17}$$

Formalizing the main result of this section, we have the following theorem:

Theorem 7.2 (The Circumgyrocenter) *Let $S = \{A_1, A_2, A_3\}$ be a pointwise independent set of three points in an Einstein gyrovector space $(\mathbb{R}_s^n, \oplus, \otimes)$. The circumgyrocenter $O \in \mathbb{R}^n$, Fig. 7.1, of gyrotriangle $A_1A_2A_3$ has the gyrobarycentric coordinate representation*

$$O = \frac{m_1\gamma_{A_1}A_1 + m_2\gamma_{A_2}A_2 + m_3\gamma_{A_3}A_3}{m_1\gamma_{A_1} + m_2\gamma_{A_2} + m_3\gamma_{A_3}} \tag{7.18}$$

with respect to the set $S = \{A_1, A_2, A_3\}$, *with gyrobarycentric coordinates* $(m_1 : m_2 : m_3)$ *given by*

$$
\begin{aligned}
m_1 &= (\gamma_{12} + \gamma_{13} - \gamma_{23} - 1)(\gamma_{23} - 1), \\
m_2 &= (\gamma_{12} - \gamma_{13} + \gamma_{23} - 1)(\gamma_{13} - 1), \\
m_3 &= (-\gamma_{12} + \gamma_{13} + \gamma_{23} - 1)(\gamma_{12} - 1),
\end{aligned}
\tag{7.19}
$$

or, equivalently, by the gyrotrigonometric gyrobarycentric coordinates

$$
\begin{aligned}
m_1 &= \sin\left(\frac{-\alpha_1 + \alpha_2 + \alpha_3}{2}\right) \sin\alpha_1, \\
m_2 &= \sin\left(\frac{\alpha_1 - \alpha_2 + \alpha_3}{2}\right) \sin\alpha_2, \\
m_3 &= \sin\left(\frac{\alpha_1 + \alpha_2 - \alpha_3}{2}\right) \sin\alpha_3.
\end{aligned}
\tag{7.20}
$$

The circumgyrocenter constant m_0 *with respect to the set* $S = \{A_1, A_2, A_3\}$ *is given by the equation*

$$
\begin{aligned}
m_0^2 &= \left\{(\gamma_{12} + \gamma_{13} + \gamma_{23} - 1)^2 - 2\left(\gamma_{12}^2 + \gamma_{13}^2 + \gamma_{23}^2 - 1\right)\right\} \\
&\quad \times \left(1 + 2\gamma_{12}\gamma_{13}\gamma_{23} - \gamma_{12}^2 - \gamma_{13}^2 - \gamma_{23}^2\right).
\end{aligned}
\tag{7.21}
$$

The circumgyrocenter lies in the ball, $O \in \mathbb{R}_s^n$, *if and only if* $m_0^2 > 0$.

7.2 Triangle Circumcenter

In this section the gyrotriangle circumgyrocenter in Fig. 7.1 will be translated into its Euclidean counterpart in Fig. 7.2.

Interestingly, the gyrobarycentric coordinate representation (7.18) with gyrotrigonometric gyrobarycentric coordinates $(m_1 : m_2 : m_3)$ given by (7.20) remains invariant in form under the Euclidean limit $s \to \infty$, so that it is valid in Euclidean geometry as well. However, for application in Euclidean geometry the representation (7.18) can be simplified owing to the fact that triangle angle sum in π.

Indeed, under the condition

$$\alpha_1 + \alpha_2 + \alpha_3 = \pi, \quad \text{(Euclidean Geometry)} \tag{7.22a}$$

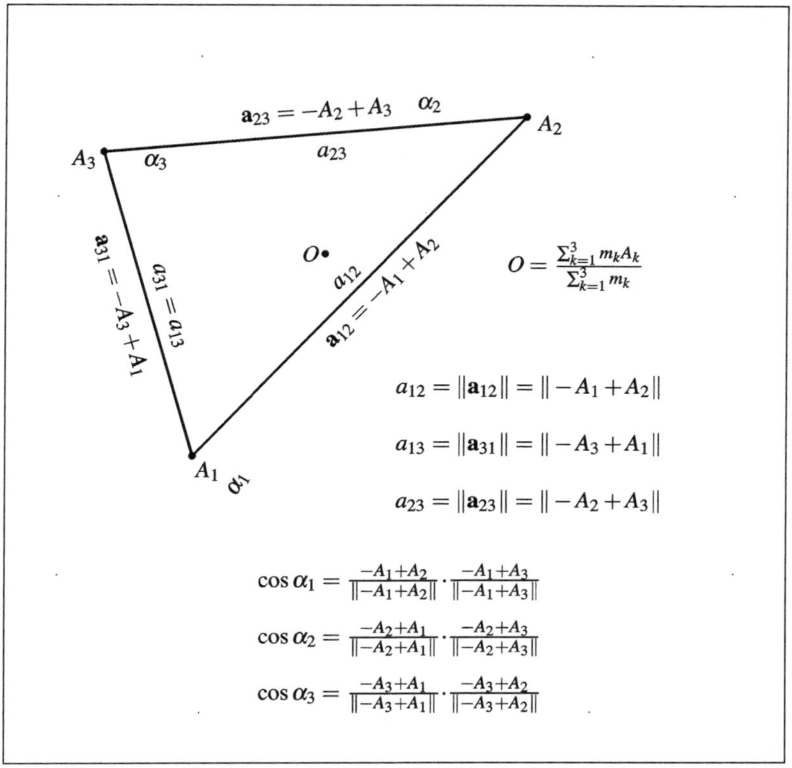

Fig. 7.2 The circumcenter O of triangle $A_1A_2A_3$ in a Euclidean vector space $\mathbb{R}^n$, $n = 2$, is shown along with its standard notation. Here $\| - A_1 + O\| = \| - A_2 + O\| = \| - A_3 + O\|$, where O is the triangle circumcenter, given by its barycentric coordinate representation (7.23) with respect to the set $S = \{A_1, A_2, A_3\}$

we have the trigonometric identities

$$
\begin{aligned}
\sin\frac{\alpha_1 - \alpha_2 - \alpha_3}{2} &= \cos\alpha_1, \\
\sin\frac{-\alpha_1 + \alpha_2 - \alpha_3}{2} &= \cos\alpha_2, \qquad \text{(Euclidean Geometry)} \\
\sin\frac{-\alpha_1 - \alpha_2 + \alpha_3}{2} &= \cos\alpha_3
\end{aligned}
\tag{7.22b}
$$

that allow m_k, $k = 1, 2, 3$, in (7.20) to be simplified.

Thus, ignoring a common factor 2, the trigonometric barycentric coordinates (7.20) of the triangle circumcenter O give rise to the following simpler trigonometric barycentric coordinates:

$$\begin{aligned} m_1 &= \sin 2\alpha_1, \\ m_2 &= \sin 2\alpha_2, \qquad \text{(Euclidean Geometry)} \\ m_3 &= \sin 2\alpha_3. \end{aligned} \tag{7.22c}$$

Hence, finally, a trigonometric barycentric coordinate representation of the circumcenter O of triangle $A_1A_2A_3$ in Fig. 7.2 with respect to the set $S = \{A_1, A_2, A_3\}$ is given by (7.23) of the following corollary of Theorem 7.2, which recovers a well-known result in Euclidean geometry [29]:

Corollary 7.3 *Let α_k, $k = 1, 2, 3$, and O be the angles and circumcenter of a triangle $A_1A_2A_3$ in a Euclidean space $\mathbb{R}^n$. Then,*

$$O = \frac{\sin 2\alpha_1 A_1 + \sin 2\alpha_2 A_2 + \sin 2\alpha_3 A_3}{\sin 2\alpha_1 + \sin 2\alpha_2 + \sin 2\alpha_3} \qquad \text{(Euclidean Geometry)} \tag{7.23}$$

Theorem 7.2 and its Corollary 7.3 form an elegant example that illustrates the result that

(i) Gyrotrigonometric gyrobarycentric coordinates of a point in an Einstein gyrovector space $\mathbb{R}^n_s$ survive unimpaired in Euclidean geometry, where they form.
(ii) Trigonometric barycentric coordinates of a point in a corresponding Euclidean vector space $\mathbb{R}^n$.
The converse is, however, not valid:
(iii) Trigonometric barycentric coordinates of a point in a Euclidean vector space $\mathbb{R}^n$ may embody the Euclidean condition the triangle angle sum in π, so that they need not survive in hyperbolic geometry.

7.3 Gyrocircle

The gyrocircle $C(r, O)$ with gyroradius r, $0 < r < s$, and gyrocenter $O \in \mathbb{R}^2_s$ in an Einstein gyrovector plane $(\mathbb{R}^2_s, \oplus, \otimes)$ is the set of all points $P \in \mathbb{R}^2_s$ such that $\|\ominus P \oplus O\| = r$; see Fig. 7.3. It is given by the equation

$$C(r, O, \theta) = O \oplus \begin{pmatrix} r\cos\theta \\ r\sin\theta \end{pmatrix} \tag{7.24}$$

for $0 \le \theta < 2\pi$. Indeed, by the left cancellation law we have

$$\|\ominus O \oplus C(r, O, \theta)\| = \left\| \begin{pmatrix} r\cos\theta \\ r\sin\theta \end{pmatrix} \right\| = r, \tag{7.25}$$

where $\|\cdot\|$ is the norm that the Einstein gyrovector plane $\mathbb{R}^2_s$ inherits from its Euclidean plane $\mathbb{R}^2$.

A sequence of gyrocircles of gyroradius $\frac{1}{4}$ in an Einstein gyrovector plane $\mathbb{R}^2_{s=1}$ with gyrocenters approaching the boundary of the open unit disc $\mathbb{R}^2_{s=1}$ is shown in

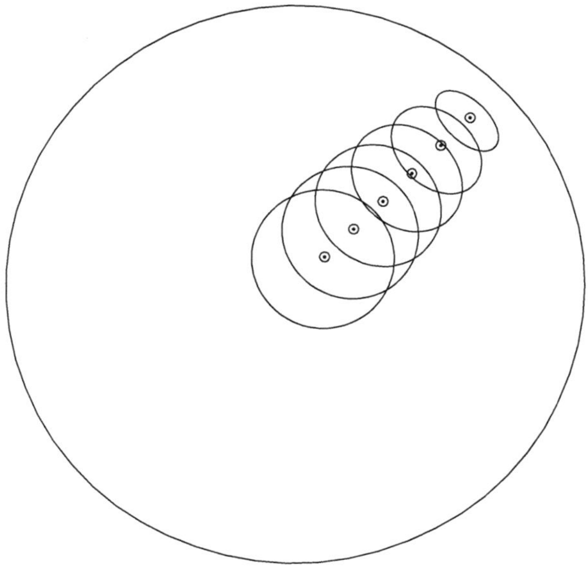

Fig. 7.3 A sequence of gyrocircles with gyroradius $\frac{1}{4}$ in an Einstein gyrovector plane $\mathbb{R}^2_{s=1}$ with gyrocenters approaching the boundary of the open unit disc $\mathbb{R}^2_{s=1}$ is shown. The center of the disc is conformal. Hence, the gyrocircle with gyrocenter at the center of the disc coincides with a Euclidean circle. The Euclidean circle is increasingly flattened as its gyrocenter approaches the boundary of the disc

Fig. 7.3. The center of the disc in Fig. 7.3 is conformal, as explained in Sect. 2.6, p. 53. Accordingly, a gyrocircle with gyrocenter at the center of the disc is identical to a Euclidean circle. This Euclidean circle is increasingly flattened in the Euclidean sense when the gyrocircle gyrocenter approaches the boundary of the disc.

The circumgyrocircle of gyrotriangle $A_1A_2A_3$ in Fig. 7.1, with circumgyrocenter at the point O, is shown in Fig. 7.4.

7.4 Gyrotriangle Circumgyroradius

In this section, we face the task of calculating the gyrotriangle circumgyroradius. The circumgyroradius R of gyrotriangle $A_1A_2A_3$ in Figs. 7.1 and 7.4 is given by

$$R = \|\ominus A_1 \oplus O\| = \|\ominus A_2 \oplus O\| = \|\ominus A_3 \oplus O\| \tag{7.26}$$

satisfying, by (7.2),

$$\gamma_R = \gamma_{\ominus A_1 \oplus O} = \frac{m_1 + m_2\gamma_{12} + m_3\gamma_{13}}{m_0}, \tag{7.27}$$

where m_1, m_2 and m_3 are given by (7.19), and where m_0 is given by (7.21).

Hence, following (7.27), (7.19) and (7.21), we have

$$\gamma_R^2 = \frac{2\gamma_{12}\gamma_{13}\gamma_{23} - (\gamma_{12}^2 + \gamma_{13}^2 + \gamma_{23}^2 - 1)}{(\gamma_{12} + \gamma_{13} + \gamma_{23} - 1)^2 - 2(\gamma_{12}^2 + \gamma_{13}^2 + \gamma_{23}^2 - 1)} \tag{7.28}$$

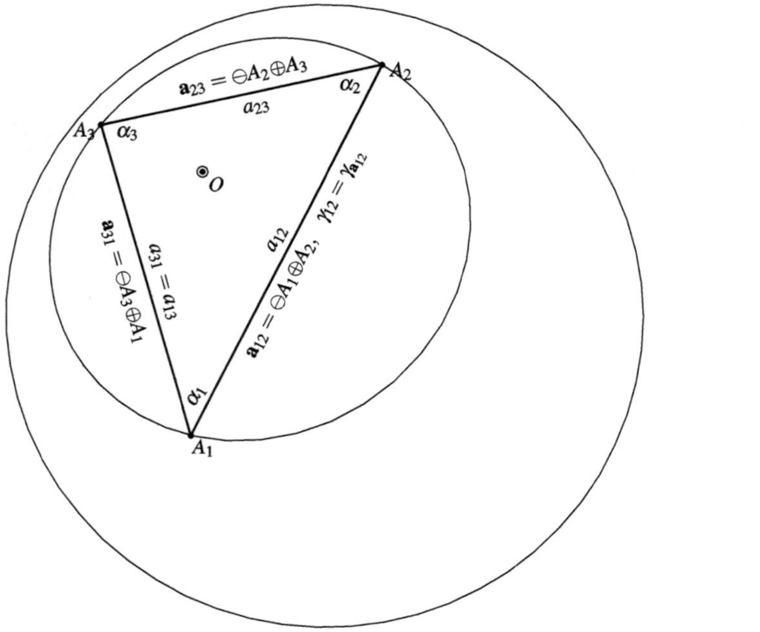

Fig. 7.4 The circumgyrocircle of gyrotriangle $A_1A_2A_3$ in an Einstein gyrovector space $(\mathbb{R}^n_s, \oplus, \otimes)$ is shown for $n = 2$. Its gyrocenter, O, is the gyrotriangle circumgyrocenter, given by its gyrobarycentric representation (7.18), p. 157, and its gyroradius r is the gyrotriangle circumgyroradius, given by each of the equations $r = \|\ominus A_k \oplus O\|$, $k = 1, 2, 3$. The gyrocircle is a flattened Euclidean circle, as shown in Fig. 7.3

so that, by (1.9), p. 5,

$$R^2 = s^2 \frac{\gamma_R^2 - 1}{\gamma_R^2} = 2s^2 \frac{(\gamma_{12}-1)(\gamma_{13}-1)(\gamma_{23}-1)}{1 + 2\gamma_{12}\gamma_{13}\gamma_{23} - \gamma_{12}^2 - \gamma_{13}^2 - \gamma_{23}^2}. \tag{7.29}$$

Hence, finally, the circumgyroradius R of gyrotriangle $A_1A_2A_3$ in Figs. 7.1 and 7.4 is given by

$$R = \sqrt{2}s \sqrt{\frac{(\gamma_{12}-1)(\gamma_{13}-1)(\gamma_{23}-1)}{1 + 2\gamma_{12}\gamma_{13}\gamma_{23} - \gamma_{12}^2 - \gamma_{13}^2 - \gamma_{23}^2}}, \tag{7.30}$$

implying

$$\sqrt{\frac{(\gamma_{12}+1)(\gamma_{13}+1)(\gamma_{23}+1)}{2}} R = s \sqrt{\frac{(\gamma_{12}^2-1)(\gamma_{13}^2-1)(\gamma_{23}^2-1)}{1 + 2\gamma_{12}\gamma_{13}\gamma_{23} - \gamma_{12}^2 - \gamma_{13}^2 - \gamma_{23}^2}}. \tag{7.31}$$

Identity (7.31) captures a remarkable analogy between the law of gyrosines and the law of sines. Indeed, following (7.31), the law of gyrosines (6.44), p. 140, for gyrotriangle $A_1A_2A_3$ in Fig. 7.4 is linked to the circumgyroradius R of the gyrotriangle by the equation

$$\frac{\gamma_{23}a_{23}}{\sin\alpha_1} = \frac{\gamma_{13}a_{13}}{\sin\alpha_2} = \frac{\gamma_{12}a_{12}}{\sin\alpha_3} = \sqrt{\frac{(\gamma_{12}+1)(\gamma_{13}+1)(\gamma_{23}+1)}{2}}R \tag{7.32}$$

called the *extended law of gyrosines*.

In the Euclidean limit of large s, $s \to \infty$, gamma factors tend to 1 and, accordingly, the extended law of gyrosines (7.32) tends to the following identity,

$$\frac{a_{23}}{\sin\alpha_1} = \frac{a_{13}}{\sin\alpha_2} = \frac{a_{12}}{\sin\alpha_3} = 2R \quad \text{(Euclidean Geometry)} \tag{7.33}$$

which is the well-known extended law of sines; see, for instance, [35, p. 87].

Identity (7.32) forms the extended law of gyrosines of a gyrotriangle $A_1A_2A_3$ with side-gyrolengths a_{23}, a_{13}, a_{12}, with gyroangles $\alpha_1, \alpha_2, \alpha_3$, and with circumgyroradius R. In full analogy, with (7.32), Identity (7.33) forms the Euclidean extended law of sines of a triangle $A_1A_2A_3$ with side-lengths a_{23}, a_{13}, a_{12}, with angles $\alpha_1, \alpha_2, \alpha_3$, and with circumradius R.

Interestingly, the gyrotriangle circumgyroradius R has an elegant representation in terms of its gyrotriangle gyroangles. Indeed, expressing the gamma factors in (7.29) in terms of the gyrotriangle gyroangles α_k, $k = 1, 2, 3$, by means of (6.33), p. 137, (7.29) takes the gyrotrigonometric form

$$\frac{R^2}{s^2} = \frac{\cos\frac{\alpha_1+\alpha_2+\alpha_3}{2}}{\cos\frac{\alpha_1-\alpha_2-\alpha_3}{2}\cos\frac{-\alpha_1+\alpha_2-\alpha_3}{2}\cos\frac{-\alpha_1-\alpha_2+\alpha_3}{2}} \tag{7.34}$$

in any Einstein gyrovector space $(\mathbb{R}_s^n, \oplus, \otimes)$. In the Euclidean limit, $s \to \infty$, each side of (7.34) tends to 0. Indeed, in that limit, the gyroangle gyrotriangle sum $\alpha_1 + \alpha_2 + \alpha_3$ tends to π so that $\cos\frac{\alpha_1+\alpha_2+\alpha_3}{2}$ tends to 0.

An important relation that results from (7.34) is formalized in the following theorem:

Theorem 7.4 *Let α_k, $k = 1, 2, 3$, and R be the gyroangles and circumgyroradius of a gyrotriangle $A_1A_2A_3$ in an Einstein gyrovector space $(\mathbb{R}_s^n \oplus, \otimes)$. Then*

$$\begin{aligned} s^2\cos\frac{\alpha_1+\alpha_2+\alpha_3}{2} &= R^2\cos\frac{\alpha_1-\alpha_2-\alpha_3}{2}\cos\frac{-\alpha_1+\alpha_2-\alpha_3}{2} \\ &\quad \times\cos\frac{-\alpha_1-\alpha_2+\alpha_3}{2}. \end{aligned} \tag{7.35}$$

Interestingly, the Euclidean limit, $s \to \infty$, of the left-hand side of Identity (7.35) of Theorem 7.4 is an indeterminate limit of type $\infty\cdot 0$, noting that in that limit $\alpha_1 + \alpha_2 + \alpha_3$ tends to π so that $\cos\frac{\alpha_1+\alpha_2+\alpha_3}{2}$ tends to 0. In contrast, the right-hand side of the identity remains invariant in form in that limit. An elegant application of Theorem 7.4 is encountered in (8.24), p. 230, where a hyperbolic geometric identity is obtained, which remains invariant in form in its transition to Euclidean geometry.

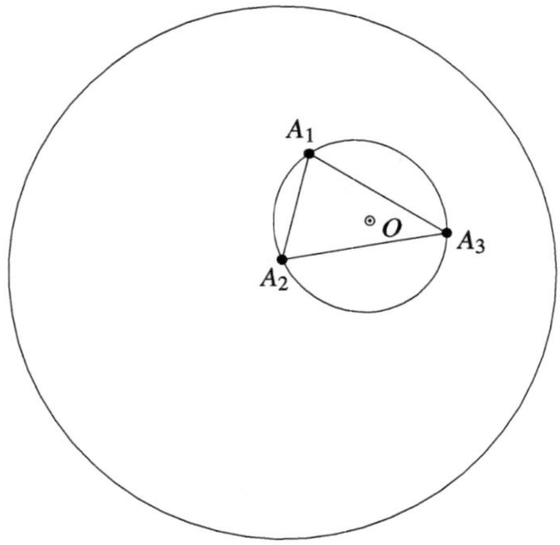

Fig. 7.5 Here A_1, A_2 and A_3 are arbitrarily selected three points of an Einstein gyrovector space $(\mathbb{R}^n_s, \oplus, \otimes)$ that satisfy the circumgyrocircle condition (7.36). Accordingly, there exists a unique gyrocircle that passes through these points

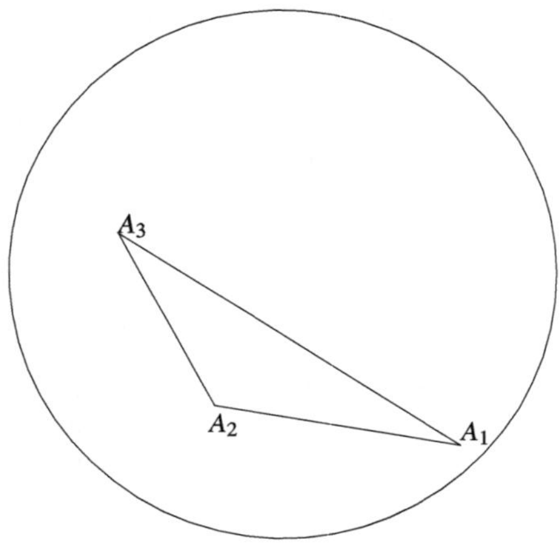

Fig. 7.6 Here A_1, A_2 and A_3 are arbitrarily selected three points of an Einstein gyrovector space $(\mathbb{R}^n_s, \oplus, \otimes)$ that do not satisfy the circumgyrocircle condition (7.36). Accordingly, there exists no gyrocircle that passes through these points

7.5 The Gyrocircle Through Three Points

In the following theorem, we use the standard gyrotriangle index notation, shown in Fig. 6.1, p. 128, and in (6.1), p. 127.

Theorem 7.5 (The Gyrocircle Through Three Points) *Let A_1, A_2 and A_3 be any three distinct points in an Einstein gyrovector space $(\mathbb{R}^n_s, \oplus, \otimes)$, see Figs.* 7.5–7.6. *There exists a unique gyrocircle that passes through these points if and only if these points obey the circumgyrocircle condition*, (7.12),

$$(\gamma_{12} + \gamma_{13} + \gamma_{23} - 1)^2 > 2\big(\gamma_{12}^2 + \gamma_{13}^2 + \gamma_{23}^2 - 1\big). \qquad (7.36)$$

When a gyrocircle exists, it is the unique gyrocircle with gyrocenter O given by, (7.18),

$$O = \frac{m_1\gamma_{A_1} A_1 + m_2\gamma_{A_2} A_2 + m_3\gamma_{A_3} A_3}{m_1\gamma_{A_1} + m_2\gamma_{A_2} + m_3\gamma_{A_3}}, \tag{7.37}$$

where

$$\begin{aligned} m_1 &= (\gamma_{12} + \gamma_{13} - \gamma_{23} - 1)(\gamma_{23} - 1), \\ m_2 &= (\gamma_{12} - \gamma_{13} + \gamma_{23} - 1)(\gamma_{13} - 1), \\ m_3 &= (-\gamma_{12} + \gamma_{13} + \gamma_{23} - 1)(\gamma_{12} - 1), \end{aligned} \tag{7.38}$$

and with gyroradius R given as in (7.30), *i.e.*,

$$R = \sqrt{2}s\sqrt{\frac{(\gamma_{12} - 1)(\gamma_{13} - 1)(\gamma_{23} - 1)}{1 + 2\gamma_{12}\gamma_{13}\gamma_{23} - \gamma_{12}^2 - \gamma_{13}^2 - \gamma_{23}^2}}. \tag{7.39}$$

Proof The gyrocircle in the theorem, if exists, is the circumgyrocircle of gyrotriangle $A_1A_2A_3$. The gyrocenter O of the gyrocircle is, therefore, given by (7.37)–(7.38), as we see from Theorem 7.2, p. 157; and the gyroradius, R, of the gyrocircle is given by (7.30), p. 162.

Finally, the circumgyrocircle of gyrotriangle $A_1A_2A_3$ exists if and only if the points A_1, A_2 and A_3 satisfy the circumgyrocircle condition (7.36), as explained in the paragraph of Inequality (7.12), p. 156. □

Example 7.6 If the three points A_1, A_2 and A_3 in Theorem 7.5 are not distinct, a gyrocircle through these points is not unique. Indeed, in this case we have

$$(\gamma_{12} + \gamma_{13} + \gamma_{23} - 1)^2 = 2(\gamma_{12}^2 + \gamma_{13}^2 + \gamma_{23}^2 - 1), \tag{7.40}$$

as one can readily check, thus violating the circumgyrocircle condition (7.36).

Example 7.7 If the three points A_1, A_2 and A_3 in Theorem 7.5 are distinct and gyrocollinear, there is no gyrocircle through these points. Hence, in this case the circumgyrocircle condition (7.36) must be violated. Hence, these points must satisfy the inequality

$$(\gamma_{12} + \gamma_{13} + \gamma_{23} - 1)^2 \le 2(\gamma_{12}^2 + \gamma_{13}^2 + \gamma_{23}^2 - 1). \tag{7.41}$$

Example 7.8 Let the three points A_1, A_2 and A_3 in Theorem 7.5 be the vertices of an equilateral gyrotriangle with side gyrolengths a. Then, $\gamma_{12} = \gamma_{13} = \gamma_{23} = \gamma_a$, so that the circumgyrocircle condition (7.36) reduces to

$$\gamma_a > 1 \tag{7.42}$$

which is satisfied by any side gyrolength a, $0 < a < s$.

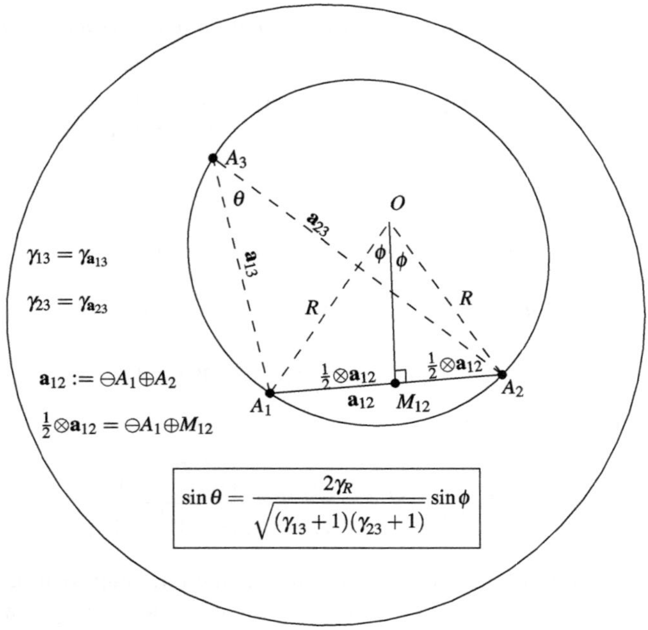

Fig. 7.7 The Inscribed Gyroangle Theorem. Illustrating the theorem, $\theta = \angle A_1 A_3 A_2$ is a gyroangle inscribed in a gyrocircle of gyroradius R (the circumgyroradius of gyrotriangle $A_1 A_2 A_3$) centered at O in an Einstein gyrovector plane $(\mathbb{R}_s^2, \oplus, \otimes)$, and $\phi = \angle A_1 O M_{12} = \angle A_2 O M_{12}$, where M_{12} is the gyromidpoint of the gyrosegment $A_1 A_2$. Accordingly, $2\phi = \angle A_1 O A_2$ is a gyrocentral gyroangle, and both θ and 2ϕ subtend on the same gyroarc on the gyrocircle. The elegant relationship between θ and ϕ, (7.43), is shown. In the Euclidean limit of large s, $s \to \infty$, gamma factors tend to 1 and, hence, the relationship between θ and ϕ in Euclidean geometry becomes $\sin\theta = \sin\phi$ or, equivalently, $\theta = \phi$

Hence, by Theorem 7.5, any equilateral gyrotriangle in an Einstein gyrovector space possesses a circumgyrocircle.

7.6 The Inscribed Gyroangle Theorem

In Fig. 7.7, we use a notation that includes the standard gyrotriangle index notation, shown in Fig. 6.1, p. 128, and in (6.1), p. 127. Fig. 7.7 presents a gyrotriangle $A_1 A_2 A_3$ and its circumgyrocircle with gyrocenter O at the gyrotriangle ingyrocenter, given by (7.18), p. 157, and with gyroradius R, given by the gyrotriangle circumgyroradius (7.30), p. 162. The gamma factor γ_R of R is given by (7.28), p. 161.

Theorem 7.9 (The Inscribed Gyroangle Theorem) *Let θ be a gyroangle inscribed in a gyrocircle of gyroradius R, and let 2ϕ be a gyrocentral gyroangle such that* *both θ and 2ϕ subtend on the same gyroarc on the gyrocircle, as shown in Fig.* 7.7. *Then, in the notation of Fig.* 7.7 *and* (6.1), *p.* 127,

$$\sin\theta = \frac{2\gamma_R}{\sqrt{(\gamma_{13}+1)(\gamma_{23}+1)}}\sin\phi. \tag{7.43}$$

Proof Under the conditions of the theorem, as described in Fig. 7.7, let M_{12} be the gyromidpoint of gyrosegment A_1A_2, implying

$$\phi := \angle A_1OM_{12} = \angle A_2OM_{12} = \frac{1}{2}\angle A_1OA_2 \tag{7.44}$$

so that 2ϕ is the gyrocentral gyroangle $\angle A_1OA_2$ shown in Fig. 7.7.

Furthermore, let

$$\mathbf{a}_{12} = \ominus A_1 \oplus A_2 \tag{7.45}$$

so that, by (4.67), p. 100,

$$\ominus A_1 \oplus M_{12} = \frac{1}{2}\otimes\mathbf{a}_{12}, \tag{7.46}$$

and hence, by (4.69), p. 100,

$$\gamma_{\frac{1}{2}\otimes\mathbf{a}_{12}}\left(\frac{1}{2}\otimes\mathbf{a}_{12}\right) = \frac{\gamma_{12}\mathbf{a}_{12}}{\sqrt{2}\sqrt{1+\gamma_{12}}}. \tag{7.47}$$

Taking magnitudes of both sides of (7.47), we have

$$\gamma_{\frac{1}{2}\otimes a_{12}}\left(\frac{1}{2}\otimes a_{12}\right) = \frac{\gamma_{12}a_{12}}{\sqrt{2}\sqrt{1+\gamma_{12}}}. \tag{7.48}$$

Applying the extended law of gyrosines (7.32), p. 163, to gyrotriangle $A_1A_2A_3$ and its circumgyrocircle in Fig. 7.7, we have

$$\frac{\gamma_{12}a_{12}}{\sin\theta} = \sqrt{\frac{(\gamma_{12}+1)(\gamma_{13}+1)(\gamma_{23}+1)}{2}}R, \tag{7.49}$$

implying

$$\sin\theta = \frac{\sqrt{2}\gamma_{12}a_{12}}{\sqrt{(1+\gamma_{12})(1+\gamma_{13})(1+\gamma_{23})}R}. \tag{7.50}$$

Applying the elementary gyrosine definition in gyrotrigonometry, (6.65), p. 146, illustrated in Fig. 6.5, p. 147, to the right gyroangled gyrotriangle $A_1M_{12}O$ in

Fig. 7.7, we obtain the first equation in (7.51),

$$\sin\phi = \frac{\gamma_{\frac{1}{2}\otimes a_{12}}(\frac{1}{2}\otimes a_{12})}{\gamma_R R} = \frac{\gamma_{12}a_{12}}{\sqrt{2}\sqrt{1+\gamma_{12}}\,\gamma_R R}. \tag{7.51}$$

The second equation in (7.51) follows from (7.48).

Finally, the desired identity (7.43) follows immediately from (7.50) and (7.51). □

7.7 Gyrotriangle Gyroangle Bisector Foot

A gyrotriangle $A_1A_2A_3$ and its gyroangle bisectors in an Einstein gyrovector space $(\mathbb{R}^n_s, \oplus, \otimes)$ are presented in Fig. 7.8, along with the gyrotriangle standard notation in Fig. 6.1, p. 128, and in (6.1), p. 127.

Let P_3 be a point on side A_1A_2 of gyrotriangle $A_1A_2A_3$ in an Einstein gyrovector space $(\mathbb{R}^n_s \oplus, \otimes)$ such that A_3P_3 is the gyroangle bisector of gyroangle $\angle A_1A_3A_2$, as shown in Fig. 7.8 for $n = 2$. Then, the point P_3 is the foot of the gyroangle bisector A_3P_3 in gyrotriangle $A_1A_2A_3$.

Let P_3 be given in terms of its gyrobarycentric coordinates $(m_1 : m_2)$ with respect to the set $S = \{A_1, A_2\}$ by the equation

$$P_3 = \frac{m_1\gamma_{A_1}A_1 + m_2\gamma_{A_2}A_2}{m_1\gamma_{A_1} + m_2\gamma_{A_2}}. \tag{7.52}$$

The gyrobarycentric coordinates m_1 and m_2 of P_3, Fig. 7.8, in (7.52) are to be determined in (7.68) below in terms of gyroangles α_1 and α_2 of the gyrotriangle $A_1A_2A_3$ and in (7.69) in terms of the side gyrolengths of the gyrotriangle.

Following the gyrocovariance of gyrobarycentric coordinate representations, Theorem 4.6, p. 90, the gyrobarycentric coordinate representation of the point P_3 in (7.52) gives rise to the identities in (7.53)–(7.55) below:

$$\ominus X\oplus P_3 = \frac{m_1\gamma_{\ominus X\oplus A_1}(\ominus X\oplus A_1) + m_2\gamma_{\ominus X\oplus A_2}(\ominus X\oplus A_2)}{m_1\gamma_{\ominus X\oplus A_1} + m_2\gamma_{\ominus X\oplus A_2}} \tag{7.53}$$

and

$$\begin{aligned} \gamma_{\ominus X\oplus P_3} &= \frac{m_1\gamma_{\ominus X\oplus A_1} + m_2\gamma_{\ominus X\oplus A_2}}{m_0}, \\ \gamma_{\ominus X\oplus P_3}(\ominus X\oplus P_3) &= \frac{m_1\gamma_{\ominus X\oplus A_1}(\ominus X\oplus A_1) + m_2\gamma_{\ominus X\oplus A_2}(\ominus X\oplus A_2)}{m_0} \end{aligned} \tag{7.54}$$

for any $X \in \mathbb{R}^n_s$, where, in the notation of Fig. 7.8 for the gamma factor γ_{12}, the constant $m_0 > 0$ in (4.28d), p. 91, specializes to

$$m_0^2 = (m_1 + m_2)^2 + 2m_1m_2(\gamma_{12} - 1) \tag{7.55}$$

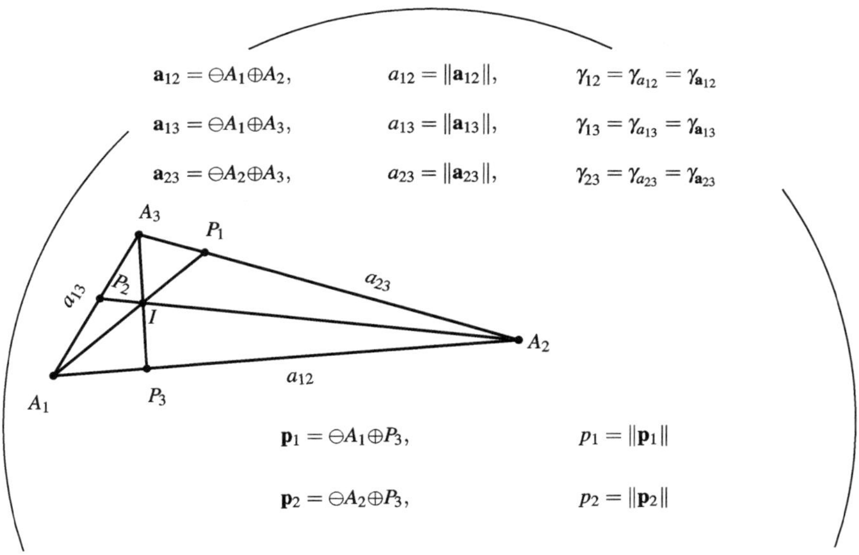

Fig. 7.8 The gyrotriangle gyroangle bisectors are concurrent. The point of concurrency, I, is called the ingyrocenter of the gyrotriangle. Let $A_1A_2A_3$ be a gyrotriangle in an Einstein gyrovector space $(\mathbb{R}^n_s, \oplus, \otimes)$. The gyroline A_kP_k is the gyroangle bisector from vertex A_k to the intersection point P_k with the opposite side, $k = 1, 2, 3$. The point P_k is the foot of the gyroangle bisector A_kP_k

in (7.54).

Using the notation in Fig. 7.8, it follows from (7.53) with $X = A_1$ that

$$\mathbf{p}_1 := \ominus A_1 \oplus P_3 = \frac{m_2 \gamma_{\ominus A_1 \oplus A_2} (\ominus A_1 \oplus A_2)}{m_1 + m_2 \gamma_{\ominus A_1 \oplus A_2}} = \frac{m_2 \gamma_{12} \mathbf{a}_{12}}{m_1 + m_2 \gamma_{12}} \tag{7.56}$$

and, similarly, with $X = A_2$,

$$\mathbf{p}_2 := \ominus A_2 \oplus P_3 = \frac{m_1 \gamma_{\ominus A_2 \oplus A_1} (\ominus A_2 \oplus A_1)}{m_1 \gamma_{\ominus A_2 \oplus A_1} + m_2} = \frac{m_1 \gamma_{21} \mathbf{a}_{21}}{m_1 \gamma_{21} + m_2}. \tag{7.57}$$

Hence, by (7.56)–(7.57), in the notation of Fig. 7.8,

$$\begin{aligned} p_1 &:= \|\mathbf{p}_1\| = \frac{m_2 \gamma_{12} a_{12}}{m_1 + m_2 \gamma_{12}}, \\ p_2 &:= \|\mathbf{p}_2\| = \frac{m_1 \gamma_{12} a_{12}}{m_1 \gamma_{12} + m_2}. \end{aligned} \tag{7.58}$$

As emphasized in (1.11)–(1.12), p. 6, one should note here that while, in general, $\mathbf{a}_{21} = \ominus A_2 \oplus A_1 \neq \ominus A_1 \oplus A_2 = \mathbf{a}_{12}$, we have $a_{21} = \|\ominus A_2 \oplus A_1\| = \|\ominus A_1 \oplus A_2\| = a_{12}$ and, hence, $\gamma_{21} = \gamma_{12}$.

Similarly, it follows from the first equation in (7.54) with $X = A_1$, and with $X = A_2$, respectively,

$$\begin{aligned}\gamma_{p_1} = \gamma_{\ominus A_1 \oplus P_3} &= \frac{m_1 + m_2\gamma_{\ominus A_1 \oplus A_2}}{m_0} = \frac{m_1 + m_2\gamma_{12}}{m_0}, \\ \gamma_{p_2} = \gamma_{\ominus A_2 \oplus P_3} &= \frac{m_1\gamma_{\ominus A_2 \oplus A_1} + m_2}{m_0} = \frac{m_1\gamma_{12} + m_2}{m_0}.\end{aligned} \tag{7.59}$$

It follows from (7.58) and (7.59) or, equivalently, from (7.52) and the second equation in (7.54) that

$$\begin{aligned}\gamma_{p_1} p_1 &= \frac{m_1 + m_2\gamma_{12}}{m_0} \frac{m_2\gamma_{12}a_{12}}{m_1 + m_2\gamma_{12}} = \frac{m_2}{m_0}\gamma_{12}a_{12}, \\ \gamma_{p_2} p_2 &= \frac{m_1\gamma_{12} + m_2}{m_0} \frac{m_1\gamma_{12}a_{12}}{m_1\gamma_{12} + m_2} = \frac{m_1}{m_0}\gamma_{12}a_{12},\end{aligned} \tag{7.60}$$

implying

$$\frac{\gamma_{p_1} p_1}{\gamma_{p_2} p_2} = \frac{m_2}{m_1}. \tag{7.61}$$

Applying the law of gyrosines (6.44), p. 140, to each of the two gyrotriangles $A_1A_3P_3$ and $A_2A_3P_3$ in Fig. 7.8, we have

$$\frac{\gamma_{P_1} P_1}{\sin\angle A_1A_3P_3} = \frac{\gamma_{13}a_{13}}{\sin\angle A_1P_3A_3} \tag{7.62}$$

and

$$\frac{\gamma_{P_2} P_2}{\sin\angle A_2A_3P_3} = \frac{\gamma_{23}a_{23}}{\sin\angle A_2P_3A_3}. \tag{7.63}$$

By the gyroangle bisector definition, $\angle A_1A_3P_3 = \angle A_2A_3P_3$, so that

$$\sin\angle A_1A_3P_3 = \sin\angle A_2A_3P_3. \tag{7.64}$$

Gyroangles $\angle A_1P_3A_3$ and $\angle A_2P_3A_3$ are supplementary (their sum is π). Hence, they have equal gyrosines,

$$\sin\angle A_1P_3A_3 = \sin\angle A_2P_3A_3. \tag{7.65}$$

If follows from (7.62)–(7.65) immediately that

$$\frac{\gamma_{p_1} p_1}{\gamma_{p_2} p_2} = \frac{\gamma_{13}a_{13}}{\gamma_{23}a_{23}}. \tag{7.66}$$

Hence, by (7.61)–(7.66), and by the law of gyrosines (6.44), p. 140,

$$\frac{m_2}{m_1} = \frac{\gamma_{13}a_{13}}{\gamma_{23}a_{23}} = \frac{\sin\alpha_2}{\sin\alpha_1} \tag{7.67}$$

so that gyrotrigonometric gyrobarycentric coordinates of point P_3 in Fig. 7.8 are given by the equation

$$(m_1 : m_2) = (\sin\alpha_1 : \sin\alpha_2). \tag{7.68}$$

It, finally, follows from (7.67) and (1.9), p. 5, that gyrobarycentric coordinates of point P_3 in Fig. 7.8 are given by the equation

$$(m_1 : m_2) = (\gamma_{23}a_{23} : \gamma_{13}a_{13}) = \left(\sqrt{\gamma_{23}^2 - 1} : \sqrt{\gamma_{13}^2 - 1}\right) \tag{7.69}$$

so that, by (7.69) and (7.52), we have

$$P_3 = \frac{\gamma_{23}a_{23}\gamma_{A_1}A_1 + \gamma_{13}a_{13}\gamma_{A_2}A_2}{\gamma_{23}a_{23}\gamma_{A_1} + \gamma_{13}a_{13}\gamma_{A_2}} = \frac{\sqrt{\gamma_{23}^2 - 1}\gamma_{A_1}A_1 + \sqrt{\gamma_{13}^2 - 1}\gamma_{A_2}A_2}{\sqrt{\gamma_{23}^2 - 1}\gamma_{A_1} + \sqrt{\gamma_{13}^2 - 1}\gamma_{A_2}}. \tag{7.70}$$

Formalizing the main result of this section, we have the following theorem:

Theorem 7.10 (Foot of a Gyrotriangle Gyroangle Bisector) *Let $S = \{A_1, A_2, A_3\}$ be a pointwise independent set of three points in an Einstein gyrovector space $(\mathbb{R}^n_s, \oplus, \otimes)$ and let P_3 be the foot of gyroangle bisector A_3P_3, Fig. 7.8, p. 169.*

Then the foot has the gyrobarycentric coordinate representation

$$P_3 = \frac{m_1\gamma_{A_1}A_1 + m_2\gamma_{A_2}A_2}{m_1\gamma_{A_1} + m_2\gamma_{A_2}} \tag{7.71}$$

with respect to the set $S = \{A_1, A_2\}$, with gyrobarycentric coordinates

$$(m_1 : m_2) = (\gamma_{23}a_{23} : \gamma_{13}a_{13}) \tag{7.72}$$

or, equivalently, with gyrotrigonometric gyrobarycentric coordinates

$$(m_1 : m_2) = (\sin\alpha_1 : \sin\alpha_2). \tag{7.73}$$

7.8 Gyrotriangle Ingyrocenter

Definition 7.11 The ingyrocircle of a gyrotriangle is the gyrocircle lying inside the gyrotriangle, tangent to each of its sides, Fig. 7.12, p. 186. The gyrocenter and the gyroradius of the ingyrocircle are called, respectively, the gyrotriangle ingyrocenter and ingyroradius.

The ingyrocenter of a gyrotriangle is the point of concurrency of the gyrotriangle gyroangle bisectors.

The three feet, P_1, P_2 and P_3 of the three gyroangle bisectors of gyrotriangle $A_1A_2A_3$ in an Einstein gyrovector space $(\mathbb{R}_s^n, \oplus, \otimes)$, shown in Fig. 7.8 for $n = 2$, are given by the equations

$$\begin{aligned} P_1 &= \frac{\gamma_{13}a_{13}\gamma_{A_2}A_2 + \gamma_{12}a_{12}\gamma_{A_3}A_3}{\gamma_{13}a_{13}\gamma_{A_2} + \gamma_{12}a_{12}\gamma_{A_3}}, \\ P_2 &= \frac{\gamma_{12}a_{12}\gamma_{A_3}A_3 + \gamma_{23}a_{23}\gamma_{A_1}A_1}{\gamma_{12}a_{12}\gamma_{A_3} + \gamma_{23}a_{23}\gamma_{A_1}}, \\ P_3 &= \frac{\gamma_{23}a_{23}\gamma_{A_1}A_1 + \gamma_{13}a_{13}\gamma_{A_2}A_2}{\gamma_{23}a_{23}\gamma_{A_1} + \gamma_{13}a_{13}\gamma_{A_2}}. \end{aligned} \tag{7.74}$$

The third equation in (7.74) is a copy from (7.70). The first and second equations in (7.70) are obtained from the third one by cyclic permutations of the vertices of gyrotriangle $A_1A_2A_3$, that is, by index permutations.

The gyroangle bisectors of gyrotriangle $A_1A_2A_3$ in an Einstein gyrovector space $(\mathbb{R}_s^n, \oplus, \otimes)$, shown in Fig. 7.8 for $n = 2$, are the gyrosegments A_1P_1, A_2P_2, and A_1P_3. Since gyrosegments in Einstein gyrovector spaces coincide with Euclidean segments, one can employ methods of linear algebra to determine the ingyrocenter, that is, the point of concurrency of the three gyroangle bisectors of gyrotriangle $A_1A_2A_3$ in Fig. 7.8.

In order to determine gyrobarycentric coordinates for the gyrotriangle ingyrocenter in Einstein gyrovector spaces, we begin with some gyroalgebraic manipulations that reduce the task we face to the task of solving a problem in linear algebra.

Let the ingyrocenter I of gyrotriangle $A_1A_2A_3$ in an Einstein gyrovector space $(\mathbb{R}_s^n, \oplus, \otimes)$, Fig. 7.8, be given in terms of its gyrobarycentric coordinate representation, (4.25), p. 90, with respect to the set $S = \{A_1, A_2, A_3\}$ of the gyrotriangle vertices by the equation

$$I = \frac{m_1\gamma_{A_1}A_1 + m_2\gamma_{A_2}A_2 + m_3\gamma_{A_3}A_3}{m_1\gamma_{A_1} + m_2\gamma_{A_2} + m_3\gamma_{A_3}}. \tag{7.75}$$

The gyrobarycentric coordinates (m_1, m_2, m_3) of I in (7.75) are to be determined in (7.103) below.

Left gyrotranslating gyrotriangle $A_1A_2A_3$ by $\ominus A_1$, the gyrotriangle becomes gyrotriangle $O(\ominus A_1 \oplus A_2)(\ominus A_1 \oplus A_3)$, where $O = \ominus A_1 \oplus A_1$ is the arbitrarily selected origin of the Einstein gyrovector space $\mathbb{R}_s^n$. The gyrotriangle gyroangle bisector feet P_1, P_2 and P_3 become, respectively, $\ominus A_1 \oplus P_1$, $\ominus A_1 \oplus P_2$ and $\ominus A_1 \oplus P_3$.

The left gyrotranslated feet are calculated in (7.76a), (7.76b), (7.76c) below by employing the Gyrobarycentric Coordinate Representation Gyrocovariance Theorem 4.6, p. 90, and the standard gyrotriangle index notation, shown in Fig. 6.1,

p. 128 and in (6.1), p. 127:

$$\ominus A_1 \oplus P_1 = \ominus A_1 \oplus \frac{\gamma_{13} a_{13} \gamma_{A_2} A_2 + \gamma_{12} a_{12} \gamma_{A_3} A_3}{\gamma_{13} a_{13} \gamma_{A_2} + \gamma_{12} a_{12} \gamma_{A_3}}$$

$$= \frac{\gamma_{13} a_{13} \gamma_{\ominus A_1 \oplus A_2} (\ominus A_1 \oplus A_2) + \gamma_{12} a_{12} \gamma_{\ominus A_1 \oplus A_3} (\ominus A_1 \oplus A_3)}{\gamma_{13} a_{13} \gamma_{\ominus A_1 \oplus A_2} + \gamma_{12} a_{12} \gamma_{\ominus A_1 \oplus A_3}}$$

$$= \frac{\gamma_{13} a_{13} \gamma_{12} \mathbf{a}_{12} + \gamma_{12} a_{12} \gamma_{13} \mathbf{a}_{13}}{\gamma_{13} a_{13} \gamma_{12} + \gamma_{12} a_{12} \gamma_{13}}, \tag{7.76a}$$

$$\ominus A_1 \oplus P_2 = \ominus A_1 \oplus \frac{\gamma_{23} a_{23} \gamma_{A_1} A_1 + \gamma_{12} a_{12} \gamma_{A_3} A_3}{\gamma_{23} a_{23} \gamma_{A_1} + \gamma_{12} a_{12} \gamma_{A_3}}$$

$$= \frac{\gamma_{12} a_{12} \gamma_{\ominus A_1 \oplus A_3} (\ominus A_1 \oplus A_3)}{\gamma_{23} a_{23} + \gamma_{12} a_{12} \gamma_{\ominus A_1 \oplus A_3}}$$

$$= \frac{\gamma_{12} a_{12} \gamma_{13} \mathbf{a}_{13}}{\gamma_{23} a_{23} + \gamma_{12} a_{12} \gamma_{13}}, \tag{7.76b}$$

$$\ominus A_1 \oplus P_3 = \ominus A_1 \oplus \frac{\gamma_{23} a_{23} \gamma_{A_1} A_1 + \gamma_{13} a_{13} \gamma_{A_2} A_2}{\gamma_{23} a_{23} \gamma_{A_1} + \gamma_{13} a_{13} \gamma_{A_2}}$$

$$= \frac{\gamma_{13} a_{13} \gamma_{\ominus A_1 \oplus A_2} (\ominus A_1 \oplus A_2)}{\gamma_{23} a_{23} + \gamma_{13} a_{13} \gamma_{\ominus A_1 \oplus A_2}}$$

$$= \frac{\gamma_{13} a_{13} \gamma_{12} \mathbf{a}_{12}}{\gamma_{23} a_{23} + \gamma_{13} a_{13} \gamma_{12}}. \tag{7.76c}$$

By Definition 4.5, p. 89, the set of points $S = \{A_1, A_2, A_3\}$ is pointwise independent in an Einstein gyrovector space $(\mathbb{R}^n_s, \oplus, \otimes)$. Hence, the two gyrovectors $\mathbf{a}_{12} = \ominus A_1 \oplus A_2$ and $\mathbf{a}_{13} = \ominus A_1 \oplus A_3$ in $\mathbb{R}^n_s \subset \mathbb{R}^n$ in (7.76a), (7.76b), (7.76c), considered as vectors in $\mathbb{R}^n$, are linearly independent in $\mathbb{R}^n$.

Similarly to the gyroalgebra in (7.76a), (7.76b), (7.76c), under a left gyrotranslation by $\ominus A_1$ the ingyrocenter I in (7.75) becomes

$$\ominus A_1 \oplus I = \ominus A_1 \oplus \frac{m_1 \gamma_{A_1} A_1 + m_2 \gamma_{A_2} A_2 + m_3 \gamma_{A_3} A_3}{m_1 \gamma_{A_1} + m_2 \gamma_{A_2} + m_3 \gamma_{A_3}}$$

$$= \frac{m_2 \gamma_{\ominus A_1 \oplus A_2} (\ominus A_1 \oplus A_2) + m_3 \gamma_{\ominus A_1 \oplus A_3} (\ominus A_1 \oplus A_3)}{m_1 + m_2 \gamma_{\ominus A_1 \oplus A_2} + m_3 \gamma_{\ominus A_1 \oplus A_3}}$$

$$= \frac{m_2\gamma_{12}\mathbf{a}_{12} + m_3\gamma_{13}\mathbf{a}_{13}}{m_1 + m_2\gamma_{12} + m_3\gamma_{13}}. \tag{7.77}$$

The gyroangle bisector of the left gyrotranslated gyrotriangle $O(\ominus A_1 \oplus A_2) \times (\ominus A_1 \oplus A_3)$ that joins the vertex

$$\ominus A_1 \oplus A_1 = O = \mathbf{0} \tag{7.78}$$

with the gyroangle bisector foot on its opposing side, $\oplus A_1 \oplus P_1$, as calculated in (7.76a),

$$\ominus A_1 \oplus P_1 = \frac{\gamma_{13}a_{13}\gamma_{12}\mathbf{a}_{12} + \gamma_{12}a_{12}\gamma_{13}\mathbf{a}_{13}}{\gamma_{13}a_{13}\gamma_{12} + \gamma_{12}a_{12}\gamma_{13}} \tag{7.79}$$

is contained in the Euclidean straight line

$$\begin{aligned} L_1 &= O + \bigl(-O + \{\ominus A_1 \oplus P_1\}\bigr)t_1 \\ &= \frac{\gamma_{13}a_{13}\gamma_{12}\mathbf{a}_{12} + \gamma_{12}a_{12}\gamma_{13}\mathbf{a}_{13}}{\gamma_{13}a_{13}\gamma_{12} + \gamma_{12}a_{12}\gamma_{13}} t_1, \end{aligned} \tag{7.80}$$

where $t_1 \in \mathbb{R}$ is the line parameter. This line passes through the point $O = \mathbf{0} \in \mathbb{R}^n_s \subset \mathbb{R}^n$ when $t_1 = 0$, and it passes through the point $\ominus A_1 \oplus P_1$ when $t_1 = 1$.

Similarly to (7.78)–(7.80), the gyroangle bisector of the left gyrotranslated gyrotriangle $O(\ominus A_1 \oplus A_2)(\ominus A_1 \oplus A_3)$ that joins the vertex

$$\ominus A_1 \oplus A_2 = \mathbf{a}_{12} \tag{7.81}$$

with the gyroangle bisector foot on its opposing side, $\oplus A_1 \oplus P_2$, as calculated in (7.76b),

$$\ominus A_1 \oplus P_2 = \frac{\gamma_{12}a_{12}\gamma_{13}\mathbf{a}_{13}}{\gamma_{23}a_{23} + \gamma_{12}a_{12}\gamma_{13}} \tag{7.82}$$

is contained in the Euclidean line

$$\begin{aligned} L_2 &= \mathbf{a}_{12} + \bigl(-\mathbf{a}_{12} + \{\ominus A_1 \oplus P_2\}\bigr)t_2 \\ &= \mathbf{a}_{12} + \left(-\mathbf{a}_{12} + \frac{\gamma_{12}a_{12}\gamma_{13}\mathbf{a}_{13}}{\gamma_{23}a_{23} + \gamma_{12}a_{12}\gamma_{13}}\right)t_2, \end{aligned} \tag{7.83}$$

where $t_2 \in \mathbb{R}$ is the line parameter. This line passes through the point $\mathbf{a}_{12} \in \mathbb{R}^n_s \subset \mathbb{R}^n$ when $t_2 = 0$, and it passes through the point $\ominus A_1 \oplus P_2$ when $t_2 = 1$.

Similarly to (7.78)–(7.80), and (7.81)–(7.83), the gyroangle bisector of the left gyrotranslated gyrotriangle $O(\ominus A_1 \oplus A_2)(\ominus A_1 \oplus A_3)$ that joins the vertex

$$\ominus A_1 \oplus A_3 = \mathbf{a}_{13} \tag{7.84}$$

with the gyroangle bisector foot on its opposing side, $\ominus A_1 \oplus P_3$, as calculated in (7.76c),

$$\ominus A_1 \oplus P_3 = \frac{\gamma_{13} a_{13} \gamma_{12} \mathbf{a}_{12}}{\gamma_{23} a_{23} + \gamma_{13} a_{13} \gamma_{12}} \tag{7.85}$$

is contained in the Euclidean line

$$\begin{aligned} L_3 &= \mathbf{a}_{13} + \left(-\mathbf{a}_{13} + \{\ominus A_1 \oplus P_3\}\right) t_3 \\ &= \mathbf{a}_{13} + \left(-\mathbf{a}_{13} + \frac{\gamma_{13} a_{13} \gamma_{12} \mathbf{a}_{12}}{\gamma_{23} a_{23} + \gamma_{13} a_{13} \gamma_{12}}\right) t_3, \end{aligned} \tag{7.86}$$

where $t_3 \in \mathbb{R}$ is the line parameter. This line passes through the point $\mathbf{a}_{13} \in \mathbb{R}^n_s \subset \mathbb{R}^n$ when $t_3 = 0$, and it passes through the point $\ominus A_1 \oplus P_3 \in \mathbb{R}^n_s \subset \mathbb{R}^n$ when $t_3 = 1$.

Hence, if the ingyrocenter I exists, its left gyrotranslated ingyrocenter, $-\ominus A_1 \oplus I$, given by (7.77), is contained in each of the three Euclidean lines L_k, $k = 1, 2, 3$, in (7.80), (7.83) and (7.86).

Formalizing, if I exists then the point P, (7.77),

$$P = \ominus A_1 \oplus I = \frac{m_2 \gamma_{12} \mathbf{a}_{12} + m_3 \gamma_{13} \mathbf{a}_{13}}{m_1 + m_2 \gamma_{12} + m_3 \gamma_{13}} \tag{7.87}$$

lies on each of the lines L_k, $k = 1, 2, 3$. Imposing the normalization condition $m_1 + m_2 + m_3 = 1$ of special gyrobarycentric coordinates, (7.87) can be simplified by means of the resulting equation $m_1 = 1 - m_2 - m_3$, obtaining

$$P = \ominus A_1 \oplus I = \frac{m_2 \gamma_{12} \mathbf{a}_{12} + m_3 \gamma_{13} \mathbf{a}_{13}}{1 + m_2(\gamma_{12} - 1) + m_3(\gamma_{13} - 1)}. \tag{7.88}$$

Since the point P lies on each of the three lines L_k, $k = 1, 2, 3$, there exist values $t_{k,0}$ of the line parameters t_k, $k = 1, 2, 3$, respectively, such that

$$\begin{aligned} P - \frac{\gamma_{13} a_{13} \gamma_{12} \mathbf{a}_{12} + \gamma_{12} a_{12} \gamma_{13} \mathbf{a}_{13}}{\gamma_{13} a_{13} \gamma_{12} + \gamma_{12} a_{12} \gamma_{13}} t_{1,0} &= 0, \\ P - \mathbf{a}_{12} - \left(-\mathbf{a}_{12} + \frac{\gamma_{12} a_{12} \gamma_{13} \mathbf{a}_{13}}{\gamma_{23} a_{23} + \gamma_{12} a_{12} \gamma_{13}}\right) t_{2,0} &= 0, \\ P - \mathbf{a}_{13} - \left(-\mathbf{a}_{13} + \frac{\gamma_{13} a_{13} \gamma_{12} \mathbf{a}_{12}}{\gamma_{23} a_{23} + \gamma_{13} a_{13} \gamma_{12}}\right) t_{3,0} &= 0. \end{aligned} \tag{7.89}$$

The kth equation in (7.89), $k = 1, 2, 3$, is equivalent to the condition that point P lies on line L_k.

The system of equations (7.89) was obtained by methods of gyroalgebra, and will be solved below by a common method of linear algebra.

Substituting P from (7.88) into (7.89), and rewriting each equation in (7.89) as a linear combination of $\mathbf{a}_{12}$ and $\mathbf{a}_{13}$ equals zero, one obtains the following homogeneous linear system of three gyrovector equations

$$\begin{aligned} c_{11}\mathbf{a}_{12} + c_{12}\mathbf{a}_{13} &= \mathbf{0}, \\ c_{21}\mathbf{a}_{12} + c_{22}\mathbf{a}_{13} &= \mathbf{0}, \\ c_{31}\mathbf{a}_{12} + c_{32}\mathbf{a}_{13} &= \mathbf{0}, \end{aligned} \tag{7.90}$$

where each coefficient c_{ij}, $i = 1, 2, 3$, $j = 1, 2$, is a function of γ_{12}, γ_{13}, γ_{23}, and the five unknowns m_2, m_3, and $t_{k,0}$, $k = 1, 2, 3$.

Since the set $S = \{A_1, A_2, A_3\}$ is pointwise independent, the two gyrovectors $\mathbf{a}_{12} = \ominus A_1 \oplus A_2$ and $\mathbf{a}_{13} = \ominus A_1 \oplus A_3$ in $\mathbb{R}^n_s$, considered as vectors in $\mathbb{R}^n$, are linearly independent in $\mathbb{R}^n$. Hence, each coefficient c_{ij} in (7.90) equals zero. Accordingly, the three gyrovector equations in (7.90) are equivalent to the following six scalar equations,

$$c_{11} = c_{12} = c_{21} = c_{22} = c_{31} = c_{32} = 0 \tag{7.91}$$

for the five unknowns m_2, m_3 and $t_{k,0}$, $k = 1, 2, 3$.

Explicitly, the six scalar equations in (7.91) are equivalent to the following six equations:

$$\begin{aligned} \gamma_{12}(a_{12} + a_{13})m_2 - (1 - m_2 - m_3 + \gamma_{12}m_2 + \gamma_{13}m_3)a_{13}t_{1,0} &= 0, \\ \gamma_{13}(a_{12} + a_{13})m_3 - (1 - m_2 - m_3 + \gamma_{12}m_2 + \gamma_{13}m_3)a_{12}t_{1,0} &= 0, \\ 1 - m_2 - m_3 + \gamma_{13}m_3 - (1 - m_2 - m_3 + \gamma_{12}m_2 + \gamma_{13}m_3)t_{2,0} &= 0, \\ (\gamma_{12}\gamma_{13}a_{12} + \gamma_{23}a_{23})m_3 - (1 - m_2 - m_3 + \gamma_{12}m_2 + \gamma_{13}m_3)\gamma_{12}a_{12}t_{2,0} &= 0, \\ (\gamma_{12}\gamma_{13}a_{13} + \gamma_{23}a_{23})m_2 - (1 - m_2 - m_3 + \gamma_{12}m_2 + \gamma_{13}m_3)\gamma_{13}a_{13}t_{3,0} &= 0, \\ 1 - m_2 - m_3 + \gamma_{12}m_2 - (1 - m_2 - m_3 + \gamma_{12}m_2 + \gamma_{13}m_3)t_{3,0} &= 0. \end{aligned} \tag{7.92}$$

The unique solution of (7.92) is given by (7.93) and (7.95) below:
The values of the line parameters are

$$\begin{aligned} t_{1,0} &= \frac{1}{D'}\gamma_{12}\gamma_{13}(a_{12} + a_{13}), \\ t_{2,0} &= \frac{1}{D'}(\gamma_{12}\gamma_{13}a_{12} + \gamma_{23}a_{23}), \\ t_{3,0} &= \frac{1}{D'}(\gamma_{12}\gamma_{13}a_{13} + \gamma_{23}a_{23}), \end{aligned} \tag{7.93}$$

where

$$D' = \gamma_{12}\gamma_{13}a_{12} + \gamma_{12}\gamma_{13}a_{13} + \gamma_{23}a_{23} > 0. \tag{7.94}$$

The special gyrobarycentric coordinates (m_1, m_2, m_3) are given by

$$\begin{aligned} m_1 &= \frac{1}{D}\gamma_{23}a_{23}, \\ m_2 &= \frac{1}{D}\gamma_{13}a_{13}, \\ m_3 &= \frac{1}{D}\gamma_{12}a_{12}, \end{aligned} \tag{7.95}$$

satisfying the normalization condition $m_1 + m_2 + m_3 = 1$, where D is given by

$$\gamma_{12}a_{12} + \gamma_{13}a_{13} + \gamma_{23}a_{23} > 0. \tag{7.96}$$

Following (7.95), convenient gyrobarycentric coordinates of the gyrotriangle ingyrocenter I are given by the equation

$$(m_1 : m_2 : m_3) = (\gamma_{23}a_{23} : \gamma_{13}a_{13} : \gamma_{12}a_{12}) \tag{7.97}$$

or, equivalently, by the equation

$$(m_1 : m_2 : m_3) = \left(\frac{\gamma_{23}a_{23}}{\gamma_{12}a_{12}} : \frac{\gamma_{13}a_{13}}{\gamma_{12}a_{12}} : 1\right) = \left(\frac{\sin\alpha_1}{\sin\alpha_3} : \frac{\sin\alpha_2}{\sin\alpha_3} : 1\right) \tag{7.98}$$

as we see from the law of gyrosines (6.44), p. 140. Hence a convenient set of gyrotrigonometric gyrobarycentric coordinates of the gyrotriangle ingyrocenter I is given by the equation

$$(m_1 : m_2 : m_3) = (\sin\alpha_1 : \sin\alpha_2 : \sin\alpha_3). \tag{7.99}$$

The gyrobarycentric coordinates in (7.99) are positive for any gyrotriangle gyroangles α_k, $k = 1, 2, 3$. Hence, by Corollary (4.10), p. 94, the gyrotriangle ingyrocenter always lies on the interior of its gyrotriangle, as shown in Fig. 7.8, p. 169.

We have thus found that the ingyrocenter of gyrotriangle $A_1A_2A_3$ lies on the interior of gyrotriangle $A_1A_2A_3$, and it has the gyrobarycentric coordinate representation with respect to the set $S = \{A_1, A_2, A_3\}$ given by each equation in the following chain of equations,

$$\begin{aligned} I &= \frac{\gamma_{23}a_{23}\gamma_{A_1}A_1 + \gamma_{13}a_{13}\gamma_{A_2}A_2 + \gamma_{12}a_{12}\gamma_{A_3}A_3}{\gamma_{23}a_{23}\gamma_{A_1} + \gamma_{13}a_{13}\gamma_{A_2} + \gamma_{12}a_{12}\gamma_{A_3}} \\ &= \frac{\sqrt{\gamma_{23}^2 - 1}\gamma_{A_1}A_1 + \sqrt{\gamma_{13}^2 - 1}\gamma_{A_2}A_2 + \sqrt{\gamma_{12}^2 - 1}\gamma_{A_3}A_3}{\sqrt{\gamma_{23}^2 - 1}\gamma_{A_1} + \sqrt{\gamma_{13}^2 - 1}\gamma_{A_2} + \sqrt{\gamma_{12}^2 - 1}\gamma_{A_3}} \\ &= \frac{\sin\alpha_1\gamma_{A_1}A_1 + \sin\alpha_2\gamma_{A_2}A_2 + \sin\alpha_3\gamma_{A_3}A_3}{\sin\alpha_1\gamma_{A_1} + \sin\alpha_2\gamma_{A_2} + \sin\alpha_3\gamma_{A_3}} \in \mathbb{R}_s^n. \end{aligned} \tag{7.100}$$

The first equation in (7.100) follows from (7.95). The second equation in (7.100) follows from the first by (1.9), p. 5, and the third equation in (7.100) follows from the first by the law of gyrosines (6.44), p. 140, according to which, by (1.9),

$$\begin{aligned}\frac{\sqrt{\gamma_{23}^2-1}}{\sqrt{\gamma_{12}^2-1}}&=\frac{\gamma_{23}a_{23}}{\gamma_{12}a_{12}}=\frac{\sin\alpha_1}{\sin\alpha_3},\\ \frac{\sqrt{\gamma_{13}^2-1}}{\sqrt{\gamma_{12}^2-1}}&=\frac{\gamma_{13}a_{13}}{\gamma_{12}a_{12}}=\frac{\sin\alpha_2}{\sin\alpha_3}.\end{aligned} \tag{7.101}$$

Formalizing the main result of this section, we obtain the following theorem:

Theorem 7.12 (The Ingyrocenter) *Let $S=\{A_1,A_2,A_3\}$ be a pointwise independent set of three points in an Einstein gyrovector space $(\mathbb{R}_s^n,\oplus,\otimes)$. The ingyrocenter $I\in\mathbb{R}_s^n$, Fig. 7.8, p. 169, of gyrotriangle $A_1A_2A_3$ has the gyrobarycentric coordinate representation*

$$I=\frac{m_1\gamma_{A_1}A_1+m_2\gamma_{A_2}A_2+m_3\gamma_{A_3}A_3}{m_1\gamma_{A_1}+m_2\gamma_{A_2}+m_3\gamma_{A_3}} \tag{7.102}$$

with respect to the set $S=\{A_1,A_2,A_3\}$, with gyrobarycentric coordinates $(m_1:m_2:m_3)$ given by each of the following three equations:

$$\begin{aligned}(m_1:m_2:m_3)&=\left(\sqrt{\gamma_{23}^2-1}:\sqrt{\gamma_{13}^2-1}:\sqrt{\gamma_{12}^2-1}\right),\\ (m_1:m_2:m_3)&=(\gamma_{23}a_{23}:\gamma_{13}a_{13}:\gamma_{12}a_{12}),\\ (m_1:m_2:m_3)&=(\sin\alpha_1:\sin\alpha_2:\sin\alpha_3).\end{aligned} \tag{7.103}$$

Interestingly, in the Euclidean limit of large s, $s\to\infty$, the three systems of gyrobarycentric coordinates $(m_1:m_2:m_3)$ in Theorem 7.12 exhibit the following different features:

The first system of gyrobarycentric coordinates of the gyrotriangle ingyrocenter in (7.103) reduces to $(m_1:m_2:m_3)=(0:0:0)$, which makes no sense in Euclidean geometry;

The second system of gyrobarycentric coordinates of the gyrotriangle ingyrocenter in (7.103) reduces to its Euclidean counterpart,

$$(m_1:m_2:m_3)=(a_{23}:a_{13}:a_{12}),\quad \text{(Euclidean Geometry)} \tag{7.104}$$

noting that in the limit of large s, $s\to\infty$, gamma factors tend to 1, and gyrolengths tend to lengths. Equation (7.104) gives a well-known barycentric coordinates of the Euclidean triangle incenter, where a_{23}, a_{13}, and a_{12}, are the side lengths of a Euclidean triangle $A_1A_2A_3$ in $\mathbb{R}^n$ [29]. We should note that a_{23},a_{13},a_{12} in (7.103) are gyrotriangle side gyrolengths while, in contrast, a_{23}, a_{13}, a_{12} in (7.104) are triangle side lengths.

The third system of gyrobarycentric coordinates of the gyrotriangle ingyrocenter in (7.103) appears in a gyrotrigonometric form. As such, it is identical, in form, with its Euclidean trigonometric counterpart. Indeed, in the limit $s \to \infty$ the third equation in (7.103), which is in a gyrotrigonometric form, remains intact in form in the transition from hyperbolic geometry to Euclidean geometry. It leads to a well-known barycentric coordinates of the Euclidean triangle incenter in a trigonometric form [29],

$$(m_1 : m_2 : m_3) = (\sin\alpha_1 : \sin\alpha_2 : \sin\alpha_3). \quad \text{(Euclidean Geometry)} \tag{7.105}$$

We should note that while the third equation in (7.103) and (7.105) are equal in form, they are different in context. The former involves gyrosines of gyrotriangle gyroangles while, in contrast, the latter involves sines of triangle angles.

By Theorem 7.12 and the ingyrocenter gyrobarycentric coordinate representation (7.75), p. 172, we obtain the following theorem:

Theorem 7.13 (The Gyrotriangle Ingyrocenter) *Let $S = \{A_1, A_2, A_3\}$ be a pointwise independent set of three points in an Einstein gyrovector space $(\mathbb{R}^n_s, \oplus, \otimes)$. The ingyrocenter I, Fig. 7.8, p. 169, of gyrotriangle $A_1A_2A_3$ has the gyrobarycentric coordinate representation*

$$I = \frac{m_1\gamma_{A_1} A_1 + m_2\gamma_{A_2} A_2 + m_3\gamma_{A_3} A_3}{m_1\gamma_{A_1} + m_2\gamma_{A_2} + m_3\gamma_{A_3}} \tag{7.106}$$

with respect to the set $S = \{A_1, A_2, A_3\}$, with gyrobarycentric coordinates given by

$$(m_1 : m_2 : m_3) = \left(\sqrt{\gamma_{23}^2 - 1} : \sqrt{\gamma_{13}^2 - 1} : \sqrt{\gamma_{12}^2 - 1}\right) \tag{7.107}$$

or, equivalently, by

$$(m_1 : m_2 : m_3) = (\gamma_{23}a_{23} : \gamma_{13}a_{13} : \gamma_{12}a_{12}) \tag{7.108}$$

or, equivalently, by the gyrotrigonometric gyrobarycentric coordinates

$$(m_1 : m_2 : m_3) = (\sin\alpha_1 : \sin\alpha_2 : \sin\alpha_3). \tag{7.109}$$

Two immediate, but interesting, corollaries of Theorem 7.13 are presented below:

Corollary 7.14 *Let $A_1A_2A_3$ be a gyrotriangle with gyroangles α_k, $k = 1, 2, 3$, in an Einstein gyrovector space $(\mathbb{R}^n_s, \oplus, \otimes)$. Then, the gyrotriangle ingyrocenter I possesses the gyrotrigonometric gyrobarycentric coordinate representation*

$$I = \frac{\sin\alpha_1\gamma_{A_1} A_1 + \sin\alpha_2\gamma_{A_2} A_2 + \sin\alpha_3\gamma_{A_3} A_3}{\sin\alpha_1\gamma_{A_1} + \sin\alpha_2\gamma_{A_2} + \sin\alpha_3\gamma_{A_3}}. \tag{7.110}$$

Corollary 7.15 *Let $A_1A_2A_3$ be a triangle with angles α_k, $k = 1, 2, 3$, in a Euclidean space $\mathbb{R}^n$. Then, the triangle incenter I possesses the trigonometric barycentric coordinate representation*

$$I = \frac{\sin\alpha_1 A_1 + \sin\alpha_2 A_2 + \sin\alpha_3 A_3}{\sin\alpha_1 + \sin\alpha_2 + \sin\alpha_3}. \tag{7.111}$$

7.9 Gyrotriangle Gyroaltitude Foot

Let A_3P_3 be the gyroaltitude of gyrotriangle $A_1A_2A_3$ drawn from vertex A_3 to its foot P_3 on its opposite side A_1A_2 in an Einstein gyrovector space $(\mathbb{R}^n_s, \oplus, \otimes)$, as shown in Figs. 7.9–7.10 for $n = 2$. Furthermore, let

$$P_3 = \frac{m_1\gamma_{A_1} A_1 + m_2\gamma_{A_2} A_2}{m_1\gamma_{A_1} + m_2\gamma_{A_2}} \tag{7.112}$$

be the gyrobarycentric coordinate representation of P_3 with respect to the set $\{A_1, A_2\}$, (4.25), p. 90, where the gyrobarycentric coordinates $(m_1 : m_2)$ are to be determined in (7.123)–(7.125) below.

Employing the Gyrobarycentric Coordinate Representation Gyrocovariance Theorem 4.6, we have from Identity (4.29b), p. 91, with $X = \ominus A_1$, using the standard

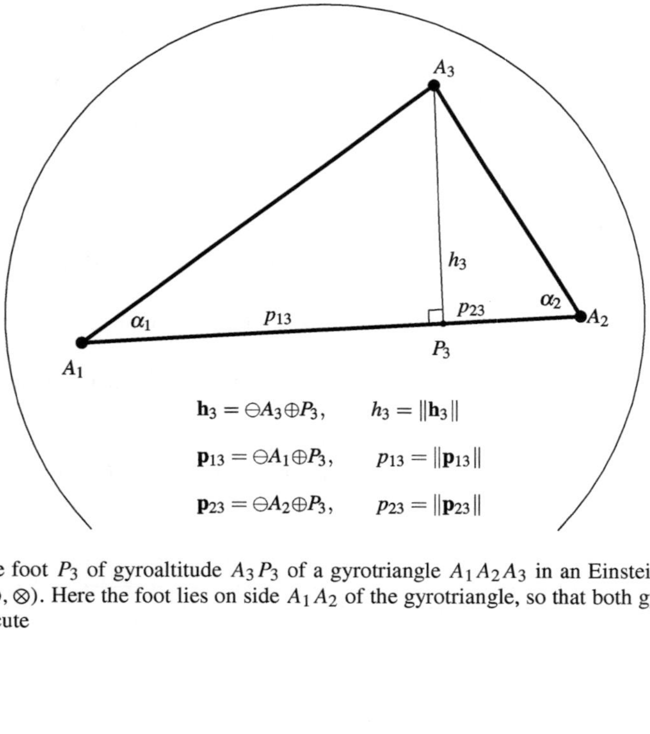

Fig. 7.9 The foot P_3 of gyroaltitude A_3P_3 of a gyrotriangle $A_1A_2A_3$ in an Einstein gyrovector space $(\mathbb{R}^n_s, \oplus, \otimes)$. Here the foot lies on side A_1A_2 of the gyrotriangle, so that both gyroangles α_1 and α_2 are acute

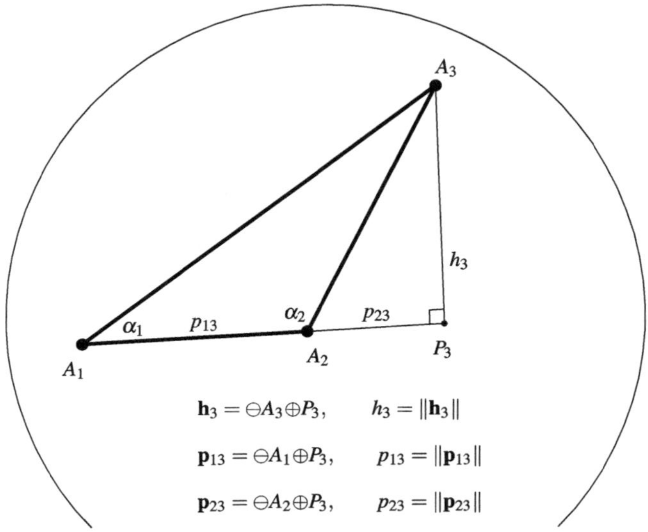

Fig. 7.10 The foot P_3 of gyroaltitude A_3P_3 of a gyrotriangle $A_1A_2A_3$ in an Einstein gyrovector space $(\mathbb{R}^n_s, \oplus, \otimes)$. Here the foot lies on the extension of side A_1A_2 of the gyrotriangle, so that gyroangle α_2 is obtuse

gyrotriangle index notation, shown in Fig. 7.9, in Fig. 6.1, p. 128, and in (6.1), p. 127,

$$\gamma_{\ominus X \oplus P_3} = \frac{m_1 \gamma_{\ominus X \oplus A_1} + m_2 \gamma_{\ominus X \oplus A_2}}{m_0}, \tag{7.113}$$

where

$$m_0^2 = m_1^2 + m_2^2 + 2m_1 m_2 \gamma_{12}. \tag{7.114}$$

Hence, for $X = A_1$, $X = A_2$ and $X = A_3$ in (7.113) we have, respectively,

$$\begin{aligned}
\gamma_{p_{13}} = \gamma_{\ominus A_1 \oplus P_3} &= \frac{m_1 + m_2 \gamma_{\ominus A_1 \oplus A_2}}{m_0} = \frac{m_1 + m_2 \gamma_{12}}{m_0}, \\
\gamma_{p_{23}} = \gamma_{\ominus A_2 \oplus P_3} &= \frac{m_1 \gamma_{\ominus A_2 \oplus A_1} + m_2}{m_0} = \frac{m_1 \gamma_{12} + m_2}{m_0}, \\
\gamma_{h_3} = \gamma_{\ominus A_3 \oplus P_3} &= \frac{m_1 \gamma_{\ominus A_3 \oplus A_1} + m_2 \gamma_{\ominus A_3 \oplus A_2}}{m_0} = \frac{m_1 \gamma_{13} + m_2 \gamma_{23}}{m_0}.
\end{aligned} \tag{7.115}$$

Applying the Einstein–Pythagoras identity (6.57), p. 144, to each of the two right gyroangled gyrotriangles $A_1P_3A_3$ and $A_2P_3A_3$ in Fig. 7.9, we have

$$\begin{aligned}\gamma_{p_{13}}\gamma_{h_3} &= \gamma_{13}, \\ \gamma_{p_{23}}\gamma_{h_3} &= \gamma_{23}.\end{aligned} \tag{7.116}$$

Substituting (7.114)–(7.115) into (7.116), we obtain a system of two equations for the two unknowns m_1 and m_2. This system does not possess a unique solution. Adding the normalization condition $m_1 + m_2 = 1$ results in the unique solution,

$$\begin{aligned}m_1 &= \frac{\gamma_{12}\gamma_{23} - \gamma_{13}}{(\gamma_{13}+\gamma_{23})(\gamma_{12}-1)}, \\ m_2 &= \frac{\gamma_{12}\gamma_{13} - \gamma_{23}}{(\gamma_{13}+\gamma_{23})(\gamma_{12}-1)},\end{aligned} \tag{7.117}$$

as one can readily check. The unique special gyrobarycentric coordinates (m_1, m_2) of the point P_3 with respect to the set $S = \{A_1, A_2\}$ in Fig. 7.9 are thus determined by (7.117).

The unique special gyrobarycentric coordinates (m_1, m_2) in (7.117) suggests the following convenient gyrobarycentric coordinates $(m'_1 : m'_2)$ of the point P_3 with respect to the set $S = \{A_1, A_2\}$,

$$\begin{aligned}m'_1 &= \gamma_{12}\gamma_{23} - \gamma_{13}, \\ m'_2 &= \gamma_{12}\gamma_{13} - \gamma_{23},\end{aligned} \tag{7.118}$$

so that a gyrobarycentric coordinate representation (7.112) of P_3 with respect to the set $S = \{A_1, A_2\}$ is given by

$$P_3 = \frac{(\gamma_{12}\gamma_{23} - \gamma_{13})\gamma_{A_1}A_1 + (\gamma_{12}\gamma_{13} - \gamma_{23})\gamma_{A_2}A_2}{(\gamma_{12}\gamma_{23} - \gamma_{13})\gamma_{A_1} + (\gamma_{12}\gamma_{13} - \gamma_{23})\gamma_{A_2}}. \tag{7.119}$$

A different convenient gyrobarycentric coordinates $(m''_1 : m''_1)$ of P_3 with respect to $\{A_1, A_2\}$ can be obtained from (7.118) by means of (6.20), p. 134,

$$\begin{aligned}(m''_1 : m''_1) &= (\gamma_{12}\gamma_{23} - \gamma_{13} : \gamma_{12}\gamma_{13} - \gamma_{23})\frac{1}{\sqrt{\gamma_{12}^2-1}\sqrt{\gamma_{13}^2-1}\sqrt{\gamma_{23}^2-1}} \\ &= \left(\frac{\cos\alpha_2}{\sqrt{\gamma_{13}^2-1}} : \frac{\cos\alpha_1}{\sqrt{\gamma_{23}^2-1}}\right).\end{aligned} \tag{7.120}$$

The advantage of the gyrobarycentric coordinates $(m''_1 : m''_1)$ of P_3 with respect to $\{A_1, A_2\}$ in (7.120) rests on the observation that the sign of m''_1 (m''_2) equals the sign of $\cos\alpha_2$ $(\cos\alpha_1)$.

Another set of convenient gyrobarycentric coordinates $(m_1''' : m_1''')$ of P_3 with respect to $\{A_1, A_2\}$ results from (7.120) and (6.39), p. 139, obtaining the following gyrotrigonometric gyrobarycentric coordinates:

$$\left(m_1''' : m_1'''\right) = (\sin\alpha_1 \cos\alpha_2 : \cos\alpha_1 \sin\alpha_2), \tag{7.121}$$

where α_k, $k = 1, 2, 3$, are the gyroangles of gyrotriangle $A_1A_2A_3$ in Fig. 7.9, in the standard gyrotriangle index notation.

Formalizing the main result of this section, we have the following theorem:

Theorem 7.16 (The Foot of a Gyrotriangle Gyroaltitude) *Let $S = \{A_1, A_2, A_3\}$ be a pointwise independent set of three points in an Einstein gyrovector space $(\mathbb{R}^n_s, \oplus, \otimes)$, and let P_3 be the foot of gyroaltitude A_3P_3, Fig. 7.9, p. 180. Then the gyroaltitude foot has the gyrobarycentric coordinate representation*

$$P_3 = \frac{m_1\gamma_{A_1} A_1 + m_2\gamma_{A_2} A_2}{m_1\gamma_{A_1} + m_2\gamma_{A_2}} \tag{7.122}$$

with respect to the set $S = \{A_1, A_2\}$, with gyrobarycentric coordinates

$$(m_1 : m_2) = (\gamma_{12}\gamma_{23} - \gamma_{13} : \gamma_{12}\gamma_{13} - \gamma_{23}) \tag{7.123}$$

or, equivalently,

$$(m_1 : m_2) = \left(\frac{\cos\alpha_2}{\sqrt{\gamma_{13}^2 - 1}} : \frac{\cos\alpha_1}{\sqrt{\gamma_{23}^2 - 1}}\right) \tag{7.124}$$

or, equivalently, with gyrotrigonometric gyrobarycentric coordinates

$$(m_1 : m_2) = (\sin\alpha_1 \cos\alpha_2 : \cos\alpha_1 \sin\alpha_2). \tag{7.125}$$

It is clear from (7.123)–(7.125) that the two gyrobarycentric coordinates m_1 and m_2 of the gyroaltitude foot P_3 are positive in Fig. 7.9, where P_3 lies on side A_1A_2 of gyrotriangle $A_1A_2A_3$, so that both gyroangles α_1 and α_2 acute. The fact that in this case the gyroaltitude foot P_3 lies on side A_1A_2 is in accordance with Corollary 4.9, p. 93.

In contrast, it is clear from (7.123)–(7.125) that the two gyrobarycentric coordinates m_1 and m_2 of the gyroaltitude foot P_3 are, respectively, negative and positive in Fig. 7.10, where P_3 lies on the extension of side A_1A_2 of gyrotriangle $A_1A_2A_3$, so that gyroangles α_1 and α_2 are, respectively, acute and obtuse. The fact that in this case the gyroaltitude foot P_3 does not lie on side A_1A_2 is in accordance with Corollary 4.9, p. 93.

7.10 Gyrotriangle Gyroaltitude

In this section, we calculate the gyrolength $h_3 = \|\ominus A_3 \oplus P_3\|$ of the gyroaltitude gyrovector $\mathbf{h}_3 = \ominus A_3 \oplus P_3$ of gyrotriangle $A_1A_2A_3$ in Figs. 7.9–7.10.

By the third equation in (7.115), we have

$$\gamma_{h_3} = \gamma_{\ominus A_3 \oplus P_3} = \frac{m_1\gamma_{13} + m_2\gamma_{23}}{m_0}, \tag{7.126}$$

where, by (7.117),

$$\begin{aligned} m_1 &= \gamma_{12}\gamma_{23} - \gamma_{13}, \\ m_2 &= \gamma_{12}\gamma_{13} - \gamma_{23}. \end{aligned} \tag{7.127}$$

Hence, following (7.127) and (7.114),

$$\begin{aligned} m_0^2 &= m_1^2 + m_2^2 + 2m_1m_2\gamma_{12} \\ &= (\gamma_{12}^2 - 1)(2\gamma_{12}\gamma_{13}\gamma_{23} - \gamma_{13}^2 - \gamma_{23}^2) \end{aligned} \tag{7.128}$$

so that, by (7.126)–(7.128),

$$\begin{aligned} \gamma_{h_3}^2 &= \frac{2\gamma_{12}\gamma_{13}\gamma_{23} - \gamma_{13}^2 - \gamma_{23}^2}{\gamma_{12}^2 - 1} \\ &= 1 + \frac{1 + 2\gamma_{12}\gamma_{13}\gamma_{23} - \gamma_{12}^2 - \gamma_{13}^2 - \gamma_{23}^2}{\gamma_{12}^2 - 1}. \end{aligned} \tag{7.129}$$

Hence, by (7.129) and (1.9), p. 5,

$$\begin{aligned} h_3^2 = s^2\frac{\gamma_{h_3}^2 - 1}{\gamma_{h_3}^2} &= s^2\frac{1 + 2\gamma_{12}\gamma_{13}\gamma_{23} - \gamma_{12}^2 - \gamma_{13}^2 - \gamma_{23}^2}{2\gamma_{12}\gamma_{13}\gamma_{23} - \gamma_{13}^2 - \gamma_{23}^2} \\ &= s^2\left(1 - \frac{\gamma_{12}^2 - 1}{2\gamma_{a_1}\gamma_{a_2}\gamma_{12} - \gamma_{a_1}^2 - \gamma_{a_2}^2}\right). \end{aligned} \tag{7.130}$$

Formalizing the results of this section, we obtain the following theorem:

Theorem 7.17 (The Gyrotriangle Gyroaltitude) *Let $A_1A_2A_3$ be a gyrotriangle in an Einstein gyrovector space $(\mathbb{R}^n_s, \oplus, \otimes)$ with a gyroaltitude A_3P_3, as shown in Figs. 7.9–7.10, 7.17, and let $h_3 = \|\ominus A_3 \oplus P_3\|$ be the gyrolength of the gyroaltitude. Then,*

$$h_3 = s\sqrt{\frac{1 + 2\gamma_{12}\gamma_{13}\gamma_{23} - \gamma_{12}^2 - \gamma_{13}^2 - \gamma_{23}^2}{2\gamma_{12}\gamma_{13}\gamma_{23} - \gamma_{13}^2 - \gamma_{23}^2}} \tag{7.131}$$

and

$$\gamma_{h_3} = \sqrt{\frac{2\gamma_{12}\gamma_{13}\gamma_{23} - \gamma_{13}^2 - \gamma_{23}^2}{\gamma_{12}^2 - 1}}. \tag{7.132}$$

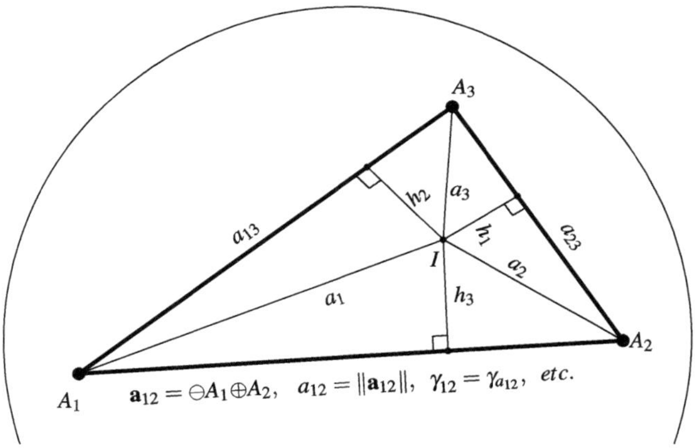

Fig. 7.11 The ingyrocenter I of a gyrotriangle $A_1A_2A_3$ in an Einstein gyrovector space $(\mathbb{R}^n_s, \oplus, \otimes)$. The gyrotriangle ingyrocenter is the point of the interior of the gyrotriangle that is equigyrodistant from the three gyrotriangle sides. Accordingly, $r := h_1 = h_2 = h_3$ is the gyrotriangle ingyroradius r

7.11 Gyrotriangle Ingyroradius

The gyrobarycentric coordinate representation, (4.25), p. 90, of the ingyrocenter I of gyrotriangle $A_1A_2A_3$ in an Einstein gyrovector space $(\mathbb{R}^n_s, \oplus, \otimes)$, Fig. 7.11, with respect to the set $S = \{A_1, A_2, A_3\}$ of the gyrotriangle vertices is given by

$$I = \frac{m_1\gamma_{A_1}A_1 + m_2\gamma_{A_2}A_2 + m_3\gamma_{A_3}A_3}{m_1\gamma_{A_1} + m_2\gamma_{A_2} + m_3\gamma_{A_3}}, \tag{7.133}$$

where, by Theorem 7.12, p. 178, the gyrobarycentric coordinates of I in (7.133) are given by

$$(m_1 : m_2 : m_3) = (\gamma_{23}a_{23} : \gamma_{13}a_{13} : \gamma_{12}a_{12}). \tag{7.134}$$

Following (7.133) and the Gyrobarycentric Coordinate Representation Gyrocovariance Theorem 4.6, p. 90, we have

$$\gamma_{X\oplus I} = \frac{m_1\gamma_{X\oplus A_1} + m_2\gamma_{X\oplus A_2} + m_3\gamma_{X\oplus A_3}}{m_0} \tag{7.135}$$

for all $X \in \mathbb{R}^n_s$, where, according to (4.28d), p. 91, m_0 is given by the equation

$$m_0^2 = m_1^2 + m_2^2 + m_3^2 + 2(m_1m_2\gamma_{12} + m_1m_3\gamma_{13} + m_2m_3\gamma_{23}). \tag{7.136}$$

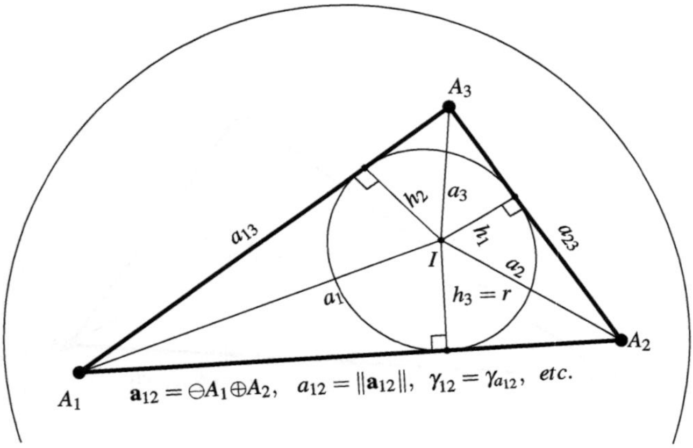

Fig. 7.12 The ingyrocircle of gyrotriangle $A_1A_2A_3$ in Fig. 7.11 is shown. Its gyrocenter, I, is the gyrotriangle ingyrocenter, and its gyroradius r, $r := h_1 = h_2 = h_3$, is the gyrotriangle ingyroradius

Substituting $X = \ominus A_1$ and $X = \ominus A_2$ into (7.135), we have, respectively,

$$\begin{aligned}\gamma_{\ominus A_1\oplus I} &= \frac{m_1 + m_2\gamma_{\ominus A_1\oplus A_2} + m_3\gamma_{\ominus A_1\oplus A_3}}{m_0},\\ \gamma_{\ominus A_2\oplus I} &= \frac{m_1\gamma_{\ominus A_2\oplus A_1} + m_2 + m_3\gamma_{\ominus A_2\oplus A_3}}{m_0}\end{aligned}\tag{7.137}$$

or, equivalently, in the notation of Fig. 7.11,

$$\begin{aligned}\gamma_{a_1} &= \frac{m_1 + m_2\gamma_{12} + m_3\gamma_{13}}{m_0},\\ \gamma_{a_2} &= \frac{m_1\gamma_{12} + m_2 + m_3\gamma_{23}}{m_0}.\end{aligned}\tag{7.138}$$

We are now in a position to apply the gyroaltitude equation (7.131) of gyrotriangle $A_1A_2A_3$ in Theorem 7.17, to gyrotriangle A_1A_2I in Fig. 7.12, obtaining in the notation of Fig. 7.12,

$$r^2 := h_3^2 = s^2\frac{1 + 2\gamma_{12}\gamma_{a_1}\gamma_{a_2} - \gamma_{12}^2 - \gamma_{a_1}^2 - \gamma_{a_2}^2}{2\gamma_{12}\gamma_{a_1}\gamma_{a_2} - \gamma_{a_1}^2 - \gamma_{a_2}^2},\tag{7.139}$$

where r is the ingyroradius of gyrotriangle $A_1A_2A_3$, shown in Fig. 7.12.

Substituting (7.138) and, subsequently, (7.136) and (7.134) into (7.139), we have

$$r^2 = \frac{s^2}{2}\frac{1 + 2\gamma_{12}\gamma_{13}\gamma_{23} - \gamma_{12}^2 - \gamma_{13}^2 - \gamma_{23}^2}{D},\tag{7.140}$$

where

$$D = \gamma_{12}\gamma_{13}\gamma_{23} - 1 + \gamma_{12}\sqrt{\gamma_{13}^2 - 1}\sqrt{\gamma_{23}^2 - 1} + \sqrt{\gamma_{12}^2 - 1}\gamma_{13}\sqrt{\gamma_{23}^2 - 1} + \sqrt{\gamma_{12}^2 - 1}\sqrt{\gamma_{13}^2 - 1}\gamma_{23}. \tag{7.141}$$

7.12 Useful Gyrotriangle Gyrotrigonometric Identities

The identities in (6.33), p. 137, and (6.43), p. 140, prove useful when expressed in the standard gyrotriangle index notation, shown in Fig. 6.1, p. 128, and in (6.1), p. 127. In this notation, these identities take the form

$$\begin{aligned}
\gamma_{12} &= \frac{\cos\alpha_3 + \cos\alpha_1\cos\alpha_2}{\sin\alpha_1\sin\alpha_2},\\
\gamma_{13} &= \frac{\cos\alpha_2 + \cos\alpha_1\cos\alpha_3}{\sin\alpha_1\sin\alpha_3},\\
\gamma_{23} &= \frac{\cos\alpha_1 + \cos\alpha_2\cos\alpha_3}{\sin\alpha_2\sin\alpha_3},
\end{aligned} \tag{7.142}$$

and

$$\begin{aligned}
\sqrt{\gamma_{12}^2 - 1} &= \gamma_{12}\frac{a_{12}}{s} = 2\frac{\sqrt{F(\alpha_1,\alpha_2,\alpha_3)}}{\sin\alpha_1\sin\alpha_2},\\
\sqrt{\gamma_{13}^2 - 1} &= \gamma_{13}\frac{a_{13}}{s} = 2\frac{\sqrt{F(\alpha_1,\alpha_2,\alpha_3)}}{\sin\alpha_1\sin\alpha_3},\\
\sqrt{\gamma_{23}^2 - 1} &= \gamma_{23}\frac{a_{23}}{s} = 2\frac{\sqrt{F(\alpha_1,\alpha_2,\alpha_3)}}{\sin\alpha_2\sin\alpha_3},
\end{aligned} \tag{7.143}$$

where

$$\begin{aligned}
F(\alpha_1,\alpha_2,\alpha_3) &= \frac{1}{4}\big(2\cos\alpha_1\cos\alpha_2\cos\alpha_3 + \cos^2\alpha_1 + \cos^2\alpha_2 + \cos^2\alpha_3 - 1\big)\\
&= \cos\frac{\alpha_1+\alpha_2+\alpha_3}{2}\cos\frac{\alpha_1-\alpha_2-\alpha_3}{2}\cos\frac{-\alpha_1+\alpha_2-\alpha_3}{2}\\
&\quad\times\cos\frac{-\alpha_1-\alpha_2+\alpha_3}{2}\\
&= \frac{1}{4}\frac{(1+2\gamma_{12}\gamma_{13}\gamma_{23} - \gamma_{12}^2 - \gamma_{13}^2 - \gamma_{23}^2)^2}{(\gamma_{12}^2-1)(\gamma_{13}^2-1)(\gamma_{23}^2-1)}.
\end{aligned} \tag{7.144}$$

The identities in (7.142) along with the common gyrotrigonometric/trigonometric addition law of the gyrocosine/cosine function,

$$\cos(\alpha \pm \beta) = \cos\alpha\cos\beta \mp \sin\alpha\sin\beta \tag{7.145}$$

for $\alpha, \beta \in \mathbb{R}$, imply

$$\begin{aligned}
\gamma_{12} \pm 1 &= \frac{\cos\alpha_3 + \cos(\alpha_1 \mp \alpha_2)}{\sin\alpha_1 \sin\alpha_2}, \\
\gamma_{13} \pm 1 &= \frac{\cos\alpha_2 + \cos(\alpha_1 \mp \alpha_3)}{\sin\alpha_1 \sin\alpha_3}, \\
\gamma_{23} \pm 1 &= \frac{\cos\alpha_1 + \cos(\alpha_2 \mp \alpha_3)}{\sin\alpha_2 \sin\alpha_3}.
\end{aligned} \tag{7.146}$$

The law of gyrocosines (6.20), p. 134, and (7.143) imply

$$\begin{aligned}
\gamma_{12}\gamma_{13} - \gamma_{23} &= \frac{4F(\alpha_1, \alpha_2, \alpha_3)\cos\alpha_1}{\sin^2\alpha_1 \sin\alpha_2 \sin\alpha_3}, \\
\gamma_{12}\gamma_{23} - \gamma_{13} &= \frac{4F(\alpha_1, \alpha_2, \alpha_3)\cos\alpha_2}{\sin\alpha_1 \sin^2\alpha_2 \sin\alpha_3}, \\
\gamma_{13}\gamma_{23} - \gamma_{12} &= \frac{4F(\alpha_1, \alpha_2, \alpha_3)\cos\alpha_3}{\sin\alpha_1 \sin\alpha_2 \sin^2\alpha_3}.
\end{aligned} \tag{7.147}$$

By (6.39), p. 139, we have

$$\gamma_{12}\gamma_{13}\gamma_{23} - 1 = \frac{4F(\alpha_1, \alpha_2, \alpha_3)(1 + \cos\alpha_1\cos\alpha_2\cos\alpha_3)}{\sin^2\alpha_1 \sin^2\alpha_2 \sin^2\alpha_3}. \tag{7.148}$$

Other elegant and useful gyrotrigonometric identities are

$$\gamma_{12} - 1 = 2\frac{\cos\frac{\alpha_1+\alpha_2+\alpha_3}{2}\cos\frac{-\alpha_1-\alpha_2+\alpha_3}{2}}{\sin\alpha_1 \sin\alpha_2}, \tag{7.149}$$

$$\gamma_{12} + 1 = 2\frac{\cos\frac{\alpha_1-\alpha_2-\alpha_3}{2}\cos\frac{-\alpha_1+\alpha_2-\alpha_3}{2}}{\sin\alpha_1 \sin\alpha_2}, \tag{7.150}$$

and

$$\gamma_{12}\sqrt{\gamma_{13}^2 - 1}\sqrt{\gamma_{23}^2 - 1} = \frac{4F(\alpha_1, \alpha_2, \alpha_3)}{\sin^2\alpha_1 \sin^2\alpha_2 \sin^2\alpha_3}(\cos\alpha_1\cos\alpha_2 + \cos\alpha_3), \tag{7.151}$$

along with their index permutations, and

$$\sqrt{1 + 2\gamma_{12}\gamma_{13}\gamma_{23} - \gamma_{12}^2 - \gamma_{13}^2 - \gamma_{23}^2} = \frac{4F(\alpha_1, \alpha_2, \alpha_3)}{\sin\alpha_1 \sin\alpha_2 \sin\alpha_3}. \tag{7.152}$$

The following resulting gyrotriangle gyrotrigonometric identities prove useful:

$$(\gamma_{12}-1)(\gamma_{13}-1)(\gamma_{23}-1)=\frac{8F(\alpha_1,\alpha_2,\alpha_3)\cos^2\frac{\alpha_1+\alpha_2+\alpha_3}{2}}{\sin^2\alpha_1\sin^2\alpha_2\sin^2\alpha_3}, \tag{7.153}$$

$$(\gamma_{12}+1)(\gamma_{13}+1)(\gamma_{23}+1)=\frac{8\cos^2\frac{\alpha_1-\alpha_2-\alpha_3}{2}\cos^2\frac{-\alpha_1+\alpha_2-\alpha_3}{2}\cos^2\frac{-\alpha_1-\alpha_2+\alpha_3}{2}}{\sin^2\alpha_1\sin^2\alpha_2\sin^2\alpha_3} \tag{7.154}$$

so that

$$\left(\gamma_{12}^2-1\right)\left(\gamma_{13}^2-1\right)\left(\gamma_{23}^2-1\right)=\frac{64F(\alpha_1,\alpha_2,\alpha_3)^3}{\sin^4\alpha_1\sin^4\alpha_4\sin^2\alpha_3}. \tag{7.155}$$

7.13 Gyrotriangle Circumgyrocenter Gyrodistance from Sides

As an application of Theorem 7.17, p. 184, we determine the gyrodistances h_1, h_2 and h_3 between a gyrotriangle circumgyrocenter, O, and its sides A_2A_3, A_1A_3 and A_1A_2, respectively, shown in Fig. 7.13.

The gyrodistance of E (note that E represents each of E_k, $k=0,1,2,3$, in Fig. 8.1) from the gyroline that passes through points A_1 and A_2, Fig. 8.1, is the gyroaltitude r_{12} of gyrotriangle A_1A_2E drawn from base A_1A_2. Hence, by Theorem 7.17, p. 184, r_{12} is given by the equation

$$\frac{h_3^2}{s^2}=\frac{1+2\gamma_{12}\gamma_{\ominus A_1\oplus O}\gamma_{\ominus A_2\oplus O}-\gamma_{12}^2-\gamma_{\ominus A_1\oplus O}^2-\gamma_{\ominus A_2\oplus O}^2}{2\gamma_{12}\gamma_{\ominus A_1\oplus O}\gamma_{\ominus A_2\oplus O}-\gamma_{\ominus A_1\oplus O}^2-\gamma_{\ominus A_2\oplus O}^2}. \tag{7.156}$$

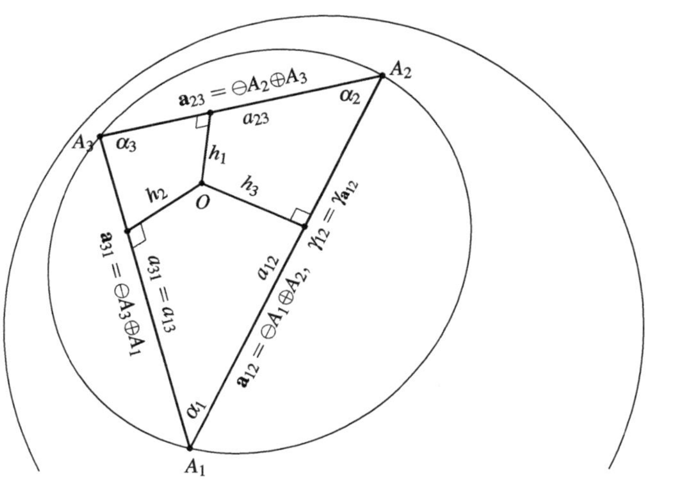

Fig. 7.13 The gyrodistances h_1, h_2 and h_3 between the circumgyrocenter of gyrotriangle $A_1A_2A_3$ and its sides in an Einstein gyrovector space $(\mathbb{R}_s^n, \oplus, \otimes)$

Since O is the gyrotriangle circumgyrocenter,

$$\|\ominus A_1 \oplus O\| = \|\ominus A_1 \oplus O\| = R, \tag{7.157}$$

where R is the circumgyroradius of gyrotriangle $A_1A_2A_3$. Hence, (7.156) can be written as

$$\frac{h_3^2}{s^2} = \frac{1 + 2\gamma_{12}\gamma_R^2 - \gamma_{12}^2 - 2\gamma_R^2}{2\gamma_{12}\gamma_R^2 - 2\gamma_R^2}. \tag{7.158}$$

Noting that $\gamma_R^2 = (1 - R^2/s^2)^{-1}$, (7.158) implies

$$\frac{h_3^2}{s^2} = \frac{1}{2}\left\{\frac{R^2}{s^2}(\gamma_{12} + 1) - (\gamma_{12} - 1)\right\}, \tag{7.159}$$

where the circumgyroradius R of gyrotriangle $A_1A_2A_3$ is given by (7.29), p. 162.

Substituting R from (7.29) into (7.159), we have

$$\frac{h_3^2}{s^2} = \frac{(\gamma_{12}^2 - 1)(\gamma_{13} - 1)(\gamma_{23} - 1)}{1 + 2\gamma_{12}\gamma_{13}\gamma_{23} - \gamma_{12}^2 - \gamma_{13}^2 - \gamma_{23}^2} - \frac{1}{2}(\gamma_{12} - 1). \tag{7.160}$$

Finally, substitutions of gyrotriangle gyrotrigonometric identities from Sect. 7.12 into (7.160) give the elegant relation

$$h_3^2 = \frac{s^2 \cos\frac{\alpha_1+\alpha_2+\alpha_3}{2} \sin\frac{\alpha_1+\alpha_2-\alpha_3}{2} \tan\frac{\alpha_1+\alpha_2-\alpha_3}{2}}{\sin\alpha_1 \sin\alpha_2}. \tag{7.161}$$

Eliminating the factor $s^2 \cos\frac{\alpha_1+\alpha_2+\alpha_3}{2}$ between (7.161) and (7.35), p. 163, we obtain the relation

$$h_3^2 = \frac{R^2 \cos\frac{\alpha_1-\alpha_2-\alpha_3}{2} \cos\frac{-\alpha_1+\alpha_2-\alpha_3}{2} \sin^2\frac{\alpha_1+\alpha_2-\alpha_3}{2}}{\sin\alpha_1 \sin\alpha_2}. \tag{7.162}$$

The ball parameter s, which appears explicitly in (7.161), disappears in (7.162). Clearly, however, its presence in (7.162) remains implicit since it involves in the measure of gyroangles. Interestingly, the elegant relation (7.162) remains invariant in form under the Euclidean limit $s \to \infty$, so that it is valid in Euclidean geometry as well.

The side A_1A_2 of gyrotriangle $A_1A_2A_3$ in Fig. 7.13 is the gyrodiameter of the gyrotriangle circumgyrocircle if and only if $h_3 = 0$. The latter, in turn, is valid if and only if $\alpha_3 = \alpha_1 + \alpha_2$, as we see from (7.162). Hence, the biggest gyroangle of a gyrotriangle has measure equal to the sum of the measures of the other two gyroangles if and only if the side opposite to the biggest gyroangle is a gyrodiameter of the circumgyrocircle. This result is known in hyperbolic geometry; see, for instance, [28, p. 133], where the result is proved synthetically.

Formalizing the results of this section, we obtain the following theorem:

Theorem 7.18 *Let $A_1A_2A_3$ be a gyrotriangle in an Einstein gyrovector space $(\mathbb{R}^n_s, \oplus, \otimes)$ and let O be its circumgyrocenter, Fig. 7.13. The gyrodistances h_k, $k = 1, 2, 3$, from O to the gyrotriangle side opposite to vertex A_k are related to the gyrotriangle circumgyrocenter R and gyroangles $\alpha_1, \alpha_2, \alpha_3$ by the equations*

$$\begin{aligned}
h_1^2 &= \frac{R^2 \cos\frac{-\alpha_1+\alpha_2-\alpha_3}{2} \cos\frac{-\alpha_1-\alpha_2+\alpha_3}{2} \sin^2\frac{-\alpha_1+\alpha_2+\alpha_3}{2}}{\sin\alpha_2 \sin\alpha_3}, \\
h_2^2 &= \frac{R^2 \cos\frac{\alpha_1-\alpha_2-\alpha_3}{2} \cos\frac{-\alpha_1-\alpha_2+\alpha_3}{2} \sin^2\frac{\alpha_1-\alpha_2+\alpha_3}{2}}{\sin\alpha_1 \sin\alpha_3}, \\
h_3^2 &= \frac{R^2 \cos\frac{\alpha_1-\alpha_2-\alpha_3}{2} \cos\frac{-\alpha_1+\alpha_2-\alpha_3}{2} \sin^2\frac{\alpha_1+\alpha_2-\alpha_3}{2}}{\sin\alpha_1 \sin\alpha_2}.
\end{aligned} \tag{7.163}$$

Proof The third equation in (7.163) is established in (7.162). The first and second equations in (7.163) are derived from the third by vertex permutations. □

Corollary 7.19 *The biggest gyroangle of a gyrotriangle in an Einstein gyrovector space has measure equal to the sum of the measures of the other two gyroangles if and only if the side opposite to the biggest gyroangle is a gyrodiameter of the gyrotriangle circumgyrocircle.*

7.14 Ingyrocircle Points of Tangency

Ingyrocircle points of tangency are associated with the gyrotriangle gyrocenter called Gergonne gyropoint G_e, shown in Fig. 7.14, and studied in Sect. 7.16

Let us consider the point of tangency F_3 in which the incircle of a gyrotriangle $A_1A_2A_3$ meets the gyrotriangle side opposite to A_3, shown in Fig. 7.14. It is the perpendicular foot of the gyrotriangle ingyrocenter I on the gyroline A_1A_2. Accordingly, F_3 is the gyroaltitude foot of gyrotriangle A_1A_2I, drawn from I, as shown in Fig. 7.14.

Hence, by Theorem 7.16, p. 183, the gyroaltitude foot F_3 possesses the gyrobarycentric coordinate representation

$$F_3 = \frac{m_1\gamma_{A_1} A_1 + m_2\gamma_{A_2} A_2}{m_1\gamma_{A_1} + m_2\gamma_{A_2}} \tag{7.164}$$

with respect to the set $S = \{A_1, A_2\}$, with gyrobarycentric coordinates

$$\begin{aligned}
m_1 &= \gamma_{\ominus A_1\oplus A_2}\gamma_{\ominus A_2\oplus I} - \gamma_{\ominus A_1\oplus I} = \gamma_{12}\gamma_{\ominus A_2\oplus I} - \gamma_{\ominus A_1\oplus I}, \\
m_2 &= \gamma_{\ominus A_1\oplus A_2}\gamma_{\ominus A_1\oplus I} - \gamma_{\ominus A_2\oplus I} = \gamma_{12}\gamma_{\ominus A_1\oplus I} - \gamma_{\ominus A_2\oplus I}.
\end{aligned} \tag{7.165}$$

The gyrobarycentric coordinates m_1 and m_2 in (7.165) involve the gamma factors $\gamma_{\ominus A_1\oplus I}$ and $\gamma_{\ominus A_2\oplus I}$, which we calculate below.

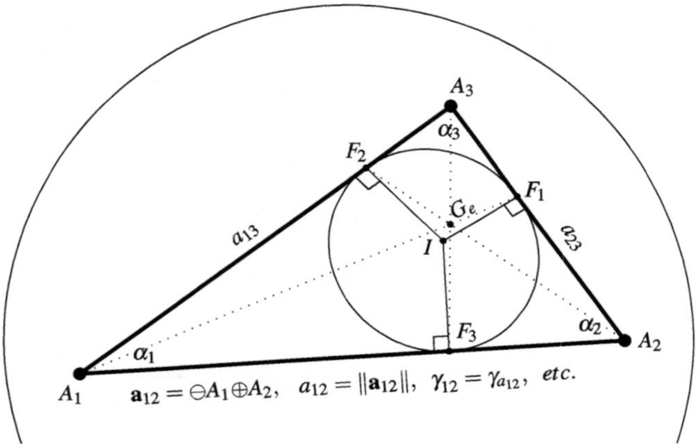

Fig. 7.14 The ingyrocircle of gyrotriangle $A_1A_2A_3$ in Fig. 7.11 is shown along with its gyrocenter I and its tangency points F_k, $k = 1, 2, 3$. The point F_k is the point in which the ingyrocircle of the gyrotriangle meets the gyrotriangle side opposite to A_k. The Gergonne gyropoint G_e of gyrotriangle $A_1A_2A_3$ is the point of concurrency of the three gyrolines A_kF_k, given by its gyrotrigonometric gyrobarycentric coordinate representation (7.202), p. 202

Being the incenter of gyrotriangle $A_1A_2A_3$, I is given by, (7.106)–(7.109), p. 179,

$$I = \frac{\gamma_{23}a_{23}\gamma_{A_1}A_1 + \gamma_{13}a_{13}\gamma_{A_2}A_2 + \gamma_{12}a_{12}\gamma_{A_3}A_3}{\gamma_{23}a_{23}\gamma_{A_1} + \gamma_{13}a_{13}\gamma_{A_2} + \gamma_{12}a_{12}\gamma_{A_3}}. \tag{7.166}$$

Hence, by Theorem 4.6, p. 90,

$$\gamma_{X\oplus I} = \frac{\gamma_{23}a_{23}\gamma_{X\oplus A_1} + \gamma_{13}a_{13}\gamma_{X\oplus A_2} + \gamma_{12}a_{12}\gamma_{X\oplus A_3}}{m_0} \tag{7.167}$$

for all $X \in \mathbb{R}^n_s$, where $m_0 > 0$ is the constant of the gyrobarycentric coordinate representation of I in (7.166). This constant need not be specified as we will see below in the transition from (7.169) to (7.170).

Following (7.167) with $X = \ominus A_k$, $k = 1, 2, 3$, we have, respectively,

$$\begin{aligned}
\gamma_{\ominus A_1\oplus I} &= \frac{\gamma_{23}a_{23} + \gamma_{13}a_{13}\gamma_{12} + \gamma_{12}a_{12}\gamma_{13}}{m_0}, \\
\gamma_{\ominus A_2\oplus I} &= \frac{\gamma_{23}a_{23}\gamma_{12} + \gamma_{13}a_{13} + \gamma_{12}a_{12}\gamma_{23}}{m_0}, \\
\gamma_{\ominus A_3\oplus I} &= \frac{\gamma_{23}a_{23}\gamma_{13} + \gamma_{13}a_{13}\gamma_{23} + \gamma_{12}a_{12}}{m_0}.
\end{aligned} \tag{7.168}$$

Substituting from (7.168) into (7.165), we have

$$\begin{aligned} m_1 &= \frac{(\gamma_{12}^2 - 1)\gamma_{23}a_{23} + (\gamma_{12}\gamma_{23} - \gamma_{13})\gamma_{12}a_{12}}{m_0}, \\ m_2 &= \frac{(\gamma_{12}^2 - 1)\gamma_{13}a_{13} + (\gamma_{12}\gamma_{13} - \gamma_{23})\gamma_{12}a_{12}}{m_0}. \end{aligned} \tag{7.169}$$

Being homogeneous, a common nonzero factor of gyrobarycentric coordinates is irrelevant, so that convenient gyrobarycentric coordinates m_1 and m_2 of the point I in (7.164) are obtained from (7.169) by removing the common denominator m_0,

$$\begin{aligned} m_1 &= (\gamma_{12}^2 - 1)\gamma_{23}a_{23} + (\gamma_{12}\gamma_{23} - \gamma_{13})\gamma_{12}a_{12}, \\ m_2 &= (\gamma_{12}^2 - 1)\gamma_{13}a_{13} + (\gamma_{12}\gamma_{13} - \gamma_{23})\gamma_{12}a_{12}. \end{aligned} \tag{7.170}$$

Substituting from (7.143)–(7.147) into (7.170), along with the abbreviation $F = F(\alpha_1, \alpha_2, \alpha_3)$, we have

$$\begin{aligned} m_1 &= \frac{8F^{3/2}}{\sin^2\alpha_1 \sin^2\alpha_2 \sin\alpha_3} \frac{1+\cos\alpha_2}{\sin\alpha_2} = \frac{8F^{3/2}}{\sin^2\alpha_1 \sin^2\alpha_2 \sin\alpha_3} \cot\frac{\alpha_2}{2}, \\ m_2 &= \frac{8F^{3/2}}{\sin^2\alpha_1 \sin^2\alpha_2 \sin\alpha_3} \frac{1+\cos\alpha_1}{\sin\alpha_1} = \frac{8F^{3/2}}{\sin^2\alpha_1 \sin^2\alpha_2 \sin\alpha_3} \cot\frac{\alpha_1}{2}. \end{aligned} \tag{7.171}$$

Since gyrobarycentric coordinates are homogeneous, a nonzero common factor of a system of gyrobarycentric coordinates is irrelevant. Hence, it follows from (7.171) that convenient gyrobarycentric coordinates for the point F_3 in (7.164) are

$$\begin{aligned} m_1 &= \cot\frac{\alpha_2}{2}, \\ m_2 &= \cot\frac{\alpha_1}{2} \end{aligned} \tag{7.172}$$

so that, by (7.164), we have

$$F_3 = \frac{\cot\frac{\alpha_2}{2}\gamma_{A_1} A_1 + \cot\frac{\alpha_1}{2}\gamma_{A_2} A_2}{\cot\frac{\alpha_2}{2}\gamma_{A_1} + \cot\frac{\alpha_1}{2}\gamma_{A_2}}. \tag{7.173}$$

We have thus obtained the following theorem:

Theorem 7.20 *Let $A_1A_2A_3$ be a gyrotriangle in an Einstein gyrovector space $\mathbb{R}_s^n$ and let F_k, $k = 1, 2, 3$, be the point in which the ingyrocircle of the gyrotriangle meets the opposite side of A_k, Fig. 7.14. A gyrotrigonometric gyrobarycentric coor-*

dinate representation of each point F_k is given by

$$
\begin{aligned}
F_1 &= \frac{\cot\frac{\alpha_3}{2}\gamma_{A_2}A_2 + \cot\frac{\alpha_2}{2}\gamma_{A_3}A_3}{\cot\frac{\alpha_3}{2}\gamma_{A_2} + \cot\frac{\alpha_2}{2}\gamma_{A_3}}, \\
F_2 &= \frac{\cot\frac{\alpha_3}{2}\gamma_{A_1}A_1 + \cot\frac{\alpha_1}{2}\gamma_{A_3}A_3}{\cot\frac{\alpha_3}{2}\gamma_{A_1} + \cot\frac{\alpha_1}{2}\gamma_{A_3}}, \\
F_3 &= \frac{\cot\frac{\alpha_2}{2}\gamma_{A_1}A_1 + \cot\frac{\alpha_1}{2}\gamma_{A_2}A_2}{\cot\frac{\alpha_2}{2}\gamma_{A_1} + \cot\frac{\alpha_1}{2}\gamma_{A_2}}.
\end{aligned}
\tag{7.174}
$$

Proof The proof of the third equation in (7.174) is given by (7.164)–(7.173). The proof of the first and the second equation in (7.174) is obtained from the first by vertex permutations. □

The three points F_k, $k = 1, 2, 3$, of Theorem 7.20 are shown in Fig. 7.14. As Fig. 7.14 indicates, the three gyrolines A_1F_1, A_2F_2 and A_3F_3 are concurrent. We show in Sect. 7.16 that this is indeed the case, giving rise to the Gergonne gyropoint.

For later convenience, we determine below the left gyrotranslation of F_k, $k = 1, 2, 3$, by $\ominus A_1$.

Applying to (7.174) the gyrocovariance of gyrobarycentric coordinate representations, Theorem 4.6, p. 90, we have from Identity (4.29a), p. 91, with $X = \ominus A_1$, using the standard gyrotriangle index notation, shown in Fig. 8.4, in Fig. 6.1, p. 128, and in (6.1), p. 127:

$$
\ominus A_1 \oplus F_1 = \ominus A_1 \oplus \frac{\cot\frac{\alpha_3}{2}\gamma_{A_2}A_2 + \cot\frac{\alpha_2}{2}\gamma_{A_3}A_3}{\cot\frac{\alpha_3}{2}\gamma_{A_2} + \cot\frac{\alpha_2}{2}\gamma_{A_3}} = \frac{\cot\frac{\alpha_3}{2}\gamma_{12}\mathbf{a}_{12} + \cot\frac{\alpha_2}{2}\gamma_{13}\mathbf{a}_{13}}{\cot\frac{\alpha_3}{2}\gamma_{12} + \cot\frac{\alpha_2}{2}\gamma_{13}},
\tag{7.175a}
$$

$$
\ominus A_1 \oplus F_2 = \ominus A_1 \oplus \frac{\cot\frac{\alpha_3}{2}\gamma_{A_1}A_1 + \cot\frac{\alpha_1}{2}\gamma_{A_3}A_3}{\cot\frac{\alpha_3}{2}\gamma_{A_1} + \cot\frac{\alpha_1}{2}\gamma_{A_3}} = \frac{\cot\frac{\alpha_1}{2}\gamma_{13}\mathbf{a}_{13}}{\cot\frac{\alpha_3}{2} + \cot\frac{\alpha_1}{2}\gamma_{13}},
\tag{7.175b}
$$

and

$$
\ominus A_1 \oplus F_3 = \ominus A_1 \oplus \frac{\cot\frac{\alpha_2}{2}\gamma_{A_1}A_1 + \cot\frac{\alpha_1}{2}\gamma_{A_2}A_2}{\cot\frac{\alpha_2}{2}\gamma_{A_1} + \cot\frac{\alpha_1}{2}\gamma_{A_2}} = \frac{\cot\frac{\alpha_1}{2}\gamma_{12}\mathbf{a}_{12}}{\cot\frac{\alpha_2}{2} + \cot\frac{\alpha_1}{2}\gamma_{12}}.
\tag{7.175c}
$$

7.15 An Unlikely Concurrence

Let $A_1A_2A_3$ be a gyrotriangle in an Einstein gyrovector space $(\mathbb{R}^n_s, \oplus, \otimes)$, and let F_k, $k = 1, 2, 3$, be the tangency points of its ingyrocircle gyrocentered at I. Furthermore, let P be the point of intersection of the gyrolines A_3I and F_1F_3, as shown in

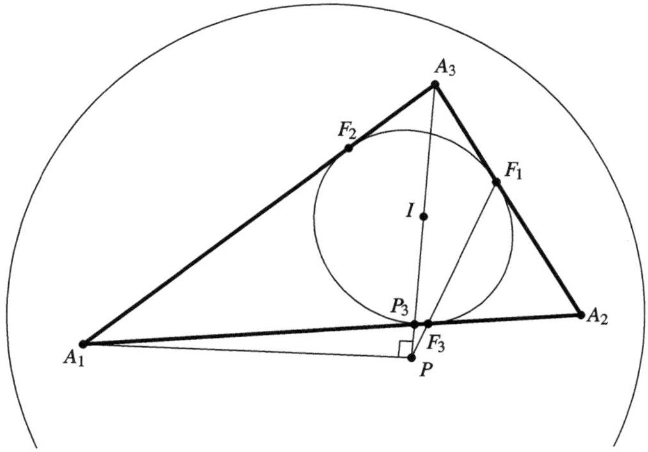

Fig. 7.15 An Unlikely Concurrence. The two gyrolines F_1F_3 and A_3I concur at P, generating the right gyrotriangle A_1PA_3. The point P_3 is the A_3-foot of the ingyrocenter of gyrotriangle $A_1A_2A_3$, so that the gyrolines A_3I and A_3P_3 coincide. F_k, $k=1,2,3$ are the points of tangency where the ingyrocircle meets the sides of the gyrotriangle

Figs. 7.15–7.16. We will show that the resulting gyrotriangle A_1PA_3 is right, with the right gyroangle $\angle A_1PA_3$.

The gyroline A_3I is the bisector of gyroangle $\angle A_1A_3A_2$ and, following (7.71) and (7.73), p. 171, its foot P_3 is given by the gyrobarycentric coordinate representation

$$P_3 = \frac{\sin\alpha_1\gamma_{A_1}A_1 + \sin\alpha_2\gamma_{A_2}A_2}{\sin\alpha_1\gamma_{A_1} + \sin\alpha_2\gamma_{A_2}} \tag{7.176}$$

with respect to the set $S=\{A_1,A_2,A_3\}$. Clearly, the gyrolines A_3I and A_3P_3 coincide.

The tangency points F_k, $K=1,3$, of the gyrotriangle ingyrocircle, Fig. 7.15, where the ingyrocircle meets the gyrotriangle sides A_2A_3 and A_1A_3 are given by, (7.174),

$$\begin{aligned} F_1 &= \frac{\cot\frac{\alpha_3}{2}\gamma_{A_2}A_2 + \cot\frac{\alpha_2}{2}\gamma_{A_3}A_3}{\cot\frac{\alpha_3}{2}\gamma_{A_2} + \cot\frac{\alpha_2}{2}\gamma_{A_3}}, \\ F_3 &= \frac{\cot\frac{\alpha_2}{2}\gamma_{A_1}A_1 + \cot\frac{\alpha_1}{2}\gamma_{A_2}A_2}{\cot\frac{\alpha_2}{2}\gamma_{A_1} + \cot\frac{\alpha_1}{2}\gamma_{A_2}}. \end{aligned} \tag{7.177}$$

Let the point of intersection, P, of gyrolines A_3I and F_1F_3 be given by its gyrobarycentric coordinate representation

$$P = \frac{m_1\gamma_{A_1} A_1 + m_2\gamma_{A_2} A_2 + m_3\gamma_{A_3} A_3}{m_1\gamma_{A_1} + m_2\gamma_{A_2} + m_3\gamma_{A_3}}, \tag{7.178}$$

where the gyrobarycentric coordinates $(m_1 : m_2 : m_3)$ are to be determined in (7.185). Clearly, P is the point of intersection of gyrolines A_3P_3 and F_1F_3 as well.

Owing to the Gyrobarycentric Coordinate Representation Gyrocovariance Theorem 4.6, p. 90, the left gyrotranslated point, $\ominus A_2 \oplus P$, of P by $\ominus A_2$ is the intersection point of the two left gyrotranslated gyrolines $(\ominus A_2 \oplus A_3)(\ominus A_2 \oplus P_3)$ and $(\ominus A_2 \oplus F_1)(\ominus A_2 \oplus F_3)$.

Following Theorem 4.6, the left gyrotranslations of points A_3, P_3, F_1, F_3, and P by $\ominus A_2$ are:

$$\begin{aligned}
\ominus A_2 \oplus A_3 &= \mathbf{a}_{23}, \\
\ominus A_2 \oplus P_3 &= \frac{\gamma_{12} \sin\alpha_1}{\gamma_{12} \sin\alpha_1 + \sin\alpha_2} \mathbf{a}_{21} =: c_1 \mathbf{a}_{21}, \\
\ominus A_2 \oplus F_1 &= \frac{\gamma_{23} \cot\frac{\alpha_2}{2}}{\cot\frac{\alpha_3}{2} + \gamma_{23} \cot\frac{\alpha_2}{2}} \mathbf{a}_{23} =: c_2 \mathbf{a}_{23}, \\
\ominus A_2 \oplus F_3 &= \frac{\gamma_{12} \cot\frac{\alpha_2}{2}}{\gamma_{12} \cot\frac{\alpha_2}{2} + \cot\frac{\alpha_1}{2}} \mathbf{a}_{21} =: c_3 \mathbf{a}_{21}, \\
\ominus A_2 \oplus P &= \frac{m_1\gamma_{12} \mathbf{a}_{21} + m_3\gamma_{23} \mathbf{a}_{23}}{m_1\gamma_{12} + m_2 + m_3\gamma_{23}} =: c_4 \mathbf{a}_{21} + c_5 \mathbf{a}_{23},
\end{aligned} \tag{7.179}$$

where we use the standard gyrotriangle index notation, shown in Fig. 6.1, p. 128, and in (6.1), p. 127.

The Euclidean straight line L_1 that passes through the points

$$\ominus A_2 \oplus A_3, \quad \ominus A_2 \oplus P_3 \in \mathbb{R}_s^n \subset \mathbb{R}^n \tag{7.180a}$$

is

$$\begin{aligned}
L_1 &= (\ominus A_2 \oplus A_3) + \big\{-(\ominus A_2 \oplus A_3) + (\ominus A_2 \oplus P_3)\big\} t_1 \\
&= \mathbf{a}_{23} + (-\mathbf{a}_{23} + c_1 \mathbf{a}_{21}) t_1
\end{aligned} \tag{7.180b}$$

where $t_1 \in \mathbb{R}$ is the line parameter.

Similarly, the Euclidean straight line L_2 that passes through the points

$$\ominus A_2 \oplus F_1, \quad \ominus A_2 \oplus F_3 \in \mathbb{R}_s^n \subset \mathbb{R}^n \tag{7.181a}$$

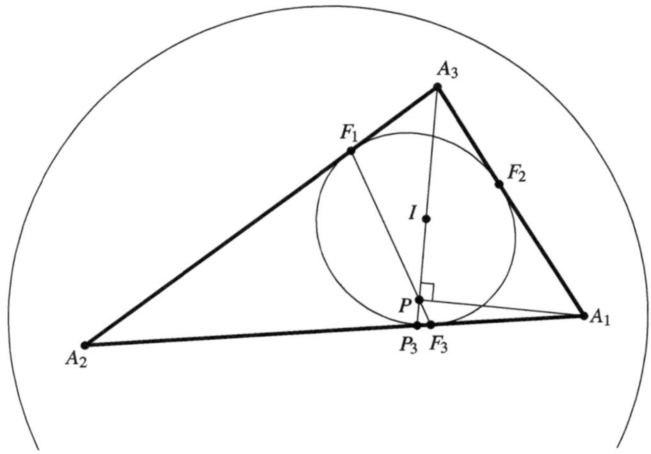

Fig. 7.16 An Unlikely Concurrence. The two points A_1 and A_2 of Fig. 7.15 are interchanged in this figure. As opposed to Fig. 7.15, here the point of concurrence, P, lies on the interior of its reference gyrotriangle $A_1A_2A_3$

is

$$\begin{aligned} L_2 &= (\ominus A_2 \oplus F_1) + \big\{-(\ominus A_2 \oplus F_1) + (\ominus A_2 \oplus F_3)\big\} t_2 \\ &= c_2 \mathbf{a}_{23} + (-c_2 \mathbf{a}_{23} + c_3 \mathbf{a}_{21}) t_2 \end{aligned} \tag{7.181b}$$

where $t_2 \in \mathbb{R}$ is the line parameter.

The point $\ominus A_2 \oplus P = c_4 \mathbf{a}_{21} + c_5 \mathbf{a}_{23}$ lies on each of the two lines L_1 and L_2, which are coplanar and nonparallel. Hence, there exist $t_{1,0}$ and $t_{2,0}$ of the line parameters t_1 and t_2, respectively, such that

$$\begin{aligned} c_4 \mathbf{a}_{21} + c_5 \mathbf{a}_{23} - \big\{\mathbf{a}_{23} + (-\mathbf{a}_{23} + c_1 \mathbf{a}_{21}) t_1\big\} &= 0, \\ c_4 \mathbf{a}_{21} + c_5 \mathbf{a}_{23} - \big\{c_2 \mathbf{a}_{23} + (-c_2 \mathbf{a}_{23} + c_3 \mathbf{a}_{21}) t_2\big\} &= 0. \end{aligned} \tag{7.182}$$

Substituting c_k, $k = 1, \ldots, 5$, from (7.179) into (7.182), we obtain two vector equations of the form

$$\begin{aligned} c_{11} \mathbf{a}_{21} + c_{12} \mathbf{a}_{23} &= \mathbf{0}, \\ c_{21} \mathbf{a}_{21} + c_{22} \mathbf{a}_{23} &= \mathbf{0}. \end{aligned} \tag{7.183}$$

The two gyrovectors $\mathbf{a}_{21}$ and $\mathbf{a}_{23}$ in $\mathbb{R}_s^n$, considered as vectors in $\mathbb{R}^n$, are linearly independent in $\mathbb{R}^n$ since the set $S = \{A_1, A_2, A_3\}$ is pointwise independent. Hence, the two vector equations in (7.183) are equivalent to the following four scalar equations,

$$c_{11} = c_{12} = c_{21} = c_{22} = 0 \tag{7.184}$$

for the five unknowns $t_{k,0}$, $k = 1, 2$, and m_k, $k = 1, 2, 3$.

Two of the resulting four equations admit a solution for $t_{1,0}$ and $t_{2,0}$. These are substituted into the remaining two equations, simplifying them into two equations for the three unknowns m_k, $k = 1, 2, 3$. The latter, in turn, determine the unknowns m_1/m_3 and m_2/m_3 uniquely. Keeping in mind that gyrobarycentric coordinates are homogeneous, by definition, resulting convenient gyrobarycentric coordinates $(m_1 : m_2 : m_3)$ for the point P in (7.178) turn out to be

$$\begin{aligned} m_1 &= \cot\frac{\alpha_3}{2}\sin\alpha_1, \\ m_2 &= \cot\frac{\alpha_3}{2}\sin\alpha_2, \\ m_3 &= \cot\frac{\alpha_2}{2}\sin\alpha_2 - \cot\frac{\alpha_1}{2}\sin\alpha_1 = 2\left(\cos^2\frac{\alpha_2}{2} - \cos^2\frac{\alpha_1}{2}\right). \end{aligned} \tag{7.185}$$

By (4.15), p. 88, the constant m_0 of the gyrobarycentric coordinate representation (7.178) of P is, in general, given by

$$m_0^2 = m_1^2 + m_2^2 + m_3^2 + 2(m_1m_2\gamma_{12} + m_1m_3\gamma_{13} + m_2m_3\gamma_{23}). \tag{7.186}$$

In order to express (7.186) gyrotrigonometrically, we substitute into (7.186) (i) the gyrobarycentric coordinates from (7.185), and (ii) the gamma factors from (7.142), p. 187, obtaining the elegant equation

$$m_0^2 = \frac{(\cos\alpha_2 + \cos\alpha_1\cos\alpha_3)^2 + \sin^2\alpha_1(1+\cos\alpha_3)^2}{\sin^2\frac{\alpha_3}{2}} \tag{7.187}$$

which demonstrates that $m_0^2 > 0$ for the gyrobarycentric coordinate representation (7.178) of $P \in \mathbb{R}^n$ with respect to any gyrotriangle $A_1A_2A_3$ in $\mathbb{R}_s^n$.

According to Corollary 4.9, p. 93, the point $P \in \mathbb{R}^n$ with the gyrobarycentric coordinate representation (7.178) lies in the ball $\mathbb{R}_s^n$ if and only if the squared constant, m_0^2, of the representation is positive. It is, therefore, interesting to realize that m_0^2 in (7.187) is always positive, so that the point P in (7.178) and in Fig. 7.15 lies in the ball (or, in the disc, if $n = 2$, as in Fig. 7.15) for any reference gyrotriangle $A_1A_2A_3$ in $\mathbb{R}_s^n$.

The gyrobarycentric coordinates m_1 and m_2 of P in (7.185) are always positive. The third gyrobarycentric coordinate, m_3, of P in (7.185) equals 0 if and only if $\alpha_1 = \alpha_2$. Otherwise, if $m_3 > 0$ for gyrotriangle $A_1A_2A_3$, as shown in Fig. 7.15, then, necessarily, $m_3 < 0$ for gyrotriangle $A_2A_1A_3$, as shown in Fig. 7.16. It is clear from (7.185) that interchanging the vertices A_1 and A_2 of gyrotriangle $A_2A_1A_3$ results in the change of the sign of m_3. Accordingly, as stated in Corollary 4.10, p. 94, The point P lies on the interior of gyrotriangle $A_1A_2A_3$ in Fig. 7.15, and on its exterior in Fig. 7.16.

Applying Theorem 4.6, p. 90, to P in (7.178) with $X = \ominus A_1$ and $X = \ominus A_2$, respectively, we obtain the following two equations:

$$\gamma_{\ominus A_1 \oplus P} = \frac{m_1 \gamma_{\ominus A_1 \oplus A_1} + m_2 \gamma_{\ominus A_1 \oplus A_2} + m_3 \gamma_{\ominus A_1 \oplus A_3}}{m_0} = \frac{m_1 + m_2 \gamma_{12} + m_3 \gamma_{13}}{m_0},$$
$$\gamma_{\ominus A_3 \oplus P} = \frac{m_1 \gamma_{\ominus A_3 \oplus A_1} + m_2 \gamma_{\ominus A_3 \oplus A_2} + m_3 \gamma_{\ominus A_3 \oplus A_3}}{m_0} = \frac{m_1 \gamma_{13} + m_2 \gamma_{23} + m_3}{m_0}, \tag{7.188}$$

where m_0 is given gyrotrigonometrically by (7.187)

In order to express (7.188) gyrotrigonometrically as well, we substitute into the numerators of the extreme right-hand sides of (7.188) (i) the gyrobarycentric coordinates from (7.185), and (ii) the gamma factors from (7.142), p. 187, obtaining the following two results, to which we add a third result taken from (7.142), p. 187:

$$\begin{aligned}
\gamma_{\ominus A_1 \oplus P} &= \frac{\tan\frac{\alpha_3}{2}}{2\sin\alpha_1} m_0, \\
\gamma_{\ominus A_3 \oplus P} &= \frac{\cos\alpha_2 + \cos\alpha_1 \cos\alpha_3}{\sin^2\frac{\alpha_3}{2}} \frac{1}{m_0}, \\
\gamma_{\ominus A_1 \oplus A_3} = \gamma_{13} &= \frac{\cos\alpha_2 + \cos\alpha_1 \cos\alpha_3}{\sin\alpha_1 \sin\alpha_3}.
\end{aligned} \tag{7.189}$$

It is clear from (7.189) that the three gamma factors in (7.189) are related by the equation

$$\gamma_{\ominus A_1 \oplus P} \gamma_{\ominus A_3 \oplus P} = \gamma_{\ominus A_1 \oplus A_3}. \tag{7.190}$$

Equation (7.190), in turn, is equivalent to the condition that gyrotriangle A_1PA_3 in Figs. 7.15–7.16 is right, with the right gyroangle at P, as indicated in Figs. 7.15–7.16 and stated in Theorem 6.13, p. 143.

The surprising coincidence described in this section, illustrated by Figs. 7.15–7.16, remains invariant in form under the Euclidean limit $s \to \infty$, so that it is valid in Euclidean geometry as well. This coincidence in Euclidean geometry is considered as an "unlikely concurrence" [26, p. 31]. Here, in the context of hyperbolic geometry, we have an additional surprise: The point of concurrence, P, in Figs. 7.15–7.16, always lies in the ball $\mathbb{R}^n_s$ unlike, for instance, the gyrotriangle circumgyrocenter and orthogyrocenter, which sometimes do not exist in the ball as shown, for instance, in Fig. 7.21, p. 212.

7.16 Gergonne Gyropoint

Definition 7.21 (Gergonne Gyropoint) Let $A_1A_2A_3$ be a gyrotriangle in an Einstein gyrovector space $(\mathbb{R}^n_s, \oplus, \otimes)$ and let F_k, $k = 1, 2, 3$, be the tangent point in which the ingyrocircle of the gyrotriangle meets the gyrotriangle side opposite to A_k, Fig. 7.14. The gyrotriangle vertices A_k and the gyrotriangle points of tangency

F_k form the three gyrolines A_kF_k that are concurrent. Owing to analogies with Euclidean geometry, this point of concurrency, G_e, is called the Gergonne gyropoint of the gyrotriangle.

Let the Gergonne gyropoint G_e of gyrotriangle $A_1A_2A_3$ in an Einstein gyrovector space $(\mathbb{R}^n_s, \oplus, \otimes)$, Fig. 7.14, be given by its gyrobarycentric coordinate representation with respect to the set $S = \{A_1, A_2, A_3\}$ of the gyrotriangle,

$$G_e = \frac{m_1\gamma_{A_1}A_1 + m_2\gamma_{A_2}A_2 + m_3\gamma_{A_3}A_3}{m_1\gamma_{A_1} + m_2\gamma_{A_2} + m_3\gamma_{A_3}}, \tag{7.191}$$

where the gyrobarycentric coordinates $(m_1 : m_2 : m_3)$ of G_e in (7.191) are to be determined.

Left gyrotranslating gyrotriangle $A_1A_2A_3$ by $\ominus A_1$, the gyrotriangle becomes gyrotriangle $O(\ominus A_1\oplus A_2)(\ominus A_1\oplus A_3)$, where $O = \ominus A_1\oplus A_1$ is the arbitrarily selected origin of the Einstein gyrovector space $(\mathbb{R}^n_s, \oplus, \otimes)$, $O = (0, 0, \dots, 0)$. The gyrotriangle tangent points F_k, $k = 1, 2, 3$, are left gyrotranslated as well, becoming, respectively, $\ominus A_1\oplus F_k$, which are given by (7.175a), (7.175b), (7.175c).

Similarly, the gyrotriangle Nagel gyropoint of the left gyrotranslated gyrotriangle becomes $P = \ominus A_1\oplus G_e$, given by

$$\begin{aligned} P &= \ominus A_1\oplus G_e \\ &= \frac{m_2\gamma_{\ominus A_1\oplus A_2}(\ominus A_1\oplus A_2) + m_3\gamma_{\ominus A_1\oplus A_3}(\ominus A_1\oplus A_3)}{m_1 + m_2\gamma_{\ominus A_1\oplus A_2} + m_3\gamma_{\ominus A_1\oplus A_3}} \\ &= \frac{m_2\gamma_{12}\mathbf{a}_{12} + m_3\gamma_{13}\mathbf{a}_{13}}{m_1 + m_2\gamma_{12} + m_3\gamma_{13}}. \end{aligned} \tag{7.192}$$

1. The tangent point $\ominus A_1\oplus F_1$ and the vertex $O = \ominus A_1\oplus A_1 = (0, 0, \dots, 0)$ of the left gyrotranslated gyrotriangle $O(\ominus A_1\oplus A_2)(\ominus A_1\oplus A_3)$ form the Euclidean line

$$L_1 = (\ominus A_1\oplus F_1)t_1 = \left(\frac{\cot\frac{\alpha_3}{2}\gamma_{12}\mathbf{a}_{12} + \cot\frac{\alpha_2}{2}\gamma_{13}\mathbf{a}_{13}}{\cot\frac{\alpha_3}{2}\gamma_{12} + \cot\frac{\alpha_2}{2}\gamma_{13}}\right)t_1 \tag{7.193}$$

as we see from (7.175a), where $t_1 \in \mathbb{R}$ is the line parameter.

2. The tangent point $\ominus A_1\oplus F_2$ and the vertex $\mathbf{a}_{12} = \ominus A_1\oplus A_2$ of the left gyrotranslated gyrotriangle $O(\ominus A_1\oplus A_2)(\ominus A_1\oplus A_3)$ form the Euclidean line

$$\begin{aligned} L_2 &= (\ominus A_1\oplus A_2) + \big(-(\ominus A_1\oplus A_2) + (\ominus A_1\oplus F_2)\big)t_2 \\ &= \mathbf{a}_{12} + \left(-\mathbf{a}_{12} + \frac{\cot\frac{\alpha_1}{2}\gamma_{13}\mathbf{a}_{13}}{\cot\frac{\alpha_3}{2} + \cot\frac{\alpha_1}{2}\gamma_{13}}\right)t_2 \end{aligned} \tag{7.194}$$

as we see from (7.175b), where $t_2 \in \mathbb{R}$ is the line parameter.

3. The tangent point $\ominus A_1 \oplus F_3$ and the vertex $\mathbf{a}_{13} = \ominus A_1 \oplus A_3$ of the left gyrotranslated gyrotriangle $O(\ominus A_1 \oplus A_2)(\ominus A_1 \oplus A_3)$ form the Euclidean line

$$\begin{aligned} L_3 &= (\ominus A_1 \oplus A_3) + \bigl(-(\ominus A_1 \oplus A_3) + (\ominus A_1 \oplus F_3)\bigr) t_3 \\ &= \mathbf{a}_{13} + \left(-\mathbf{a}_{13} + \frac{\cot \frac{\alpha_1}{2} \gamma_{12} \mathbf{a}_{12}}{\cot \frac{\alpha_2}{2} + \cot \frac{\alpha_1}{2} \gamma_{12}}\right) t_3 \end{aligned} \tag{7.195}$$

as we see from (7.175c), where $t_3 \in \mathbb{R}$ is the line parameter.

Since the point P lies on each of the three lines L_k, $k = 1, 2, 3$, there exist values $t_{k,0}$ of the line parameters t_k, $k = 1, 2, 3$, respectively, such that

$$\begin{aligned} P - \left(\frac{\cot \frac{\alpha_3}{2} \gamma_{12} \mathbf{a}_{12} + \cot \frac{\alpha_2}{2} \gamma_{13} \mathbf{a}_{13}}{\cot \frac{\alpha_3}{2} \gamma_{12} + \cot \frac{\alpha_2}{2} \gamma_{13}}\right) t_{1,0} &= 0, \\ P - \mathbf{a}_{12} - \left(-\mathbf{a}_{12} + \frac{\cot \frac{\alpha_1}{2} \gamma_{13} \mathbf{a}_{13}}{\cot \frac{\alpha_3}{2} + \cot \frac{\alpha_1}{2} \gamma_{13}}\right) t_{2,0} &= 0, \\ P - \mathbf{a}_{13} - \left(-\mathbf{a}_{13} + \frac{\cot \frac{\alpha_1}{2} \gamma_{12} \mathbf{a}_{12}}{\cot \frac{\alpha_2}{2} + \cot \frac{\alpha_1}{2} \gamma_{12}}\right) t_{3,0} &= 0, \end{aligned} \tag{7.196}$$

where P is given by (7.192).

The system of equations (7.196) was obtained by methods of gyroalgebra, and will be solved below by a common method of linear algebra.

Substituting P from (7.192) into (7.196), and rewriting each of the resulting equations as a linear combination of $\mathbf{a}_{12}$ and $\mathbf{a}_{13}$ equals zero, one obtains the following homogeneous linear system of three gyrovector equations

$$\begin{aligned} c_{11}\mathbf{a}_{12} + c_{12}\mathbf{a}_{13} &= \mathbf{0}, \\ c_{21}\mathbf{a}_{12} + c_{22}\mathbf{a}_{13} &= \mathbf{0}, \\ c_{31}\mathbf{a}_{12} + c_{32}\mathbf{a}_{13} &= \mathbf{0}, \end{aligned} \tag{7.197}$$

where each coefficient c_{ij}, $i = 1, 2, 3$, $j = 1, 2$, is a function of the gyrotriangle parameters $\gamma_{12}, \gamma_{13}, \gamma_{23}$ and α_k, and the six unknowns $t_{k,0}$ and m_k, $k = 1, 2, 3$.

Since the set $S = \{A_1, A_2, A_3\}$ is pointwise independent, the two gyrovectors $\mathbf{a}_{12} = \ominus A_1 \oplus A_2$ and $\mathbf{a}_{13} = \ominus A_1 \oplus A_3$ in $\mathbb{R}^n_s$, considered as vectors in $\mathbb{R}^n$, are linearly independent in $\mathbb{R}^n$. Hence, each coefficient c_{ij} in (7.197) equals zero. Accordingly, the three gyrovector equations in (7.197) are equivalent to the following six scalar equations,

$$c_{11} = c_{12} = c_{21} = c_{22} = c_{31} = c_{32} = 0 \tag{7.198}$$

for the six unknowns $t_{k,0}$ and m_k, $k = 1, 2, 3$.

An explicit presentation of the resulting system (7.198) reveals that it is slightly nonlinear. However, it is linear in the unknowns $t_{1,0}, t_{2,0}, t_{3,0}$. Solving three equations of the system for $t_{1,0}, t_{2,0}, t_{3,0}$, and substituting these into the remaining equations of the system we obtain a system that determines the ratios m_1/m_3 and m_2/m_3 uniquely, from which convenient (homogeneous) gyrobarycentric coordinates $(m_1 : m_2 : m_3)$ are obtained. The unique determination of m_1/m_3 and m_2/m_3 turns out to be

$$\begin{aligned}\frac{m_1}{m_3} &= \frac{\cot\frac{\alpha_3}{2}}{\cot\frac{\alpha_1}{2}},\\ \frac{m_2}{m_3} &= \frac{\cot\frac{\alpha_3}{2}}{\cot\frac{\alpha_2}{2}}\end{aligned} \tag{7.199}$$

from which two convenient gyrobarycentric coordinates result. These are:

$$(m_1 : m_2 : m_3) = \left(\cot\frac{\alpha_2}{2}\cot\frac{\alpha_3}{2} : \cot\frac{\alpha_1}{2}\cot\frac{\alpha_3}{2} : \cot\frac{\alpha_1}{2}\cot\frac{\alpha_2}{2}\right) \tag{7.200}$$

and, equivalently,

$$(m_1 : m_2 : m_3) = \left(\tan\frac{\alpha_1}{2} : \tan\frac{\alpha_2}{2} : \tan\frac{\alpha_3}{2}\right). \tag{7.201}$$

Formalizing the main result of this section, we have the following theorem:

Theorem 7.22 *Let $A_1A_2A_3$ be a gyrotriangle in an Einstein gyrovector space $(\mathbb{R}_s^n, \oplus, \otimes)$. A gyrotrigonometric gyrobarycentric coordinate representation of the gyrotriangle Gergonne gyropoint G_e, Fig.* 7.14, *p.* 192, *is given by*

$$G_e = \frac{\tan\frac{\alpha_1}{2}\gamma_{A_1} A_1 + \tan\frac{\alpha_2}{2}\gamma_{A_2} A_2 + \tan\frac{\alpha_3}{2}\gamma_{A_3} A_3}{\tan\frac{\alpha_1}{2}\gamma_{A_1} + \tan\frac{\alpha_2}{2}\gamma_{A_2} + \tan\frac{\alpha_3}{2}\gamma_{A_3}}. \tag{7.202}$$

Proof The proof follows immediately from (7.191) and (7.201). □

7.17 Gyrotriangle Orthogyrocenter

The hyperbolic triangle orthocenter, H, shown in Fig. 7.17, is called in gyrolanguage a gyrotriangle orthogyrocenter.

Definition 7.23 The orthogyrocenter, H, of a gyrotriangle is the point of concurrency of the gyrotriangle gyroaltitudes.

The three feet, P_1, P_2 and P_3 of the three gyroaltitudes of gyrotriangle $A_1A_2A_3$ in an Einstein gyrovector space $(\mathbb{R}^n_s, \oplus, \otimes)$, shown in Fig. 7.17 for $n = 2$, are given by

$$
\begin{aligned}
P_1 &= \frac{(\gamma_{13}\gamma_{23} - \gamma_{12})\gamma_{A_2} A_2 + (\gamma_{12}\gamma_{23} - \gamma_{13})\gamma_{A_3} A_3}{(\gamma_{13}\gamma_{23} - \gamma_{12})\gamma_{A_2} + (\gamma_{12}\gamma_{23} - \gamma_{13})\gamma_{A_3}}, \\
P_2 &= \frac{(\gamma_{13}\gamma_{23} - \gamma_{12})\gamma_{A_1} A_1 + (\gamma_{12}\gamma_{13} - \gamma_{23})\gamma_{A_3} A_3}{(\gamma_{13}\gamma_{23} - \gamma_{12})\gamma_{A_1} + (\gamma_{12}\gamma_{13} - \gamma_{23})\gamma_{A_3}}, \\
P_3 &= \frac{(\gamma_{12}\gamma_{23} - \gamma_{13})\gamma_{A_1} A_1 + (\gamma_{12}\gamma_{13} - \gamma_{23})\gamma_{A_2} A_2}{(\gamma_{12}\gamma_{23} - \gamma_{13})\gamma_{A_1} + (\gamma_{12}\gamma_{13} - \gamma_{23})\gamma_{A_2}}.
\end{aligned}
\tag{7.203}
$$

The third equation in (7.203) is a copy of (7.119). The first and second equations in (7.203) are obtained from the third one by cyclic permutations of the vertices of gyrotriangle $A_1A_2A_3$.

Gyrotriangle gyroaltitudes are concurrent. The gyroaltitudes of gyrotriangle $A_1A_2A_3$ in an Einstein gyrovector space $(\mathbb{R}^n_s, \oplus, \otimes)$, shown in Fig. 7.17 for $n = 2$, are the gyrosegments A_1P_1, A_2P_2, and A_1P_3. Since gyrosegments in Einstein gyrovector spaces coincide with Euclidean segments, one can employ methods of linear algebra to determine the point of concurrency, that is, the orthogyrocenter, of the three gyroaltitudes of gyrotriangle $A_1A_2A_3$ in Fig. 7.17.

In order to determine the gyrobarycentric coordinates of the gyrotriangle orthogyrocenter in Einstein gyrovector spaces we begin with some gyroalgebraic manipulations that reduce the task we face to a problem in linear algebra.

Let the orthogyrocenter H of gyrotriangle $A_1A_2A_3$ in an Einstein gyrovector space $(\mathbb{R}^n_s, \oplus, \otimes)$, Fig. 7.17, be given in terms of its gyrobarycentric coordinate representation with respect to the set $S = \{A_1, A_2, A_3\}$ of the gyrotriangle vertices by the equation

$$
H = \frac{m_1\gamma_{A_1} A_1 + m_2\gamma_{A_2} A_2 + m_3\gamma_{A_3} A_3}{m_1\gamma_{A_1} + m_2\gamma_{A_2} + m_3\gamma_{A_3}}
\tag{7.204}
$$

where the gyrobarycentric coordinates $(m_1 : m_2 : m_3)$ of H in (7.204) are to be determined.

Left gyrotranslating gyrotriangle $A_1A_2A_3$ by $\ominus A_1$, the gyrotriangle becomes gyrotriangle $O(\ominus A_1 \oplus A_2)(\ominus A_1 \oplus A_3)$, where $O = \ominus A_1 \oplus A_1$ is the arbitrarily selected origin of the Einstein gyrovector space $\mathbb{R}^n_s$. The gyrotriangle gyroaltitude feet P_1, P_2 and P_3 become, respectively, $\ominus A_1 \oplus P_1$, $\ominus A_1 \oplus P_2$ and $\ominus A_1 \oplus P_3$. These are calculated in (7.205a), (7.205b), (7.205c) below. Employing the Gyrobarycentric Coordinate Representation Gyrocovariance Theorem 4.6, p. 90, we have from Identity (4.29a), p. 91, with $X = \ominus A_1$, using the standard gyrotriangle index notation,

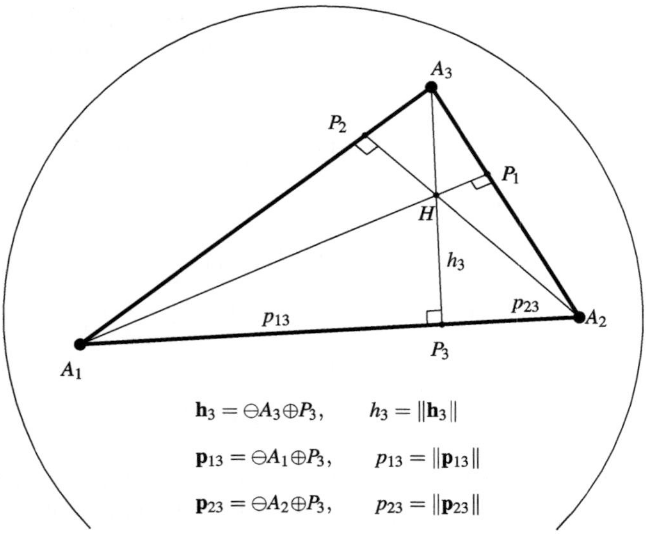

Fig. 7.17 The orthogyrocenter H of a gyrotriangle $A_1A_2A_3$ in an Einstein gyrovector space $(\mathbb{R}^n_s, \oplus, \otimes)$. Here the orthogyrocenter lies inside its gyrotriangle. There are gyrotriangles with their orthogyrocenter lying out of their gyrotriangles, and there are gyrotriangles that possess no orthogyrocenter, as shown in Figs. 7.18–7.21

shown in Fig. 7.17, in Fig. 6.1, p. 128, and in (6.1), p. 127:

$$\begin{aligned}
&\ominus A_1 \oplus P_1 \\
&\quad = \ominus A_1 \oplus \frac{(\gamma_{13}\gamma_{23} - \gamma_{12})\gamma_{A_2} A_2 + (\gamma_{12}\gamma_{23} - \gamma_{13})\gamma_{A_3} A_3}{(\gamma_{13}\gamma_{23} - \gamma_{12})\gamma_{A_2} + (\gamma_{12}\gamma_{23} - \gamma_{13})\gamma_{A_3}} \\
&\quad = \frac{(\gamma_{13}\gamma_{23} - \gamma_{12})\gamma_{\ominus A_1 \oplus A_2}(\ominus A_1 \oplus A_2) + (\gamma_{12}\gamma_{23} - \gamma_{13})\gamma_{\ominus A_1 \oplus A_3}(\ominus A_1 \oplus A_3)}{(\gamma_{13}\gamma_{23} - \gamma_{12})\gamma_{\ominus A_1 \oplus A_2} + (\gamma_{12}\gamma_{23} - \gamma_{13})\gamma_{\ominus A_1 \oplus A_3}} \\
&\quad = \frac{(\gamma_{13}\gamma_{23} - \gamma_{12})\gamma_{12}\mathbf{a}_{12} + (\gamma_{12}\gamma_{23} - \gamma_{13})\gamma_{13}\mathbf{a}_{13}}{(\gamma_{13}\gamma_{23} - \gamma_{12})\gamma_{12} + (\gamma_{12}\gamma_{23} - \gamma_{13})\gamma_{13}},
\end{aligned} \tag{7.205a}$$

$$\begin{aligned}
&\ominus A_1 \oplus P_2 \\
&\quad = \ominus A_1 \oplus \frac{(\gamma_{13}\gamma_{23} - \gamma_{12})\gamma_{A_1} A_1 + (\gamma_{12}\gamma_{13} - \gamma_{23})\gamma_{A_3} A_3}{(\gamma_{13}\gamma_{23} - \gamma_{12})\gamma_{A_1} + (\gamma_{12}\gamma_{13} - \gamma_{23})\gamma_{A_3}} \\
&\quad = \frac{(\gamma_{12}\gamma_{13} - \gamma_{23})\gamma_{\ominus A_1 \oplus A_3}(\ominus A_1 \oplus A_3)}{(\gamma_{13}\gamma_{23} - \gamma_{12}) + (\gamma_{12}\gamma_{13} - \gamma_{23})\gamma_{\ominus A_1 \oplus A_3}}
\end{aligned}$$

$$= \frac{(\gamma_{12}\gamma_{13} - \gamma_{23})\gamma_{13}\mathbf{a}_{13}}{(\gamma_{13}\gamma_{23} - \gamma_{12}) + (\gamma_{12}\gamma_{13} - \gamma_{23})\gamma_{13}}, \tag{7.205b}$$

$$\ominus A_1 \oplus P_3$$

$$= \ominus A_1 \oplus \frac{(\gamma_{12}\gamma_{23} - \gamma_{13})\gamma_{A_1} A_1 + (\gamma_{12}\gamma_{13} - \gamma_{23})\gamma_{A_2} A_2}{(\gamma_{12}\gamma_{23} - \gamma_{13})\gamma_{A_1} + (\gamma_{12}\gamma_{13} - \gamma_{23})\gamma_{A_2}}$$

$$= \frac{(\gamma_{12}\gamma_{13} - \gamma_{23})\gamma_{\ominus A_1 \oplus A_2}(\ominus A_1 \oplus A_2)}{(\gamma_{12}\gamma_{23} - \gamma_{13}) + (\gamma_{12}\gamma_{13} - \gamma_{23})\gamma_{\ominus A_1 \oplus A_2}}$$

$$= \frac{(\gamma_{12}\gamma_{13} - \gamma_{23})\gamma_{12}\mathbf{a}_{12}}{(\gamma_{12}\gamma_{23} - \gamma_{13}) + (\gamma_{12}\gamma_{13} - \gamma_{23})\gamma_{12}}. \tag{7.205c}$$

Note that, by Definition 4.5, p. 89, the set of points $S = \{A_1, A_2, A_3\}$ is pointwise independent in an Einstein gyrovector space $(\mathbb{R}_s^n, \oplus, \otimes)$. Hence, the two gyrovectors $\mathbf{a}_{12} = \ominus A_1 \oplus A_2$ and $\mathbf{a}_{13} = \ominus A_1 \oplus A_3$ in $\mathbb{R}_s^n \subset \mathbb{R}^n$ in (7.205a), (7.205b), (7.205c) are linearly independent in $\mathbb{R}^n$.

Similarly to the gyroalgebra in (7.205a), (7.205b), (7.205c), under a left gyrotranslation by $\ominus A_1$ the orthogyrocenter H in (7.204) becomes

$$\begin{aligned}\ominus A_1 \oplus H &= \frac{m_2\gamma_{\ominus A_1 \oplus A_2}(\ominus A_1 \oplus A_2) + m_3\gamma_{\ominus A_1 \oplus A_3}(\ominus A_1 \oplus A_3)}{m_1 + m_2\gamma_{\ominus A_1 \oplus A_2} + m_3\gamma_{\ominus A_1 \oplus A_3}} \\ &= \frac{m_2\gamma_{12}\mathbf{a}_{12} + m_3\gamma_{13}\mathbf{a}_{13}}{m_1 + m_2\gamma_{12} + m_3\gamma_{13}}.\end{aligned} \tag{7.206}$$

The gyroaltitude of the left gyrotranslated gyrotriangle $O(\ominus A_1 \oplus A_2)(\ominus A_1 \oplus A_3)$ that joins the vertex

$$\ominus A_1 \oplus A_1 = O = \mathbf{0} \tag{7.207}$$

with the gyroaltitude foot on its opposing side, P_1, as calculated in (7.205a),

$$\ominus A_1 \oplus P_1 = \frac{(\gamma_{13}\gamma_{23} - \gamma_{12})\gamma_{12}\mathbf{a}_{12} + (\gamma_{12}\gamma_{23} - \gamma_{13})\gamma_{13}\mathbf{a}_{13}}{(\gamma_{13}\gamma_{23} - \gamma_{12})\gamma_{12} + (\gamma_{12}\gamma_{23} - \gamma_{13})\gamma_{13}} \tag{7.208}$$

is contained in the Euclidean line

$$\begin{aligned}L_1 &= O + \big(-O + \{\ominus A_1 \oplus P_1\}\big)t_1 \\ &= \frac{(\gamma_{13}\gamma_{23} - \gamma_{12})\gamma_{12}\mathbf{a}_{12} + (\gamma_{12}\gamma_{23} - \gamma_{13})\gamma_{13}\mathbf{a}_{13}}{(\gamma_{13}\gamma_{23} - \gamma_{12})\gamma_{12} + (\gamma_{12}\gamma_{23} - \gamma_{13})\gamma_{13}} t_1,\end{aligned} \tag{7.209}$$

where $t_1 \in \mathbb{R}$ is the line parameter. This line passes through the point $O = \mathbf{0} \in \mathbb{R}_s^n \subset \mathbb{R}^n$ when $t_1 = 0$, and it passes through the point $\ominus A_1 \oplus P_1$ when $t_1 = 1$.

Similarly to (7.207)–(7.209), the gyroaltitude of the left gyrotranslated gyrotriangle $O(\ominus A_1 \oplus A_2)(\ominus A_1 \oplus A_3)$ that joins the vertex

$$\ominus A_1 \oplus A_2 = \mathbf{a}_{12} \tag{7.210}$$

with the gyroaltitude foot on its opposing side, P_2, as calculated in (7.205b),

$$\ominus A_1 \oplus P_2 = \frac{(\gamma_{12}\gamma_{13} - \gamma_{23})\gamma_{13}\mathbf{a}_{13}}{(\gamma_{13}\gamma_{23} - \gamma_{12}) + (\gamma_{12}\gamma_{13} - \gamma_{23})\gamma_{13}} \tag{7.211}$$

is contained in the Euclidean line

$$\begin{aligned} L_2 &= \mathbf{a}_{12} + \left(-\mathbf{a}_{12} + \{\ominus A_1 \oplus P_2\}\right)t_2 \\ &= \mathbf{a}_{12} + \left(-\mathbf{a}_{12} + \frac{(\gamma_{12}\gamma_{13} - \gamma_{23})\gamma_{13}\mathbf{a}_{13}}{(\gamma_{13}\gamma_{23} - \gamma_{12}) + (\gamma_{12}\gamma_{13} - \gamma_{23})\gamma_{13}}\right)t_2, \end{aligned} \tag{7.212}$$

where $t_2 \in \mathbb{R}$ is the line parameter. This line passes through the point $\mathbf{a}_{12} \in \mathbb{R}^n_s \subset \mathbb{R}^n$ when $t_2 = 0$, and it passes through the point $\ominus A_1 \oplus P_2$ when $t_2 = 1$.

Similarly to (7.207)–(7.209), and similarly to (7.210)–(7.212), the gyroaltitude of the left gyrotranslated gyrotriangle $O(\ominus A_1 \oplus A_2)(\ominus A_1 \oplus A_3)$ that joins the vertex

$$\ominus A_1 \oplus A_3 = \mathbf{a}_{13} \tag{7.213}$$

with the gyroaltitude foot on its opposing side, P_3, as calculated in (7.205c),

$$\ominus A_1 \oplus P_3 = \frac{(\gamma_{12}\gamma_{13} - \gamma_{23})\gamma_{12}\mathbf{a}_{12}}{(\gamma_{12}\gamma_{23} - \gamma_{13}) + (\gamma_{12}\gamma_{13} - \gamma_{23})\gamma_{12}} \tag{7.214}$$

is contained in the Euclidean line

$$\begin{aligned} L_3 &= \mathbf{a}_{13} + \left(-\mathbf{a}_{13} + \{\ominus A_1 \oplus P_3\}\right)t_3 \\ &= \mathbf{a}_{13} + \left(-\mathbf{a}_{13} + \frac{(\gamma_{12}\gamma_{13} - \gamma_{23})\gamma_{12}\mathbf{a}_{12}}{(\gamma_{12}\gamma_{23} - \gamma_{13}) + (\gamma_{12}\gamma_{13} - \gamma_{23})\gamma_{12}}\right)t_3, \end{aligned} \tag{7.215}$$

where $t_3 \in \mathbb{R}$ is the line parameter. This line passes through the point $\mathbf{a}_{13} \in \mathbb{R}^n_s \subset \mathbb{R}^n$ when $t_3 = 0$, and it passes through the point $\ominus A_1 \oplus P_3 \in \mathbb{R}^n_s \subset \mathbb{R}^n$ when $t_3 = 1$.

Hence, if the orthogyrocenter H exists, its left gyrotranslated orthogyrocenter, $-\ominus A_1 \oplus H$, given by (7.206), is contained in each of the three Euclidean lines L_k, $k = 1, 2, 3$, in (7.209), (7.212) and (7.215). Formalizing, if H exists then the point P, (7.206),

$$P = \ominus A_1 \oplus H = \frac{m_2\gamma_{12}\mathbf{a}_{12} + m_3\gamma_{13}\mathbf{a}_{13}}{m_1 + m_2\gamma_{12} + m_3\gamma_{13}} \tag{7.216}$$

lies on each of the lines L_k, $k = 1, 2, 3$. Imposing the normalization condition $m_1 + m_2 + m_3 = 1$ of gyrobarycentric coordinates, (7.216) can be simplified by means of the resulting equation $m_1 = 1 - m_2 - m_3$, obtaining

$$P = \ominus A_1 \oplus H = \frac{m_2 \gamma_{12} \mathbf{a}_{12} + m_3 \gamma_{13} \mathbf{a}_{13}}{1 + m_2(\gamma_{12} - 1) + m_3(\gamma_{13} - 1)}. \tag{7.217}$$

Since the point P lies on each of the three lines L_k, $k = 1, 2, 3$, there exist values $t_{k,0}$ of the line parameters t_k, $k = 1, 2, 3$, respectively, such that

$$\begin{aligned}
&P - \frac{(\gamma_{13}\gamma_{23} - \gamma_{12})\gamma_{12}\mathbf{a}_{12} + (\gamma_{12}\gamma_{23} - \gamma_{13})\gamma_{13}\mathbf{a}_{13}}{(\gamma_{13}\gamma_{23} - \gamma_{12})\gamma_{12} + (\gamma_{12}\gamma_{23} - \gamma_{13})\gamma_{13}} t_{1,0} = 0, \\
&P - \mathbf{a}_{12} - \left(-\mathbf{a}_{12} + \frac{(\gamma_{12}\gamma_{13} - \gamma_{23})\gamma_{13}\mathbf{a}_{13}}{(\gamma_{13}\gamma_{23} - \gamma_{12}) + (\gamma_{12}\gamma_{13} - \gamma_{23})\gamma_{13}}\right) t_{2,0} = 0, \\
&P - \mathbf{a}_{13} - \left(-\mathbf{a}_{13} + \frac{(\gamma_{12}\gamma_{13} - \gamma_{23})\gamma_{12}\mathbf{a}_{12}}{(\gamma_{12}\gamma_{23} - \gamma_{13}) + (\gamma_{12}\gamma_{13} - \gamma_{23})\gamma_{12}}\right) t_{3,0} = 0.
\end{aligned} \tag{7.218}$$

The kth equation in (7.218), $k = 1, 2, 3$, is equivalent to the condition that point P lies on line L_k.

The system of equations (7.218) was obtained by methods of gyroalgebra, and will be solved below by a common method of linear algebra.

Substituting P from (7.217) into (7.218), and rewriting each equation in (7.218) as a linear combination of $\mathbf{a}_{12}$ and $\mathbf{a}_{13}$ equals zero, one obtains the following linear homogeneous system of three gyrovector equations

$$\begin{aligned}
c_{11}\mathbf{a}_{12} + c_{12}\mathbf{a}_{13} &= \mathbf{0}, \\
c_{21}\mathbf{a}_{12} + c_{22}\mathbf{a}_{13} &= \mathbf{0}, \\
c_{31}\mathbf{a}_{12} + c_{32}\mathbf{a}_{13} &= \mathbf{0},
\end{aligned} \tag{7.219}$$

where each coefficient c_{ij}, $i = 1, 2, 3$, $j = 1, 2$, is a function of γ_{12}, γ_{13}, γ_{23}, and the five unknowns m_2, m_3, and $t_{k,0}$, $k = 1, 2, 3$.

Since the set $S = \{A_1, A_2, A_3\}$ is pointwise independent, the two gyrovectors $\mathbf{a}_{12} = \ominus A_1 \oplus A_2$ and $\mathbf{a}_{13} = \ominus A_1 \oplus A_3$ in $\mathbb{R}_s^n$, considered as vectors in $\mathbb{R}^n$, are linearly independent in $\mathbb{R}^n$. Hence, each coefficient c_{ij} in (7.219) equals zero. Accordingly, the three gyrovector equations in (7.219) are equivalent to the following six scalar equations,

$$c_{11} = c_{12} = c_{21} = c_{22} = c_{31} = c_{32} = 0 \tag{7.220}$$

for the five unknowns m_2, m_3 and $t_{k,0}$, $k = 1, 2, 3$.

Explicitly, the six scalar equations in (7.220) are equivalent to the following six equations:

$$\begin{aligned}
&(2\gamma_{12}\gamma_{13}\gamma_{23} - \gamma_{12}^2 - \gamma_{13}^2)m_2 \\
&\quad - (\gamma_{13}\gamma_{23} - \gamma_{12})(1 - m_2 - m_3 + \gamma_{12}m_2 + \gamma_{13}m_3)t_{1,0} = 0, \\
&(2\gamma_{12}\gamma_{13}\gamma_{23} - \gamma_{12}^2 - \gamma_{13}^2)m_3 \\
&\quad - (\gamma_{12}\gamma_{23} - \gamma_{13})(1 - m_2 - m_3 + \gamma_{12}m_2 + \gamma_{13}m_3)t_{1,0} = 0, \\
&1 - m_2 - m_3 + \gamma_{13}m_3 - (1 - m_2 - m_3 + \gamma_{12}m_2 + \gamma_{13}m_3)t_{2,0} = 0, \\
&\gamma_{12}(\gamma_{13}^2 - 1)m_3 - (\gamma_{12}\gamma_{13} - \gamma_{23})(1 - m_2 - m_3 + \gamma_{12}m_2 + \gamma_{13}m_3)t_{2,0} = 0, \\
&\gamma_{13}(\gamma_{12}^2 - 1)m_2 - (\gamma_{12}\gamma_{13} - \gamma_{23})(1 - m_2 - m_3 + \gamma_{12}m_2 + \gamma_{13}m_3)t_{3,0} = 0, \\
&1 - m_2 - m_3 + \gamma_{12}m_2 - (1 - m_2 - m_3 + \gamma_{12}m_2 + \gamma_{13}m_3)t_{3,0} = 0.
\end{aligned} \tag{7.221}$$

The system (7.221) is slightly nonlinear. It is, however, linear in the unknowns $t_{1,0}, t_{2,0}, t_{3,0}$. Solving three equations of the system for $t_{1,0}, t_{2,0}, t_{3,0}$, and substituting these into the remaining equations of the system determine the ratios m_2/m_1 and m_3/m_1 uniquely, from which convenient (homogeneous) gyrobarycentric coordinates $(m_1 : m_2 : m_3)$ are obtained. A solution of (7.221) is given by (7.222) and (7.224) below:

The values of the line parameters are

$$\begin{aligned}
t_{1,0} &= \frac{(\gamma_{12}\gamma_{13} - \gamma_{23})(2\gamma_{12}\gamma_{13}\gamma_{23} - \gamma_{12}^2 - \gamma_{13}^2)}{\gamma_{12}\gamma_{13}(1 + 2\gamma_{12}\gamma_{13}\gamma_{23} - \gamma_{12}^2 - \gamma_{13}^2)}, \\
t_{2,0} &= \frac{(\gamma_{13}^2 - 1)(\gamma_{12}\gamma_{23} - \gamma_{13})}{\gamma_{13}(1 + 2\gamma_{12}\gamma_{13}\gamma_{23} - \gamma_{12}^2 - \gamma_{13}^2)}, \\
t_{3,0} &= \frac{(\gamma_{12}^2 - 1)(\gamma_{13}\gamma_{23} - \gamma_{12})}{\gamma_{12}(1 + 2\gamma_{12}\gamma_{13}\gamma_{23} - \gamma_{12}^2 - \gamma_{13}^2)},
\end{aligned} \tag{7.222}$$

where

$$1 + 2\gamma_{12}\gamma_{13}\gamma_{23} - \gamma_{12}^2 - \gamma_{13}^2 > 0 \tag{7.223}$$

by (6.28), p. 136.

The gyrobarycentric coordinates (m_1, m_2, m_3) are given by

$$\begin{aligned}
m_1 &= \frac{1}{D}(\gamma_{12}\gamma_{23} - \gamma_{13})(\gamma_{13}\gamma_{23} - \gamma_{12}), \\
m_2 &= \frac{1}{D}(\gamma_{12}\gamma_{13} - \gamma_{23})(\gamma_{13}\gamma_{23} - \gamma_{12}), \\
m_3 &= \frac{1}{D}(\gamma_{12}\gamma_{23} - \gamma_{13})(\gamma_{12}\gamma_{23} - \gamma_{13}),
\end{aligned} \tag{7.224}$$

satisfying the normalization condition $m_1+m_2+m_3=1$, where D is the determinant

$$D=\begin{vmatrix}\gamma_{12}\gamma_{23}-\gamma_{13} & -(\gamma_{13}\gamma_{23}-\gamma_{12}) \\ \gamma_{12}\gamma_{13}-\gamma_{23} & (\gamma_{12}\gamma_{13}-\gamma_{23})+(\gamma_{13}\gamma_{23}-\gamma_{12})\end{vmatrix} \tag{7.225}$$

or, equivalently,

$$\begin{aligned} D &= (\gamma_{13}\gamma_{23}-\gamma_{12})(\gamma_{12}\gamma_{23}-\gamma_{13})+(\gamma_{13}\gamma_{23}-\gamma_{12})(\gamma_{12}\gamma_{13}-\gamma_{23}) \\ &\quad +(\gamma_{12}\gamma_{23}-\gamma_{13})(\gamma_{12}\gamma_{13}-\gamma_{23}). \end{aligned} \tag{7.226}$$

Following (7.224), convenient gyrobarycentric coordinates of the gyrotriangle orthogyrocenter H are given by the equation

$$(m_1:m_2:m_3)=(C_{12}C_{13}:C_{12}C_{23}:C_{13}C_{23}) \tag{7.227}$$

or, equivalently, by the equation

$$(m_1:m_2:m_3)=\left(\frac{C_{12}}{C_{23}}:\frac{C_{12}}{C_{13}}:1\right) \tag{7.228}$$

where

$$\begin{aligned} C_{12} &= \gamma_{13}\gamma_{23}-\gamma_{12}, \\ C_{13} &= \gamma_{12}\gamma_{23}-\gamma_{13}, \\ C_{23} &= \gamma_{12}\gamma_{13}-\gamma_{23}. \end{aligned} \tag{7.229}$$

Accordingly, the gyrobarycentric coordinate representation of the orthogyrocenter H of gyrotriangle $A_1A_2A_3$ with respect to the set of the gyrotriangle vertices is given by the equation

$$H=\frac{C_{12}C_{13}\gamma_{A_1}A_1+C_{12}C_{23}\gamma_{A_2}A_2+C_{13}C_{23}\gamma_{A_3}A_3}{C_{12}C_{13}\gamma_{A_1}+C_{12}C_{23}\gamma_{A_2}+C_{13}C_{23}\gamma_{A_3}}. \tag{7.230}$$

Substituting from (7.147), p. 188, into (7.229), we have

$$\begin{aligned} \frac{C_{12}}{C_{23}} &= \frac{\tan\alpha_1}{\tan\alpha_3}, \\ \frac{C_{12}}{C_{13}} &= \frac{\tan\alpha_2}{\tan\alpha_3}. \end{aligned} \tag{7.231}$$

Hence, the gyrobarycentric coordinates of H in (7.228) can be written as

$$(m_1:m_2:m_3)=\left(\frac{\tan\alpha_1}{\tan\alpha_3}:\frac{\tan\alpha_2}{\tan\alpha_3}:1\right) \tag{7.232}$$

which are, in turn, equivalent to the gyrobarycentric coordinates

$$(m_1 : m_2 : m_3) = (\tan\alpha_1 : \tan\alpha_2 : \tan\alpha_3). \tag{7.233}$$

Interestingly, the gyrotrigonometric gyrobarycentric coordinates (7.233) of the gyrotriangle orthogyrocenter H are identical in form with trigonometric barycentric coordinates of the triangle orthocenter in Euclidean geometry, as we see from [29].

Following (7.227) and the definition, Definition (4.5), p. 89, of the constant m_0, (4.27), of a point P with a gyrobarycentric representation, the constant m_0 of the gyrotriangle orthogyrocenter H in (7.230) with respect to the set of the gyrotriangle vertices is given by the equation

$$\begin{aligned} m_0^2 &= m_1^2 + m_2^2 + m_3^2 + 2m_1m_2\gamma_{12} + 2m_1m_3\gamma_{13} + 2m_2m_3\gamma_{23} \\ &= \frac{1}{2} f_1\left(f_1^2 + f_2\right), \end{aligned} \tag{7.234}$$

where f_1 and f_2 are factors given by

$$\begin{aligned} f_1 &= 1 + 2\gamma_{12}\gamma_{13}\gamma_{23} - \gamma_{12}^2 - \gamma_{13}^2 - \gamma_{23}^2 \\ &= 2(\gamma_{12}\gamma_{13}\gamma_{23} - 1) - (\gamma_{12}^2 - 1) - (\gamma_{13}^2 - 1) - (\gamma_{23}^2 - 1), \\ f_2 &= 2(\gamma_{12}\gamma_{13}\gamma_{23} - 1)^2 - (\gamma_{12}^2 - 1)^2 - (\gamma_{13}^2 - 1)^2 - (\gamma_{23}^2 - 1)^2. \end{aligned} \tag{7.235}$$

Since $f_1 > 0$, by (6.23), p. 135, the constant m_0^2 in (7.234) is positive, zero, or negative if and only if $f_1^2 + f_2$ is positive, zero, or negative, respectively. Hence, equivalently, the constant m_0^2 in (7.234) is positive, zero, or negative if and only if $f_1^4 - f_2^2$ is positive, zero, or negative, respectively. Expressing of gamma factors of sides of gyrotriangle $A_1A_2A_3$ in terms of the gyrotriangle gyroangles by (7.142), p. 187, we have

$$\begin{aligned} f_1^4 - f_2^2 = \big\{&\cos^2\alpha_1\cos^2\alpha_2 + \cos^2\alpha_1\cos^2\alpha_3 + \cos^2\alpha_2\cos^2\alpha_3 \\ &+ 2\cos\alpha_1\cos\alpha_2\cos\alpha_3\left(\cos^2\alpha_1 + \cos^2\alpha_2 + \cos^2\alpha_3\right) \\ &+ 3\cos^2\alpha_1\cos^2\alpha_2\cos^2\alpha_3\big\}Q, \end{aligned} \tag{7.236}$$

where Q is a positive valued function of $\cos\alpha_k$, $k = 1, 2, 3$.

Hence, the constant m_0^2 in (7.234) is positive, zero, or negative if and only if f_3 is positive, zero, or negative, respectively, where

$$\begin{aligned} f_3 = {}&\cos^2\alpha_1\cos^2\alpha_2 + \cos^2\alpha_1\cos^2\alpha_3 + \cos^2\alpha_2\cos^2\alpha_3 \\ &+ 2\cos\alpha_1\cos\alpha_2\cos\alpha_3\left(\cos^2\alpha_1 + \cos^2\alpha_2 + \cos^2\alpha_3\right) \\ &+ 3\cos^2\alpha_1\cos^2\alpha_2\cos^2\alpha_3. \end{aligned} \tag{7.237}$$

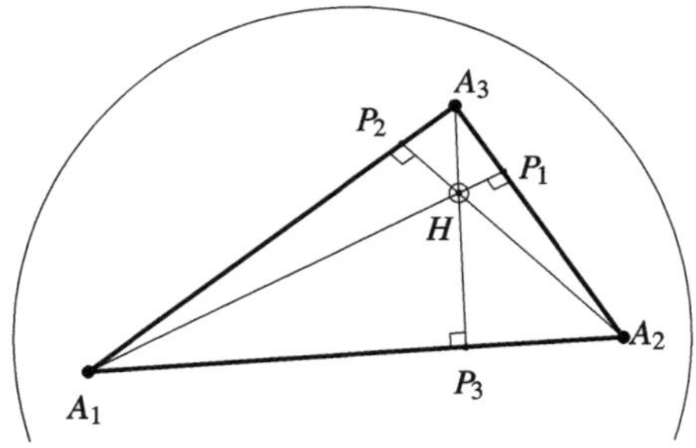

Fig. 7.18 The gyroaltitudes, and the orthogyrocenter H, of a gyrotriangle $A_1A_2A_3$ in an Einstein gyrovector space. Case I: The orthogyrocenter H lies inside the acute gyrotriangle. Gyrobarycentric coordinates $(m_1 : m_2 : m_3)$ of the orthogyrocenter H relative to the set $\{A_1, A_2, A_3\}$ of the gyrotriangle vertices, given by (7.233), are all positive so that $m_0^2 > 0$ in (7.234), in agreement with Corollary 4.10, p. 94

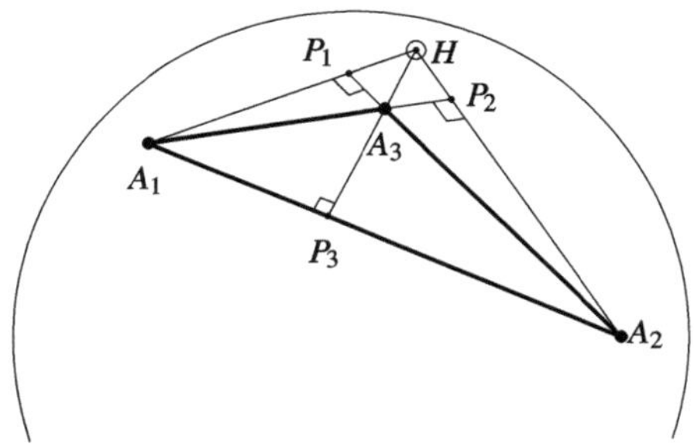

Fig. 7.19 The gyroaltitudes, and the orthogyrocenter H, of a gyrotriangle $A_1A_2A_3$ in an Einstein gyrovector space. Case II: The orthogyrocenter H lies outside the obtuse gyrotriangle. One of the gyrobarycentric coordinates $(m_1 : m_2 : m_3)$ of the orthogyrocenter H relative to the set $\{A_1, A_2, A_3\}$ of the gyrotriangle vertices, given by (7.233), is negative and the other two are positive, in agreement with Corollary 4.9

According to Corollary 4.9, p. 93, if $m_0^2 > 0$ then gyrotriangle $A_1A_2A_3$ possesses a orthogyrocenter H. The orthogyrocenter H lies in the interior of gyrotriangle $A_1A_2A_3$ if and only if gyrobarycentric coordinates of H are all positive or all negative. The gyrotriangle $A_1A_2A_3$ does not have a orthogyrocenter H when $m_0^2 \leq 0$ in (7.234). When $m_0^2 = 0$, the point H lies on the boundary of the ball $\mathbb{R}_s^n$, and when $m_0^2 < 0$ the point H lies outside of the ball, as shown in Figs. 7.18–7.21. Indeed,

1. $f_3 > 0$ for gyrotriangle $A_1A_2A_3$ in Figs. 7.17, p. 204, and 7.18–7.19
2. $f_3 = 0$ for gyrotriangle $A_1A_2A_3$ in Fig. 7.20, p. 212; and
3. $f_3 < 0$ for gyrotriangle $A_1A_2A_3$ in Fig. 7.21, p. 212

Formalizing the main result of this section, we have the following theorem:

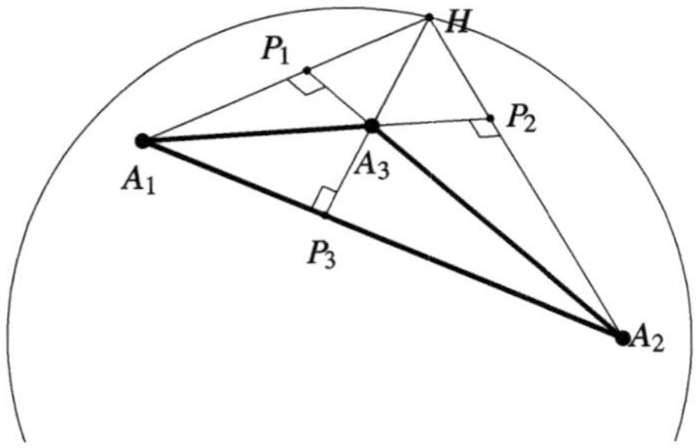

Fig. 7.20 A gyrotriangle $A_1A_2A_3$ that does not possess an orthogyrocenter H in an Einstein gyrovector plane $(\mathbb{R}^2_s, \oplus, \otimes)$. The point $H \in \mathbb{R}^n$ lies on the boundary of the ball $\mathbb{R}^n_s \subset \mathbb{R}^n$. Accordingly $m_0^2 = 0$ in (7.234), in agreement with Corollary 4.9, p. 93

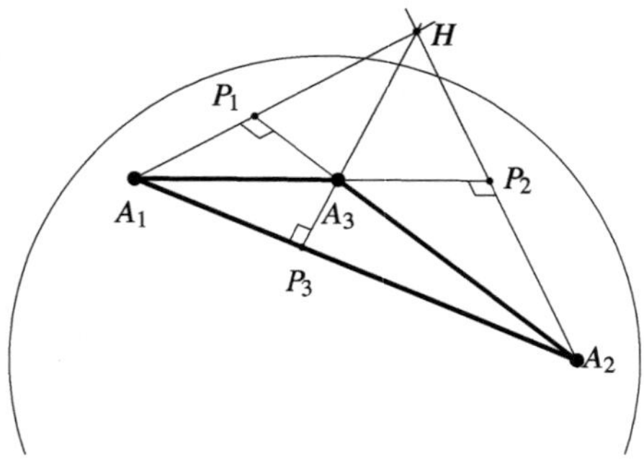

Fig. 7.21 A gyrotriangle $A_1A_2A_3$ that does not possess an orthogyrocenter H in an Einstein gyrovector plane $(\mathbb{R}^2_s, \oplus, \otimes)$. The point $H \in \mathbb{R}^n$ lies outside of the ball $\mathbb{R}^n_s \subset \mathbb{R}^n$. Accordingly $m_0^2 < 0$ in (7.234), in agreement with Corollary 4.9, p. 93

Theorem 7.24 (The Orthogyrocenter) *Let $S = \{A_1, A_2, A_3\}$ be a pointwise independent set of three points in an Einstein gyrovector space $(\mathbb{R}^n_s, \oplus, \otimes)$. The orthogyrocenter H, see Figs.* 7.18–7.21, *of gyrotriangle $A_1A_2A_3$ has the gyrobarycentric coordinate representation*

$$H = \frac{m_1\gamma_{A_1} A_1 + m_2\gamma_{A_2} A_2 + m_3\gamma_{A_3} A_3}{m_1\gamma_{A_1} + m_2\gamma_{A_2} + m_3\gamma_{A_3}} \tag{7.238}$$

with respect to the set $S = \{A_1, A_2, A_3\}$, with gyrotrigonometric gyrobarycentric coordinates given by each of the two equations

$$(m_1 : m_2 : m_3) = (\tan\alpha_1 : \tan\alpha_2 : \tan\alpha_3) \tag{7.239}$$

and

$$\begin{aligned} m_1 &= (\gamma_{12}\gamma_{23} - \gamma_{13})(\gamma_{13}\gamma_{23} - \gamma_{12}), \\ m_2 &= (\gamma_{12}\gamma_{13} - \gamma_{23})(\gamma_{13}\gamma_{23} - \gamma_{12}), \\ m_3 &= (\gamma_{12}\gamma_{23} - \gamma_{13})(\gamma_{12}\gamma_{23} - \gamma_{13}). \end{aligned} \tag{7.240}$$

The existence of the gyrotriangle orthogyrocenter H is determined by the squared orthogyrocenter constant m_0^2 with respect to the set of the gyrotriangle vertices,

$$\begin{aligned} m_0^2 &= \tan^2\alpha_1 + \tan^2\alpha_2 + \tan^2\alpha_3 \\ &\quad + 2(\tan\alpha_1\tan\alpha_2\gamma_{12} + \tan\alpha_1\tan\alpha_3\gamma_{13} + \tan\alpha_2\tan\alpha_3\gamma_{23}) \\ &= \bigl\{\cos^2\alpha_1\cos^2\alpha_2 + \cos^2\alpha_1\cos^2\alpha_3 + \cos^2\alpha_2\cos^2\alpha_3 \\ &\quad + 2\cos\alpha_1\cos\alpha_2\cos\alpha_3\bigl(\cos^2\alpha_1 + \cos^2\alpha_2 + \cos^2\alpha_3\bigr) \\ &\quad + 3\cos^2\alpha_1\cos^2\alpha_2\cos^2\alpha_3\bigr\}\frac{1}{\cos^2\alpha_1\cos^2\alpha_2\cos^2\alpha_3}. \end{aligned} \tag{7.241}$$

The gyrotriangle orthogyrocenter H exists if and only if $m_0^2 > 0$. Furthermore, the gyrotriangle orthogyrocenter H lies on the interior of its gyrotriangle if and only if $\tan\alpha_1 > 0$, $\tan\alpha_2 > 0$ *and* $\tan\alpha_3 > 0$ *or, equivalently, if and only if the gyrotriangle is acute, see Figs.* 7.18–7.21.

The gyrotrigonometric gyrobarycentric coordinates (7.239) remain invariant in form under the Euclidean limit $s \to \infty$, resulting in the following corollary of Theorem 7.24:

Corollary 7.25 (The Orthocenter) *Let $S = \{A_1, A_2, A_3\}$ be a pointwise independent set of three points in a Euclidean vector space $\mathbb{R}^n$. The orthocenter H of triangle $A_1A_2A_3$ has the barycentric coordinate representation*

$$H = \frac{m_1A_1 + m_2A_2 + m_3A_3}{m_1 + m_2 + m_3} \tag{7.242}$$

with respect to the set $S = \{A_1, A_2, A_3\}$, with trigonometric barycentric coordinates given by

$$(m_1 : m_2 : m_3) = (\tan\alpha_1 : \tan\alpha_2 : \tan\alpha_3). \tag{7.243}$$

7.18 The Gyrodistance Between *O* and *I*

Let O and I be the circumgyrocenter and ingyrocenter of a gyrotriangle $A_1A_2A_3$ in an Einstein gyrovector space $(\mathbb{R}^n_s, \oplus, \otimes)$. Their gyrobarycentric coordinate representations with respect to the set $S = \{A_1, A_2, A_3\}$ are, by (7.18) and (7.20), p. 158,

$$O = \frac{m'_1\gamma_{A_1}A_1 + m'_2\gamma_{A_2}A_2 + m'_3\gamma_{A_3}A_3}{m'_1\gamma_{A_1} + m'_2\gamma_{A_2} + m'_3\gamma_{A_3}}, \tag{7.244a}$$

where the gyrobarycentric coordinates m'_k, $k = 1, 2, 3$, are given by

$$
\begin{aligned}
m'_1 &= \sin\left(\frac{-\alpha_1 + \alpha_2 + \alpha_3}{2}\right) \sin\alpha_1, \\
m'_2 &= \sin\left(\frac{\alpha_1 - \alpha_2 + \alpha_3}{2}\right) \sin\alpha_2, \\
m'_3 &= \sin\left(\frac{\alpha_1 + \alpha_2 - \alpha_3}{2}\right) \sin\alpha_3,
\end{aligned}
\tag{7.244b}
$$

and, by (7.75), p. 172, and Theorem 7.12, p. 178,

$$
I = \frac{m_1\gamma_{A_1} A_1 + m_2\gamma_{A_2} A_2 + m_3\gamma_{A_3} A_3}{m_1\gamma_{A_1} + m_2\gamma_{A_2} + m_3\gamma_{A_3}},
\tag{7.245a}
$$

where the gyrobarycentric coordinates m_k, $k = 1, 2, 3$, are given by

$$
\begin{aligned}
m_1 &= \sin\alpha_1, \\
m_2 &= \sin\alpha_2, \\
m_3 &= \sin\alpha_3.
\end{aligned}
\tag{7.245b}
$$

Hence, by (4.121), p. 113,

$$
\begin{aligned}
\gamma_{\ominus O \oplus I} = \frac{1}{m_0 m'_0} \{ &(m_1 m'_2 + m'_1 m_2)\gamma_{12} + (m_1 m'_3 + m'_1 m_3)\gamma_{13} \\
&+ (m_2 m'_3 + m'_2 m_3)\gamma_{23} + m_1 m'_1 + m_2 m'_2 + m_3 m'_3 \},
\end{aligned}
\tag{7.246}
$$

where, by (4.118b) and (4.119b), p. 112, $m_0 > 0$ and $m'_0 > 0$ are given by

$$
\begin{aligned}
m_0^2 &= m_1^2 + m_2^2 + m_3^2 + 2(m_1 m_2 \gamma_{12} + m_1 m_3 \gamma_{13} + m_2 m_3 \gamma_{23}), \\
(m'_0)^2 &= (m'_1)^2 + (m'_2)^2 + (m'_3)^2 + 2(m'_1 m'_2 \gamma_{12} + m'_1 m'_3 \gamma_{13} + m'_2 m'_3 \gamma_{23}),
\end{aligned}
\tag{7.247}
$$

noting that always $m_0^2 > 0$; and that $(m'_0)^2 > 0$ if and only if gyrotriangle $A_1A_2A_3$ possesses a circumgyrocenter.

Substituting (7.244b) and (7.245b) into (7.246) and squaring, one obtains $\gamma^2_{\ominus O \oplus I}$ expressed in terms of the gyrotriangle gyroangles α_k, $k = 1, 2, 3$. Substituting the latter, in turn, into the identity, (1.9), p. 5,

$$
\|\ominus O \oplus I\|^2 = s^2 \frac{\gamma^2_{\ominus O \oplus I} - 1}{\gamma^2_{\ominus O \oplus I}},
\tag{7.248}
$$

we finally obtain the desired gyrodistance,

$$
\begin{aligned}
&\|\ominus O \oplus I\|^2 \\
&= \frac{4s^2 \cos\frac{\alpha_1+\alpha_2+\alpha_3}{2} \cos\frac{\alpha_1}{2} \cos\frac{\alpha_2}{2} \cos\frac{\alpha_3}{2}}{(\sin\alpha_1 + \sin\alpha_2 + \sin\alpha_3)^2 \cos\frac{\alpha_1-\alpha_2-\alpha_3}{2} \cos\frac{-\alpha_1+\alpha_2-\alpha_3}{2} \cos\frac{-\alpha_1-\alpha_2+\alpha_3}{2}} \\
&\quad \times \left\{ \cos\frac{\alpha_1-\alpha_2-\alpha_3}{2} + \cos\frac{-\alpha_1+\alpha_2-\alpha_3}{2} + \cos\frac{-\alpha_1-\alpha_2+\alpha_3}{2} \right. \\
&\quad \left. - \cos\frac{3\alpha_1-\alpha_2-\alpha_3}{2} - \cos\frac{-\alpha_1+3\alpha_2-\alpha_3}{2} - \cos\frac{-\alpha_1-\alpha_2+3\alpha_3}{2} \right\}.
\end{aligned}
\tag{7.249}
$$

Eliminating the factor $s^2 \cos\frac{\alpha_1+\alpha_2+\alpha_3}{2}$ between (7.249) and (7.35), p. 163, we obtain the result (7.250) of the following theorem:

Theorem 7.26 *Let α_k, $k = 1, 2, 3$, O and I be the gyroangles, circumgyrocenter and ingyrocenter of a gyrotriangle $A_1A_2A_3$ in an Einstein gyrovector space $(\mathbb{R}^n_s \oplus, \otimes)$. Then,*

$$
\begin{aligned}
\|\ominus O \oplus I\|^2 &= \frac{4R^2 \cos\frac{\alpha_1}{2} \cos\frac{\alpha_2}{2} \cos\frac{\alpha_3}{2}}{(\sin\alpha_1 + \sin\alpha_2 + \sin\alpha_3)^2} \\
&\quad \times \left\{ \cos\frac{\alpha_1-\alpha_2-\alpha_3}{2} + \cos\frac{-\alpha_1+\alpha_2-\alpha_3}{2} + \cos\frac{-\alpha_1-\alpha_2+\alpha_3}{2} \right. \\
&\quad - \cos\frac{3\alpha_1-\alpha_2-\alpha_3}{2} - \cos\frac{-\alpha_1+3\alpha_2-\alpha_3}{2} \\
&\quad \left. - \cos\frac{-\alpha_1-\alpha_2+3\alpha_3}{2} \right\}.
\end{aligned}
\tag{7.250}
$$

Interestingly, Equation (7.250) remains invariant in form under the Euclidean limit $s \to \infty$, so that the equation is valid in Euclidean geometry as well. However, for application in Euclidean geometry (7.250) can be simplified, owing to the fact that triangle angle sum in π.

Indeed, under the condition

$$\alpha_1 + \alpha_2 + \alpha_3 = \pi, \tag{7.251a}$$

we have the trigonometric identities similar to (7.22b), p. 159,

$$
\begin{aligned}
\cos\frac{\alpha_1-\alpha_2-\alpha_3}{2} &= \sin\alpha_1, \\
\cos\frac{3\alpha_1-\alpha_2-\alpha_3}{2} &= \sin 2\alpha_1, \qquad \text{etc.}
\end{aligned}
\tag{7.251b}
$$

Hence, we obtain the following corollary of Theorem 7.26:

Corollary 7.27 *Let α_k, $k = 1, 2, 3$, O and I be the angles, circumcenter and incenter of a triangle $A_1A_2A_3$ in a Euclidean space $\mathbb{R}^n$. Then,*

$$\|-O+I\|^2 = \frac{4R^2 \cos\frac{\alpha_1}{2} \cos\frac{\alpha_2}{2} \cos\frac{\alpha_3}{2}}{(\sin\alpha_1 + \sin\alpha_2 + \sin\alpha_3)^2} \times \{\sin\alpha_1 + \sin\alpha_2 + \sin\alpha_3 - \sin 2\alpha_1 - \sin 2\alpha_2 - \sin 2\alpha_3\}. \tag{7.252}$$

7.19 Problems

Problem 7.1 The constant of a Gyrobarycentric Coordinate Representation:
Derive (7.11), p. 156, from (7.10) and (7.3).

Problem 7.2 Gyrotrigonometric Substitutions:
Substitute from (7.13) into (7.10) to obtain (7.14), p. 156.

Problem 7.3 Gyrotrigonometric Substitutions:
Derive the gyrotrigonometric representation (7.34), p. 163, of the gyrotriangle circumgyroradius R by expressing the gamma factors in (7.29), p. 162, in terms of the gyrotriangle gyroangles α_k, $k = 1, 2, 3$, by means of (6.33), p. 137.

Remarkably, this task in gyrotrigonometry can straightforwardly be performed by *Mathematica*, a software for computer algebra, using commands that manipulate common trigonometric identities like `TrigToExp`, `ExpToTrig`, `TrigReduce` and `TrigFactor`.

Problem 7.4 Show that (7.40), p. 165, holds when the three points A_1, A_2 and A_3 in Theorem 7.5 are not distinct.

Problem 7.5 A Gyrotriangle Gyroangle Inequality:
Employ (7.13), p. 156, to derive Inequality (7.17), p. 157, from Inequality (7.12), p. 156.

Problem 7.6 Linear Algebra:
Provide the missing technical details in the derivation of (7.95), p. 177, from (7.89), p. 175.

Problem 7.7 Gyrotriangle Gyrotrigonometric Identities:
Verify the gyrotriangle gyrotrigonometric identities in (7.149)–(7.151), p. 188.

Problem 7.8 Gyrotrigonometric Substitutions:
Derive (7.187), p. 198, by substitutions from (7.185), p. 198, and from (7.142), p. 187.

Problem 7.9 Gyrotrigonometric Substitutions:
Derive the first two equations in (7.189), p. 199, from (7.188).

Problem 7.10 Derive (7.190), p. 199, from (7.189).

Problem 7.11 Gyrotriangle Orthogyrocenter:
Solve the system (7.221), p. 208, and hence derive the gyrobarycentric coordinates (7.224).

Problem 7.12 Gyrotrigonometric Substitutions:
By substitutions from (6.33), p. 137, derive the gyrotrigonometric condition f_3 in (7.237), p. 210, that determines whether the orthogyrocenter H of gyrotriangle $A_1A_2A_3$ exists.

Problem 7.13 Gyrotrigonometric Substitutions:
By substitutions from (7.244b) and (7.245b) into (7.246) and squaring, express $\gamma^2_{\ominus O \oplus I}$ in terms of the gyroangles α_k, $k = 1, 2, 3$, of the reference gyrotriangle $A_1A_2A_3$. Furthermore, substitute the latter into (7.248) to obtain the squared gyrodistance $\|\ominus O \oplus I\|^2$ in (7.249), p. 214.

Chapter 8
Gyrotriangle Exgyrocircles

Abstract Hyperbolic triangle excircles are called, in gyrolanguage, gyrotriangle exgyrocircles. These are determined in this chapter in terms of their gyrocenters and gyroradii. Their gyrocenters, in turn, are determined in terms of their gyrobarycentric coordinate representations with respect to the reference gyrotriangle. Moreover, relationships between the exgyroradii of a gyrotriangle exgyrocircles, and the gyrotriangle ingyroradius and circumgyroradius are obtained.

8.1 Introduction

Each gyrotriangle has an ingyrocircle, defined in Definition 7.11, p. 171, and up to three exgyrocircles the definition of which follows.

Definition 8.1 An exgyrocircle of a gyrotriangle is a gyrocircle lying outside the gyrotriangle, tangent to one of its sides and tangent to the extensions of the other two, see Figs. 8.1–8.2. The gyrocenter and the gyroradius of an exgyrocircle of a gyrotriangle are called the gyrotriangle exgyrocenter and exgyroradius.

Two additional gyrotriangle gyrocenters that are associated with ingyrocircles and exgyrocircles are the Nagel gyropoint and the Gergonne gyropoint, which are studied in this chapter.

8.2 Gyrotriangle Exgyrocircles and Ingyrocircles

In this section, we obtain gyrobarycentric coordinates for gyrotriangle exgyrocenters. Strikingly, as a byproduct we obtain gyrobarycentric coordinates for gyrotriangle ingyrocenters as well.

A.A. Ungar, *Hyperbolic Triangle Centers*,
Fundamental Theories of Physics 166,
DOI 10.1007/978-90-481-8637-2_8, © Springer Science+Business Media B.V. 2010

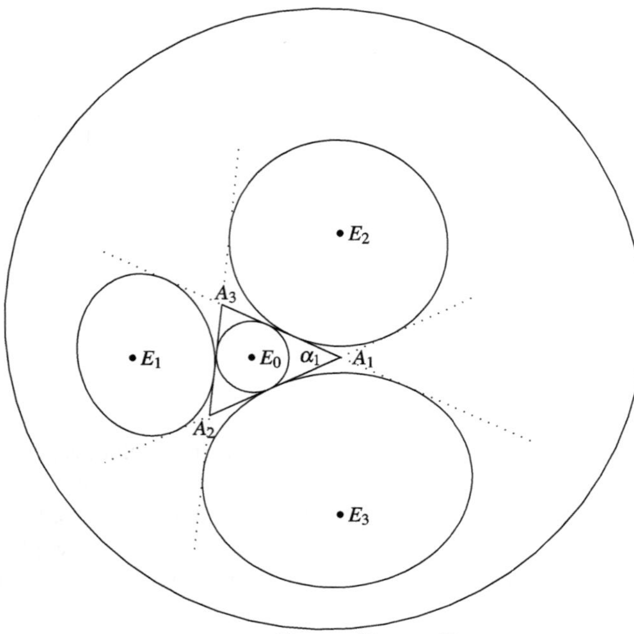

$$E_1: \quad e_1 := -\cos\tfrac{\alpha_1}{2} + \sin\tfrac{\alpha_2}{2} + \sin\tfrac{\alpha_3}{2} = 0.1320$$

$$E_2: \quad e_2 := -\cos\tfrac{\alpha_2}{2} + \sin\tfrac{\alpha_1}{2} + \sin\tfrac{\alpha_3}{2} = 0.1176$$

$$E_3: \quad e_3 := -\cos\tfrac{\alpha_3}{2} + \sin\tfrac{\alpha_1}{2} + \sin\tfrac{\alpha_2}{2} = 0.0626$$

Fig. 8.1 The gyrotriangle exgyrocircles, and exgyrocenters E_k, $k = 1, 2, 3$, in an Einstein gyrovector space $(\mathbb{R}^n_s, \oplus, \otimes)$. Exgyrocircle with exgyrocenter E_k exists if and only if $e_k > 0$, $k = 1, 2, 3$. When e_k tends to 0, E_k approaches the boundary of the ball $\mathbb{R}^n_s$, as shown here for $n = 2$

Let E be a generic exgyrocenter or ingyrocenter of a gyrotriangle $A_1A_2A_3$ in an Einstein gyrovector space $(\mathbb{R}^n_s, \oplus, \otimes)$, shown in Figs. 8.1–8.3. Furthermore, let

$$E = \frac{m_1\gamma_{A_1} A_1 + m_2\gamma_{A_2} A_2 + m_3\gamma_{A_3} A_3}{m_1\gamma_{A_1} + m_2\gamma_{A_2} + m_3\gamma_{A_3}} \tag{8.1}$$

be the gyrobarycentric coordinate representation of E with respect to the set $S = \{A_1, A_2, A_3\}$ of the gyrotriangle vertices, where the gyrobarycentric coordinates $(m_1 : m_2 : m_3)$ of E in (8.2) are to be determined in (8.13a), (8.13b), (8.13c), (8.13d), p. 224.

By Identity (4.29c), p. 91, with $X = \ominus A_k$ $k = 1, 2, 3$, we have, respectively,

$$\gamma_{\ominus A_1 \oplus E} = \frac{m_1\gamma_{\ominus A_1\oplus A_1} + m_2\gamma_{\ominus A_1\oplus A_2} + m_3\gamma_{\ominus A_1\oplus A_3}}{m_0} = \frac{m_1 + m_2\gamma_{12} + m_3\gamma_{13}}{m_0},$$

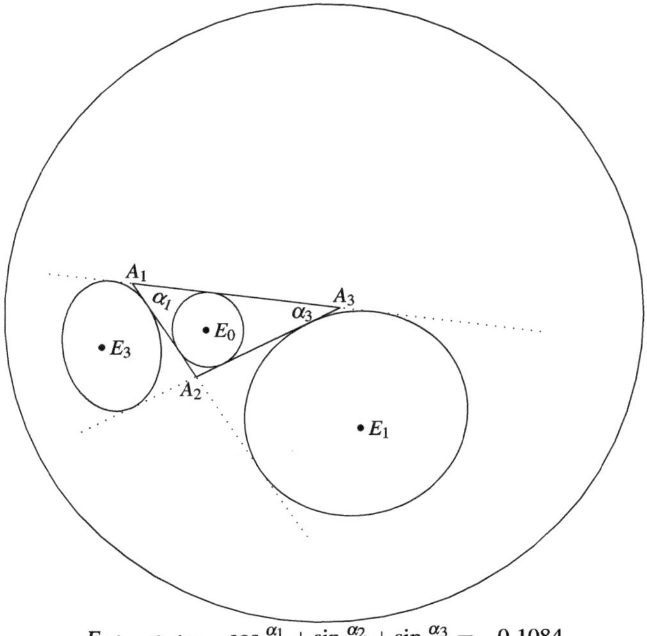

$$
\begin{aligned}
E_1: &\quad e_1 := -\cos\tfrac{\alpha_1}{2} + \sin\tfrac{\alpha_2}{2} + \sin\tfrac{\alpha_3}{2} = 0.1084 \\
E_2: &\quad e_2 := -\cos\tfrac{\alpha_2}{2} + \sin\tfrac{\alpha_1}{2} + \sin\tfrac{\alpha_3}{2} = -0.0091 \\
E_3: &\quad e_3 := -\cos\tfrac{\alpha_3}{2} + \sin\tfrac{\alpha_1}{2} + \sin\tfrac{\alpha_2}{2} = 0.1583
\end{aligned}
$$

Fig. 8.2 Exgyrocircle with exgyrocenter E_k exists if and only if $e_k > 0$, $k = 1, 2, 3$. When e_k tends to 0, E_k approaches the boundary of the ball $\mathbb{R}^n_s$, as shown here for $n = 2$. Indeed, $e_2 < 0$ and, accordingly, exgyrocircle with exgyrocenter E_2 does not exist

$$
\begin{aligned}
\gamma_{\ominus A_2 \oplus E} &= \frac{m_1 \gamma_{\ominus A_2 \oplus A_1} + m_2 \gamma_{\ominus A_2 \oplus A_2} + m_3 \gamma_{\ominus A_2 \oplus A_3}}{m_0} = \frac{m_1 \gamma_{12} + m_2 + m_3 \gamma_{23}}{m_0}, \\
\gamma_{\ominus A_3 \oplus E} &= \frac{m_1 \gamma_{\ominus A_3 \oplus A_1} + m_2 \gamma_{\ominus A_3 \oplus A_2} + m_3 \gamma_{\ominus A_3 \oplus A_3}}{m_0} = \frac{m_1 \gamma_{13} + m_2 \gamma_{23} + m_3}{m_0},
\end{aligned}
\tag{8.2}
$$

where we use the standard gyrotriangle index notation, shown in Fig. 6.1, p. 128, and in (6.1), p. 127, and where m_0 is the constant of the gyrobarycentric coordinate representation of E in (8.1) which, according to (4.29d), p. 91, is given by the equation

$$
m_0^2 = m_1^2 + m_2^2 + m_3^2 + 2m_1 m_2 \gamma_{12} + 2m_1 m_3 \gamma_{13} + 2m_2 m_3 \gamma_{23}. \tag{8.3}
$$

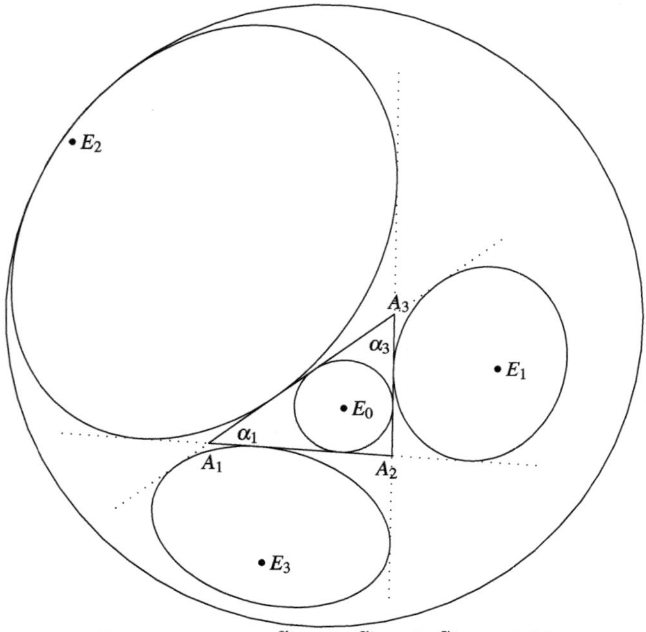

$$E_1: \quad e_1 := -\cos\tfrac{\alpha_1}{2} + \sin\tfrac{\alpha_2}{2} + \sin\tfrac{\alpha_3}{2} = 0.1502$$

$$E_2: \quad e_2 := -\cos\tfrac{\alpha_2}{2} + \sin\tfrac{\alpha_1}{2} + \sin\tfrac{\alpha_3}{2} = 0.0003$$

$$E_3: \quad e_3 := -\cos\tfrac{\alpha_3}{2} + \sin\tfrac{\alpha_1}{2} + \sin\tfrac{\alpha_2}{2} = 0.0694$$

Fig. 8.3 Exgyrocircle with exgyrocenter E_k exists if and only if $e_k > 0$, $k = 1, 2, 3$. When e_k tends to 0, E_k approaches the boundary of the ball $\mathbb{R}_s^n$, as shown here for $n = 2$. Indeed, $e_2 \approx 0$ and, accordingly, exgyrocenter E_2 lies close to the boundary of the disk

1. The gyrodistance of E (note that E represents each of E_k, $k = 0, 1, 2, 3$, in Fig. 8.1) from the gyroline that passes through points A_1 and A_2, Fig. 8.1, is the gyroaltitude r_{12} of gyrotriangle A_1A_2E drawn from base A_1A_2. Hence, by Theorem 7.17, p. 184, r_{12} is given by the equation

$$\gamma_{r_{12}}^2 = \frac{2\gamma_{12}\gamma_{\ominus A_1 \oplus E}\gamma_{\ominus A_2 \oplus E} - \gamma_{\ominus A_1 \oplus E}^2 - \gamma_{\ominus A_2 \oplus E}^2}{\gamma_{12}^2 - 1}. \tag{8.4a}$$

2. The gyrodistance of E from the gyroline that passes through points A_1 and A_3, Fig. 8.1, is the gyroaltitude r_{13} of gyrotriangle A_1A_3E drawn from base A_1A_3. Hence, by Theorem 7.17, p. 184, r_{13} is given by the equation

$$\gamma_{r_{13}}^2 = \frac{2\gamma_{13}\gamma_{\ominus A_1 \oplus E}\gamma_{\ominus A_3 \oplus E} - \gamma_{\ominus A_1 \oplus E}^2 - \gamma_{\ominus A_3 \oplus E}^2}{\gamma_{13}^2 - 1}. \tag{8.4b}$$

3. The gyrodistance of E from the gyroline that passes through points A_2 and A_3, Fig. 8.1, is the gyroaltitude r_{23} of gyrotriangle A_2A_3E drawn from base A_2A_3. Hence, by Theorem 7.17, p. 184, r_{23} is given by the equation

$$\gamma_{r_{23}}^2 = \frac{2\gamma_{23}\gamma_{\ominus A_2\oplus E}\gamma_{\ominus A_3\oplus E} - \gamma_{\ominus A_2\oplus E}^2 - \gamma_{\ominus A_3\oplus E}^2}{\gamma_{23}^2 - 1}. \tag{8.4c}$$

The gyrodistances from E to each of the three gyrolines A_1A_2, A_1A_3 and A_2A_3 are equal, implying

$$\begin{aligned} \gamma_{r_{12}}^2 &= \gamma_{r_{13}}^2, \\ \gamma_{r_{12}}^2 &= \gamma_{r_{23}}^2. \end{aligned} \tag{8.5}$$

Substituting successively (8.4a), (8.4b), (8.4c) and (8.2) into (8.5), along with the convenient normalization condition

$$m_1^2 + m_2^2 + m_3^2 = 1, \tag{8.6}$$

we obtain from (8.5)–(8.6) the following system of three equations for the three unknowns m_1^2, m_2^2 and m_3^2:

$$\begin{aligned} m_1^2(\gamma_{12}^2 - 1) - m_3^2(\gamma_{23}^2 - 1) &= 0, \\ m_2^2(\gamma_{12}^2 - 1) - m_3^2(\gamma_{13}^2 - 1) &= 0, \\ m_1^2 + m_2^2 + m_3^2 &= 1. \end{aligned} \tag{8.7}$$

The system (8.7) can be written as a matrix equation,

$$\begin{pmatrix} 1 & 1 & 1 \\ \gamma_{12}^2 - 1 & 0 & -(\gamma_{23}^2 - 1) \\ 0 & \gamma_{12}^2 - 1 & -(\gamma_{13}^2 - 1) \end{pmatrix} \begin{pmatrix} m_1^2 \\ m_2^2 \\ m_3^2 \end{pmatrix} = \begin{pmatrix} 1 \\ 0 \\ 0 \end{pmatrix} \tag{8.8}$$

so that

$$\begin{aligned} \begin{pmatrix} m_1^2 \\ m_2^2 \\ m_3^2 \end{pmatrix} &= \begin{pmatrix} 1 & 1 & 1 \\ \gamma_{12}^2 - 1 & 0 & -(\gamma_{23}^2 - 1) \\ 0 & \gamma_{12}^2 - 1 & -(\gamma_{13}^2 - 1) \end{pmatrix}^{-1} = \begin{pmatrix} 1 \\ 0 \\ 0 \end{pmatrix} \\ &= \frac{1}{D} \begin{pmatrix} (\gamma_{12}^2 - 1)(\gamma_{23}^2 - 1) & (\gamma_{12}^2 - 1)(\gamma_{13}^2 - 1) & -(\gamma_{23}^2 - 1) \\ (\gamma_{12}^2 - 1)(\gamma_{33}^2 - 1) & -(\gamma_{13}^2 - 1) & (\gamma_{12}^2 - 1)(\gamma_{23}^2 - 1) \\ (\gamma_{12}^2 - 1)^2 & -(\gamma_{12}^2 - 1) & (\gamma_{12}^2 - 1) \end{pmatrix} \begin{pmatrix} 1 \\ 0 \\ 0 \end{pmatrix}, \end{aligned} \tag{8.9}$$

where D is the determinant of the coefficient matrix in (8.8), given by

$$D = \{(\gamma_{12}^2 - 1) + (\gamma_{13}^2 - 1) + (\gamma_{23}^2 - 1)\}(\gamma_{12}^2 - 1) > 0. \tag{8.10}$$

Hence, by (8.9),

$$\begin{aligned} m_1^2 &= \frac{1}{D}(\gamma_{12}^2 - 1)(\gamma_{23}^2 - 1), \\ m_2^2 &= \frac{1}{D}(\gamma_{12}^2 - 1)(\gamma_{13}^2 - 1), \\ m_3^2 &= \frac{1}{D}(\gamma_{12}^2 - 1)^2, \end{aligned} \tag{8.11}$$

where the gyrobarycentric coordinates are normalized by (8.6).

However, while the normalization (8.6) was temporarily convenient, it is irrelevant since gyrobarycentric coordinates are homogeneous by definition, Definition 4.2, p. 86. Hence, for the sake of simplicity, we drop the common factor $(\gamma_{12}^2 - 1)/D$ in (8.11), obtaining the following equations for gyrobarycentric coordinates m_k, $k = 1, 2, 3$, of E in (8.1) in which we employ Identity (1.9), p. 5, and introduce the common factor s^2 for convenience:

$$\begin{aligned} m_1^2 &= s^2(\gamma_{23}^2 - 1) = \gamma_{23}^2 a_{23}^2, \\ m_2^2 &= s^2(\gamma_{13}^2 - 1) = \gamma_{13}^2 a_{13}^2, \\ m_3^2 &= s^2(\gamma_{12}^2 - 1) = \gamma_{12}^2 a_{12}^2. \end{aligned} \tag{8.12}$$

Finally, in order to determine gyrobarycentric coordinates for the exgyrocenter E in (8.1) it remains to determine the signs of the gyrobarycentric coordinates m_k, $k = 1, 2, 3$, in (8.12). Being homogeneous, the two selections of a positive sign for each m_k and a negative sign for each m_k are indistinguishable. Similarly, the two selections of one positive and two negative signs for the m_k's and two positive and one negative signs for the m_k's are indistinguishable.

Hence, there are precisely four distinct sign selections for the m_k's, which turn out to correspond to the three exgyrocenters and the ingyrocenter of gyrotriangle $A_1A_2A_3$. These are:

$$E_1: \quad (m_1 : m_2 : m_3) = (-\gamma_{23}a_{23} : \gamma_{13}a_{13} : \gamma_{12}a_{12}), \tag{8.13a}$$

$$E_2: \quad (m_1 : m_2 : m_3) = (\gamma_{23}a_{23} : -\gamma_{13}a_{13} : \gamma_{12}a_{12}), \tag{8.13b}$$

$$E_3: \quad (m_1 : m_2 : m_3) = (\gamma_{23}a_{23} : \gamma_{13}a_{13} : -\gamma_{12}a_{12}), \tag{8.13c}$$

$$E_0: \quad (m_1 : m_2 : m_3) = (\gamma_{23}a_{23} : \gamma_{13}a_{13} : \gamma_{12}a_{12}). \tag{8.13d}$$

1. Substituting (8.13a) into (8.1), the point E of (8.1) becomes E_1, $E = E_1$, where

$$E_1 = \frac{-\gamma_{23}a_{23}\gamma_{A_1}A_1 + \gamma_{13}a_{13}\gamma_{A_2}A_2 + \gamma_{12}a_{12}\gamma_{A_3}A_3}{-\gamma_{23}a_{23}\gamma_{A_1} + \gamma_{13}a_{13}\gamma_{A_2} + \gamma_{12}a_{12}\gamma_{A_3}}. \tag{8.14a}$$

   The point E_1 is the exgyrocenter of gyrotriangle $A_1A_2A_3$ opposite to vertex A_1, as shown in Fig. 8.1.
2. Substituting (8.13b) into (8.1), the point E of (8.1) becomes E_2, $E = E_2$, where

$$E_2 = \frac{\gamma_{23}a_{23}\gamma_{A_1}A_1 - \gamma_{13}a_{13}\gamma_{A_2}A_2 + \gamma_{12}a_{12}\gamma_{A_3}A_3}{\gamma_{23}a_{23}\gamma_{A_1} - \gamma_{13}a_{13}\gamma_{A_2} + \gamma_{12}a_{12}\gamma_{A_3}}. \tag{8.14b}$$

   The point E_2 is the exgyrocenter of gyrotriangle $A_1A_2A_3$ opposite to vertex A_2, as shown in Fig. 8.1.
3. Substituting (8.13c) into (8.1), the point E of (8.1) becomes E_3, $E = E_3$, where

$$E_3 = \frac{\gamma_{23}a_{23}\gamma_{A_1}A_1 + \gamma_{13}a_{13}\gamma_{A_2}A_2 - \gamma_{12}a_{12}\gamma_{A_3}A_3}{\gamma_{23}a_{23}\gamma_{A_1} + \gamma_{13}a_{13}\gamma_{A_2} - \gamma_{12}a_{12}\gamma_{A_3}}. \tag{8.14c}$$

   The point E_3 is the exgyrocenter of gyrotriangle $A_1A_2A_3$ opposite to vertex A_3, as shown into Fig. 8.1.
4. Substituting (8.13d) into (8.1), the point E of (8.1) becomes E_0, $E = E_0$, where

$$E_0 = \frac{\gamma_{23}a_{23}\gamma_{A_1}A_1 + \gamma_{13}a_{13}\gamma_{A_2}A_2 + \gamma_{12}a_{12}\gamma_{A_3}A_3}{\gamma_{23}a_{23}\gamma_{A_1} + \gamma_{13}a_{13}\gamma_{A_2} + \gamma_{12}a_{12}\gamma_{A_3}}. \tag{8.14d}$$

   The point E_0 is the ingyrocenter of gyrotriangle $A_1A_2A_3$, shown in Fig. 8.1.

The gyrobarycentric coordinates of E_0 with respect to the set $S = \{A_1, \dots, A_N\}$ in (8.14d) are all positive, so that m_0^2 in (8.3) is positive. Hence, by Corollary 4.9, p. 93, the point E_0 lies on the interior of gyrotriangle $A_1A_2A_3$. Accordingly, E_0 is the ingyrocenter of gyrotriangle $A_1A_2A_3$.

The gyrobarycentric coordinate representation (8.14d) of the ingyrocenter of gyrotriangle $A_1A_2A_3$ is obtained in (7.103), p. 178, by a different method. Here, in contrast, it is obtained as a byproduct of the exgyrocenters determination.

8.3 Existence of Gyrotriangle Exgyrocircles

According to Corollary 4.9, p. 93, a point with a gyrobarycentric coordinate representation exists if and only if its constant m_0 is real, that is, if and only if $m_0^2 > 0$. Let us, therefore, calculate the constant m_0^2 of the gyrobarycentric coordinate representation of each of the gyrotriangle exgyrocenters and its ingyrocenter.

Following (4.27), p. 90, of Definition 4.5, the constant m_0 of the gyrobarycentric coordinate representation of the exgyrocenter E in (8.1), p. 220, is given by the equation

$$m_0^2 = m_1^2 + m_2^2 + m_3^2 + 2m_1m_2\gamma_{12} + 2m_1m_3\gamma_{13} + 2m_2m_3\gamma_{23}\,. \tag{8.15}$$

1. Substituting the gyrobarycentric coordinates $(m_1 : m_2 : m_3)$ from (8.13a) into (8.15), we obtain the constant m_0^2 of exgyrocenter E_1 in (8.14a),

$$m_0^2 = (\gamma_{12}^2 - 1) + (\gamma_{13}^2 - 1) + (\gamma_{23}^2 - 1) + 2\gamma_{12}\gamma_{13}\gamma_{23}(a_{12}a_{13} - a_{12}a_{13} - a_{12}a_{13}). \tag{8.16a}$$

2. Substituting the gyrobarycentric coordinates $(m_1 : m_2 : m_3)$ from (8.13b) into (8.15), we obtain the constant m_0^2 of exgyrocenter E_2 in (8.14b),

$$m_0^2 = (\gamma_{12}^2 - 1) + (\gamma_{13}^2 - 1) + (\gamma_{23}^2 - 1) + 2\gamma_{12}\gamma_{13}\gamma_{23}(-a_{12}a_{13} + a_{12}a_{13} - a_{12}a_{13}). \tag{8.16b}$$

3. Substituting the gyrobarycentric coordinates $(m_1 : m_2 : m_3)$ from (8.13c) into (8.15), we obtain the constant m_0^2 of exgyrocenter E_3 in (8.14c),

$$m_0^2 = (\gamma_{12}^2 - 1) + (\gamma_{13}^2 - 1) + (\gamma_{23}^2 - 1) + 2\gamma_{12}\gamma_{13}\gamma_{23}(-a_{12}a_{13} - a_{12}a_{13} + a_{12}a_{13}). \tag{8.16c}$$

4. Substituting the gyrobarycentric coordinates $(m_1 : m_2 : m_3)$ from (8.13d) into (8.15), we obtain the constant m_0^2 of ingyrocenter E_0 in (8.14d),

$$m_0^2 = (\gamma_{12}^2 - 1) + (\gamma_{13}^2 - 1) + (\gamma_{23}^2 - 1) + 2\gamma_{12}\gamma_{13}\gamma_{23}(a_{12}a_{13} + a_{12}a_{13} + a_{12}a_{13}). \tag{8.16d}$$

Hence,

1. The exgyrocircle with exgyrocenter E_1 of gyrotriangle $A_1A_2A_3$, opposite to vertex A_1, Fig. 8.1, exists if and only if

$$(\gamma_{12}^2 - 1) + (\gamma_{13}^2 - 1) + (\gamma_{23}^2 - 1) > 2\gamma_{12}\gamma_{13}\gamma_{23}(-a_{12}a_{13} + a_{12}a_{13} + a_{12}a_{13}), \tag{8.17a}$$

as we see from (8.16a).

2. The exgyrocircle with exgyrocenter E_2 of gyrotriangle $A_1A_2A_3$, opposite to vertex A_2, Fig. 8.1, exists if and only if

$$(\gamma_{12}^2 - 1) + (\gamma_{13}^2 - 1) + (\gamma_{23}^2 - 1) > 2\gamma_{12}\gamma_{13}\gamma_{23}(a_{12}a_{13} - a_{12}a_{13} + a_{12}a_{13}), \tag{8.17b}$$

as we see from (8.16b).

3. The exgyrocircle with exgyrocenter E_3 of gyrotriangle $A_1A_2A_3$, opposite to vertex A_3, Fig. 8.1, exists if and only if

$$(\gamma_{12}^2 - 1) + (\gamma_{13}^2 - 1) + (\gamma_{23}^2 - 1) > 2\gamma_{12}\gamma_{13}\gamma_{23}(a_{12}a_{13} + a_{12}a_{13} - a_{12}a_{13}), \tag{8.17c}$$

as we see from (8.16c).

4. The ingyrocircle with ingyrocenter E_0 of gyrotriangle $A_1A_2A_3$, Fig. 8.1, exists if and only if

$$(\gamma_{12}^2-1)+(\gamma_{13}^2-1)+(\gamma_{23}^2-1)>2\gamma_{12}\gamma_{13}\gamma_{23}(-a_{12}a_{13}-a_{12}a_{13}-a_{12}a_{13}), \tag{8.17d}$$

as we see from (8.16d). But, Inequality (8.17d) is valid for any gyrotriangle $A_1A_2A_3$. Hence, the ingyrocenter E_0 exists for any gyrotriangle $A_1A_2A_3$, as expected.

8.4 Exgyroradius and Ingyroradius

Following (8.4a) and (1.9), p. 5, we have

$$r_{12}^2=s^2\frac{\gamma_{r_{12}}^2-1}{\gamma_{r_{12}}^2}. \tag{8.18}$$

1. Substituting, successively, (8.4a), (8.2), and (8.13a) into (8.18), r_{12} becomes r_1, $r_{12}=r_1$, where r_1 is given by the equation

$$r_1^2=\frac{s^2}{2}\frac{1+2\gamma_{12}\gamma_{13}\gamma_{23}-\gamma_{12}^2-\gamma_{13}^2-\gamma_{23}^2}{D_1}, \tag{8.19a}$$

where

$$D_1=\gamma_{12}\gamma_{13}\gamma_{23}-1-\gamma_{12}\sqrt{\gamma_{13}^2-1}\sqrt{\gamma_{23}^2-1}-\sqrt{\gamma_{12}^2-1}\gamma_{13}\sqrt{\gamma_{23}^2-1}$$
$$+\sqrt{\gamma_{12}^2-1}\sqrt{\gamma_{13}^2-1}\gamma_{23}.$$

2. Substituting, successively, (8.4a), (8.2), and (8.13b) into (8.18), r_{12} becomes r_2, $r_{12}=r_2$, where r_2 is given by the equation

$$r_2^2=\frac{s^2}{2}\frac{1+2\gamma_{12}\gamma_{13}\gamma_{23}-\gamma_{12}^2-\gamma_{13}^2-\gamma_{23}^2}{D_2}, \tag{8.19b}$$

where

$$D_2=\gamma_{12}\gamma_{13}\gamma_{23}-1-\gamma_{12}\sqrt{\gamma_{13}^2-1}\sqrt{\gamma_{23}^2-1}+\sqrt{\gamma_{12}^2-1}\gamma_{13}\sqrt{\gamma_{23}^2-1}$$
$$-\sqrt{\gamma_{12}^2-1}\sqrt{\gamma_{13}^2-1}\gamma_{23}.$$

3. Substituting, successively, (8.4a), (8.2), and (8.13c) into (8.18), r_{12} becomes r_3, $r_{12}=r_3$, where r_3 is given by the equation

$$r_3^2=\frac{s^2}{2}\frac{1+2\gamma_{12}\gamma_{13}\gamma_{23}-\gamma_{12}^2-\gamma_{13}^2-\gamma_{23}^2}{D_3}, \tag{8.19c}$$

where

$$D_3 = \gamma_{12}\gamma_{13}\gamma_{23} - 1 + \gamma_{12}\sqrt{\gamma_{13}^2 - 1}\sqrt{\gamma_{23}^2 - 1} - \sqrt{\gamma_{12}^2 - 1}\gamma_{13}\sqrt{\gamma_{23}^2 - 1}$$
$$- \sqrt{\gamma_{12}^2 - 1}\sqrt{\gamma_{13}^2 - 1}\gamma_{23}.$$

4. Substituting, successively, (8.4a), (8.2), and (8.13d) into (8.18), r_{12} becomes r_0, $r_{12} = r_0$, where r_0 is given by the equation

$$r_0^2 = \frac{s^2}{2}\frac{1 + 2\gamma_{12}\gamma_{13}\gamma_{23} - \gamma_{12}^2 - \gamma_{13}^2 - \gamma_{23}^2}{D_0}, \tag{8.19d}$$

where

$$D_0 = \gamma_{12}\gamma_{13}\gamma_{23} - 1 + \gamma_{12}\sqrt{\gamma_{13}^2 - 1}\sqrt{\gamma_{23}^2 - 1} + \sqrt{\gamma_{12}^2 - 1}\gamma_{13}\sqrt{\gamma_{23}^2 - 1}$$
$$+ \sqrt{\gamma_{12}^2 - 1}\sqrt{\gamma_{13}^2 - 1}\gamma_{23}.$$

Formalizing the results of Sects. 8.2–8.4, we have the following theorem:

Theorem 8.2 (The Gyrotriangle Ingyrocircle and Exgyrocircles Theorem) *Let $A_1A_2A_3$ be a gyrotriangle in an Einstein gyrovector space $(\mathbb{R}_s^n, \oplus, \otimes)$, let E_0 and r_0 be the gyrotriangle ingyrocenter and ingyroradius, and let E_k and r_k be the gyrotriangle exgyrocenter and exgyroradius of the exgyrocircle opposite to vertex A_k, $k = 1, 2, 3$.*

Then, in the standard gyrotriangle index notation, shown in Fig. 6.1, *p.* 128, *and in* (6.1), *p.* 127,

0. *Ingyrocenter E_0 of gyrotriangle $A_1A_2A_3$ always exists. Equivalently, ingyrocenter E_0 of gyrotriangle $A_1A_2A_3$ exists if and only if*

$$(\gamma_{12}^2 - 1) + (\gamma_{13}^2 - 1) + (\gamma_{23}^2 - 1) > 2\gamma_{12}\gamma_{13}\gamma_{23}(-a_{12}a_{13} - a_{12}a_{13} - a_{12}a_{13}).$$

1. *Exgyrocenter E_1 of gyrotriangle $A_1A_2A_3$ exists if and only if*

$$(\gamma_{12}^2 - 1) + (\gamma_{13}^2 - 1) + (\gamma_{23}^2 - 1) > 2\gamma_{12}\gamma_{13}\gamma_{23}(-a_{12}a_{13} + a_{12}a_{13} + a_{12}a_{13}).$$

2. *Exgyrocenter E_2 of gyrotriangle $A_1A_2A_3$ exists if and only if*

$$(\gamma_{12}^2 - 1) + (\gamma_{13}^2 - 1) + (\gamma_{23}^2 - 1) > 2\gamma_{12}\gamma_{13}\gamma_{23}(a_{12}a_{13} - a_{12}a_{13} + a_{12}a_{13}).$$

3. *Exgyrocenter E_3 of gyrotriangle $A_1A_2A_3$ exists if and only if*

$$(\gamma_{12}^2 - 1) + (\gamma_{13}^2 - 1) + (\gamma_{23}^2 - 1) > 2\gamma_{12}\gamma_{13}\gamma_{23}(a_{12}a_{13} + a_{12}a_{13} - a_{12}a_{13}).$$

The ingyrocenter E_0, and each exgyrocenter E_k, when exists, $k = 1, 2, 3$, are given by

$$
\begin{aligned}
E_0 &= \frac{\gamma_{23}a_{23}\gamma_{A_1}A_1 + \gamma_{13}a_{13}\gamma_{A_2}A_2 + \gamma_{12}a_{12}\gamma_{A_3}A_3}{\gamma_{23}a_{23}\gamma_{A_1} + \gamma_{13}a_{13}\gamma_{A_2} + \gamma_{12}a_{12}\gamma_{A_3}}, \\
E_1 &= \frac{-\gamma_{23}a_{23}\gamma_{A_1}A_1 + \gamma_{13}a_{13}\gamma_{A_2}A_2 + \gamma_{12}a_{12}\gamma_{A_3}A_3}{-\gamma_{23}a_{23}\gamma_{A_1} + \gamma_{13}a_{13}\gamma_{A_2} + \gamma_{12}a_{12}\gamma_{A_3}}, \\
E_2 &= \frac{\gamma_{23}a_{23}\gamma_{A_1}A_1 - \gamma_{13}a_{13}\gamma_{A_2}A_2 + \gamma_{12}a_{12}\gamma_{A_3}A_3}{\gamma_{23}a_{23}\gamma_{A_1} - \gamma_{13}a_{13}\gamma_{A_2} + \gamma_{12}a_{12}\gamma_{A_3}}, \\
E_3 &= \frac{\gamma_{23}a_{23}\gamma_{A_1}A_1 + \gamma_{13}a_{13}\gamma_{A_2}A_2 - \gamma_{12}a_{12}\gamma_{A_3}A_3}{\gamma_{23}a_{23}\gamma_{A_1} + \gamma_{13}a_{13}\gamma_{A_2} - \gamma_{12}a_{12}\gamma_{A_3}}.
\end{aligned}
\tag{8.20}
$$

The ingyroradius r_0, and each exgyroradius r_k that exists, $k = 1, 2, 3$, are given by

$$
r_k^2 = \frac{s^2}{2}\frac{1 + 2\gamma_{12}\gamma_{13}\gamma_{23} - \gamma_{12}^2 - \gamma_{13}^2 - \gamma_{23}^2}{\gamma_{12}\gamma_{13}\gamma_{23} - 1 + G_k}, \tag{8.21}
$$

$k = 0, 1, 2, 3$, *where*

$$
\begin{aligned}
G_0 &= \gamma_{12}\sqrt{\gamma_{13}^2 - 1}\sqrt{\gamma_{23}^2 - 1} + \sqrt{\gamma_{12}^2 - 1}\gamma_{13}\sqrt{\gamma_{23}^2 - 1} + \sqrt{\gamma_{12}^2 - 1}\sqrt{\gamma_{13}^2 - 1}\gamma_{23}, \\
G_1 &= -\gamma_{12}\sqrt{\gamma_{13}^2 - 1}\sqrt{\gamma_{23}^2 - 1} - \sqrt{\gamma_{12}^2 - 1}\gamma_{13}\sqrt{\gamma_{23}^2 - 1} + \sqrt{\gamma_{12}^2 - 1}\sqrt{\gamma_{13}^2 - 1}\gamma_{23}, \\
G_2 &= -\gamma_{12}\sqrt{\gamma_{13}^2 - 1}\sqrt{\gamma_{23}^2 - 1} + \sqrt{\gamma_{12}^2 - 1}\gamma_{13}\sqrt{\gamma_{23}^2 - 1} - \sqrt{\gamma_{12}^2 - 1}\sqrt{\gamma_{13}^2 - 1}\gamma_{23}, \\
G_3 &= \gamma_{12}\sqrt{\gamma_{13}^2 - 1}\sqrt{\gamma_{23}^2 - 1} - \sqrt{\gamma_{12}^2 - 1}\gamma_{13}\sqrt{\gamma_{23}^2 - 1} - \sqrt{\gamma_{12}^2 - 1}\sqrt{\gamma_{13}^2 - 1}\gamma_{23}.
\end{aligned}
\tag{8.22}
$$

Clearly,

$$
\sum_{k=0}^{3} G_k = 0, \tag{8.23}
$$

as we see from (8.22) of Theorem 8.2.

Hence, by (8.21), and by (7.148) and (7.152), p. 188, and by (7.35), p. 163,

$$\begin{aligned}
&\frac{1}{r_0^2}+\frac{1}{r_1^2}+\frac{1}{r_2^2}+\frac{1}{r_3^2}\\
&\quad =\frac{8}{s^2}\frac{\gamma_{12}\gamma_{13}\gamma_{23}-1}{1+2\gamma_{12}\gamma_{13}\gamma_{23}-\gamma_{12}^2-\gamma_{13}^2-\gamma_{23}^2}\\
&\quad =2\frac{1+\cos\alpha_1\cos\alpha_2\cos\alpha_3}{s^2F}\\
&\quad =2\frac{1+\cos\alpha_1\cos\alpha_2\cos\alpha_3}{s^2\cos\frac{\alpha_1+\alpha_2+\alpha_3}{2}\cos\frac{\alpha_1-\alpha_2-\alpha_3}{2}\cos\frac{-\alpha_1+\alpha_2-\alpha_3}{2}\cos\frac{-\alpha_1-\alpha_2+\alpha_3}{2}}\\
&\quad =2\frac{1+\cos\alpha_1\cos\alpha_2\cos\alpha_3}{R^2\cos^2\frac{\alpha_1-\alpha_2-\alpha_3}{2}\cos^2\frac{-\alpha_1+\alpha_2-\alpha_3}{2}\cos^2\frac{-\alpha_1-\alpha_2+\alpha_3}{2}}
\end{aligned}\tag{8.24}$$

where F is given by (7.144), p. 187.

8.5 In-Exgyroradii Relations

By substituting gyrotriangle gyrotrigonometric identities from Sect. 7.12, p. 187, into (8.21)–(8.22) and employing the result of Theorem 7.4, p. 163, we obtain gyrotrigonometric relations for the gyrotriangle ingyroradius r_0, exgyroradii r_k, $k=1,2,3$, and circumradius R, which are presented in the following theorem:

Theorem 8.3 (The In-Exgyroradii Theorem) *Let $A_1A_2A_3$ be a gyrotriangle in an Einstein gyrovector space $(\mathbb{R}_s^n,\oplus,\otimes)$ with ingyroradius r_0, exgyroradii r_k, $k=1,2,3$, and circumgyroradius R. Then in the gyrotriangle index notation in Fig. 6.1, p. 128,*

$$\begin{aligned}
r_0&=\frac{1}{2}R\frac{\cos\frac{\alpha_1-\alpha_2-\alpha_3}{2}}{\cos\frac{\alpha_1}{2}}\frac{\cos\frac{-\alpha_1+\alpha_2-\alpha_3}{2}}{\cos\frac{\alpha_2}{2}}\frac{\cos\frac{-\alpha_1-\alpha_2+\alpha_3}{2}}{\cos\frac{\alpha_3}{2}},\\
r_1&=\frac{1}{2}R\frac{\cos\frac{\alpha_1-\alpha_2-\alpha_3}{2}}{\cos\frac{\alpha_1}{2}}\frac{\cos\frac{-\alpha_1+\alpha_2-\alpha_3}{2}}{\sin\frac{\alpha_2}{2}}\frac{\cos\frac{-\alpha_1-\alpha_2+\alpha_3}{2}}{\sin\frac{\alpha_3}{2}},\\
r_2&=\frac{1}{2}R\frac{\cos\frac{\alpha_1-\alpha_2-\alpha_3}{2}}{\sin\frac{\alpha_1}{2}}\frac{\cos\frac{-\alpha_1+\alpha_2-\alpha_3}{2}}{\cos\frac{\alpha_2}{2}}\frac{\cos\frac{-\alpha_1-\alpha_2+\alpha_3}{2}}{\sin\frac{\alpha_3}{2}},\\
r_3&=\frac{1}{2}R\frac{\cos\frac{\alpha_1-\alpha_2-\alpha_3}{2}}{\sin\frac{\alpha_1}{2}}\frac{\cos\frac{-\alpha_1+\alpha_2-\alpha_3}{2}}{\sin\frac{\alpha_2}{2}}\frac{\cos\frac{-\alpha_1-\alpha_2+\alpha_3}{2}}{\cos\frac{\alpha_3}{2}},
\end{aligned}\tag{8.25}$$

and

$$\frac{1}{r_0} - \left\{\frac{1}{r_1} + \frac{1}{r_2} + \frac{1}{r_3}\right\} = \frac{2}{R} \frac{\cos\frac{\alpha_1+\alpha_2+\alpha_3}{2}}{\cos\frac{\alpha_1-\alpha_2-\alpha_3}{2}\cos\frac{-\alpha_1+\alpha_2-\alpha_3}{2}\cos\frac{-\alpha_1-\alpha_2+\alpha_3}{2}} \tag{8.26}$$

and

$$\begin{aligned} &r_1 + r_2 + r_3 - (4R + r_0) \\ &\quad = R\frac{\cos\frac{\alpha_1+\alpha_2+\alpha_3}{2}}{\sin\alpha_1 \sin\alpha_2 \sin\alpha_3}\left\{3\sin\frac{\alpha_1+\alpha_2+\alpha_3}{2} + \sin\frac{3\alpha_1-\alpha_2-\alpha_3}{2}\right. \\ &\quad\quad \left. + \sin\frac{-\alpha_1+3\alpha_2-\alpha_3}{2} + \sin\frac{-\alpha_1-\alpha_2+3\alpha_3}{2}\right\}. \end{aligned} \tag{8.27}$$

Proof The results of the theorem follow straightforwardly by substitutions of gyrotriangle gyrotrigonometric identities from Sect. 7.12, p. 187, into (8.21)–(8.22) and employing the result of Theorem 7.4, p. 163.

Equation (8.26) follows from (8.25) straightforwardly owing to the elegant trigonometric/gyrotrigonometric identity

$$\begin{aligned} &\cos\frac{\alpha_1}{2}\cos\frac{\alpha_2}{2}\cos\frac{\alpha_3}{2} - \cos\frac{\alpha_1}{2}\sin\frac{\alpha_2}{2}\sin\frac{\alpha_3}{2} \\ &\quad - \sin\frac{\alpha_1}{2}\cos\frac{\alpha_2}{2}\sin\frac{\alpha_3}{2} - \sin\frac{\alpha_1}{2}\sin\frac{\alpha_2}{2}\cos\frac{\alpha_3}{2} \\ &= \cos\frac{\alpha_1+\alpha_2+\alpha_3}{2}. \end{aligned} \tag{8.28}$$

Similarly, (8.27) follows from (8.25) straightforwardly; see Problem 8.6, p. 266. □

In the Euclidean limit, $s \to \infty$, gyrolengths and gyroangles tend to corresponding lengths and angles. Accordingly, gyroangle gyrotriangle sum tends to angle triangle sum, π, so that the gyrocosine function of a gyroangle,

$$\cos\frac{\alpha_1+\alpha_2+\alpha_3}{2}, \tag{8.29a}$$

tends to the corresponding cosine function of an angle,

$$\cos\frac{\pi}{2} = 0. \tag{8.29b}$$

Hence, the elegant relations (8.26)–(8.27) for gyrotriangles reduce in that limit to the familiar relations

$$\frac{1}{r_1} + \frac{1}{r_2} + \frac{1}{r_3} = \frac{1}{r_0} \quad \text{(Euclidean Geometry)} \tag{8.30}$$

and

$$r_1 + r_2 + r_3 = 4R + r_0 \quad \text{(Euclidean Geometry)} \tag{8.31}$$

for triangles in Euclidean geometry.

Equation (8.30) is found, for instance, in [9, p. 13], and (8.31) is found, for instance, in [8, p. 13].

The incircle and the three excircles, each touching all the three sides of their reference triangle, are called the four *tritangent circles* of the triangle.

8.6 In-Exradii Relations

Interestingly, the elegant relations (8.25) remain invariant in form under the Euclidean limit $s \to \infty$, so that they are valid in Euclidean geometry as well. However, for application in Euclidean geometry the relations (8.25) can be simplified owing to the fact that triangle angle sum in π.

Indeed, under the condition

$$\alpha_1 + \alpha + \alpha_3 = \pi, \tag{8.32a}$$

we have the following trigonometric identities for triangle angles:

$$\begin{aligned}
\cos\frac{\alpha_1 - \alpha - \alpha_3}{2} &= \sin\alpha_1 = 2\sin\frac{\alpha_1}{2}\cos\frac{\alpha_1}{2}, \\
\cos\frac{-\alpha_1 + \alpha_2 - \alpha_3}{2} &= \sin\alpha_2 = 2\sin\frac{\alpha_2}{2}\cos\frac{\alpha_2}{2}, \\
\cos\frac{-\alpha_1 - \alpha_2 + \alpha_3}{2} &= \sin\alpha_3 = 2\sin\frac{\alpha_3}{2}\cos\frac{\alpha_3}{2}.
\end{aligned} \tag{8.32b}$$

Substituting the simplifications offered by (8.32a), (8.32b) into (8.25) we obtain the following corollary of Theorem 8.3:

Corollary 8.4 *Let $A_1A_2A_3$ be a triangle with angles α_k, exradii r_k, $k = 1, 2, 3$, inradius r_0 and circumradius R in a Euclidean space $\mathbb{R}^n$. Then,*

$$\begin{aligned}
r_0 &= 4R\sin\frac{\alpha_1}{2}\sin\frac{\alpha_2}{2}\sin\frac{\alpha_3}{2}, \\
r_1 &= 4R\sin\frac{\alpha_1}{2}\cos\frac{\alpha_2}{2}\cos\frac{\alpha_3}{2}, \\
r_2 &= 4R\cos\frac{\alpha_1}{2}\sin\frac{\alpha_2}{2}\cos\frac{\alpha_3}{2}, \\
r_3 &= 4R\cos\frac{\alpha_1}{2}\cos\frac{\alpha_2}{2}\sin\frac{\alpha_3}{2},
\end{aligned} \tag{8.33}$$

$$r_1 + r_2 + r_3 = 4R + r_0, \tag{8.34}$$

and

$$\frac{1}{r_1}+\frac{1}{r_2}+\frac{1}{r_3}=\frac{1}{r_0}. \tag{8.35}$$

Proof Substituting the simplifications offered by (8.32a), (8.32b) into (8.25) we obtain (8.33).

Equation (8.34) follows from (8.33) straightforwardly by means of the elegant trigonometric/gyrotrigonometric identity

$$\begin{aligned}
&\sin\frac{\alpha_1}{2}\sin\frac{\alpha_2}{2}\sin\frac{\alpha_3}{2}-\sin\frac{\alpha_1}{2}\cos\frac{\alpha_2}{2}\cos\frac{\alpha_3}{2}\\
&\quad-\cos\frac{\alpha_1}{2}\sin\frac{\alpha_2}{2}\cos\frac{\alpha_3}{2}-\cos\frac{\alpha_1}{2}\cos\frac{\alpha_2}{2}\sin\frac{\alpha_3}{2}\\
&=\sin\frac{\alpha_1+\alpha_2+\alpha_3}{2},
\end{aligned} \tag{8.36}$$

and (8.35) is established in (8.30). □

8.7 In-Exgyrocenter Gyrotrigonometric Gyrobarycentric Representations

It is useful to express the gyrobarycentric coordinate representations of the ingyrocenter and the exgyrocenters of a gyrotriangle $A_1A_2A_3$ in an Einstein gyrovector space $(\mathbb{R}^n_s, \oplus, \otimes)$ gyrotrigonometrically. Being homogeneous, and following the law of gyrosines (6.44), p. 140, the gyrobarycentric coordinates in (8.13a), (8.13b), (8.13c), (8.13d), p. 224, are, respectively, equivalent to

$$\begin{aligned}
E_0:&\quad (m_1:m_2:m_3)=\left(\frac{\gamma_{23}a_{23}}{\gamma_{12}a_{12}}:\frac{\gamma_{13}a_{13}}{\gamma_{12}a_{12}}:1\right)=\left(\frac{\sin\alpha_1}{\sin\alpha_3}:\frac{\sin\alpha_2}{\sin\alpha_3}:1\right),\\
E_1:&\quad (m_1:m_2:m_3)=\left(-\frac{\gamma_{23}a_{23}}{\gamma_{12}a_{12}}:\frac{\gamma_{13}a_{13}}{\gamma_{12}a_{12}}:1\right)=\left(-\frac{\sin\alpha_1}{\sin\alpha_3}:\frac{\sin\alpha_2}{\sin\alpha_3}:1\right),\\
E_2:&\quad (m_1:m_2:m_3)=\left(\frac{\gamma_{23}a_{23}}{\gamma_{12}a_{12}}:-\frac{\gamma_{13}a_{13}}{\gamma_{12}a_{12}}:1\right)=\left(\frac{\sin\alpha_1}{\sin\alpha_3}:-\frac{\sin\alpha_2}{\sin\alpha_3}:1\right),\\
E_3:&\quad (m_1:m_2:m_3)=\left(\frac{\gamma_{23}a_{23}}{\gamma_{12}a_{12}}:\frac{\gamma_{13}a_{13}}{\gamma_{12}a_{12}}:-1\right)=\left(\frac{\sin\alpha_1}{\sin\alpha_3}:\frac{\sin\alpha_2}{\sin\alpha_3}:-1\right).
\end{aligned}$$

These, in turn, give rise to the following gyrotrigonometric gyrobarycentric coordinates for the exgyrocenters E_k, $k=1,2,3$, and the ingyrocenter E_0 of gyrotriangle $A_1A_2A_3$ in an Einstein gyrovector space $(\mathbb{R}^n_s, \oplus, \otimes)$:

$$E_0:\quad (m_1:m_2:m_3)=(\sin\alpha_1:\sin\alpha_2:\sin\alpha_3), \tag{8.37a}$$

$$E_1: \quad (m_1 : m_2 : m_3) = (-\sin\alpha_1 : \sin\alpha_2 : \sin\alpha_3), \tag{8.37b}$$

$$E_2: \quad (m_1 : m_2 : m_3) = (\sin\alpha_1 : -\sin\alpha_2 : \sin\alpha_3), \tag{8.37c}$$

$$E_3: \quad (m_1 : m_2 : m_3) = (\sin\alpha_1 : \sin\alpha_2 : -\sin\alpha_3). \tag{8.37d}$$

We have thus obtained the following theorem:

Theorem 8.5 (In-Exgyrocenters) *Let $S = \{A_1, A_2, A_3\}$ be a pointwise independent set of three points in an Einstein gyrovector space $(\mathbb{R}^n_s, \oplus, \otimes)$. The ingyrocenter E_0 and the exgyrocenters E_k, $k = 1, 2, 3$, of gyrotriangle $A_1A_2A_3$ have the following gyrotrigonometric gyrobarycentric coordinate representations*:

$$E_0 = \frac{\sin\alpha_1\gamma_{A_1}A_1 + \sin\alpha_2\gamma_{A_2}A_2 + \sin\alpha_3\gamma_{A_3}A_3}{\sin\alpha_1\gamma_{A_1} + \sin\alpha_2\gamma_{A_2} + \sin\alpha_3\gamma_{A_3}} \tag{8.38a}$$

$$E_1 = \frac{-\sin\alpha_1\gamma_{A_1}A_1 + \sin\alpha_2\gamma_{A_2}A_2 + \sin\alpha_3\gamma_{A_3}A_3}{-\sin\alpha_1\gamma_{A_1} + \sin\alpha_2\gamma_{A_2} + \sin\alpha_3\gamma_{A_3}}, \tag{8.38b}$$

$$E_2 = \frac{\sin\alpha_1\gamma_{A_1}A_1 - \sin\alpha_2\gamma_{A_2}A_2 + \sin\alpha_3\gamma_{A_3}A_3}{\sin\alpha_1\gamma_{A_1} - \sin\alpha_2\gamma_{A_2} + \sin\alpha_3\gamma_{A_3}}, \tag{8.38c}$$

$$E_3 = \frac{\sin\alpha_1\gamma_{A_1}A_1 + \sin\alpha_2\gamma_{A_2}A_2 - \sin\alpha_3\gamma_{A_3}A_3}{\sin\alpha_1\gamma_{A_1} + \sin\alpha_2\gamma_{A_2} - \sin\alpha_3\gamma_{A_3}}. \tag{8.38d}$$

Following (8.15), p. 226, and the gyrotrigonometric gyrobarycentric coordinates m_k, $k = 1, 2, 3$, in (8.37a), (8.37b), (8.37c), (8.37d), the constant m_0^2 of each of the gyrotrigonometric gyrobarycentric coordinate representations of E_k, $k = 1, 2, 3, 0$, in (8.38a), (8.38b), (8.38c), (8.38d) is, respectively,

$$\begin{aligned}
E_0: \quad m_0^2 &= \sin^2\alpha_1 + \sin^2\alpha_2 + \sin^2\alpha_3 \\
&\quad + 2(\cos\alpha_1\cos\alpha_2 + \cos\alpha_1\cos\alpha_3 + \cos\alpha_2\cos\alpha_3) \\
&\quad + 2(\cos\alpha_1 + \cos\alpha_2 + \cos\alpha_3) \\
&= 4\left\{\cos^2\frac{\alpha_1}{2} - \left(\cos\frac{\alpha_2}{2} - \cos\frac{\alpha_3}{2}\right)^2\right\} \\
&\quad \times \left\{-\cos^2\frac{\alpha_1}{2} + \left(\cos\frac{\alpha_2}{2} + \cos\frac{\alpha_3}{2}\right)^2\right\},
\end{aligned} \tag{8.39a}$$

$$\begin{aligned}
E_1: \quad m_0^2 &= \sin^2\alpha_1 + \sin^2\alpha_2 + \sin^2\alpha_3 \\
&\quad + 2(-\cos\alpha_1\cos\alpha_2 - \cos\alpha_1\cos\alpha_3 + \cos\alpha_2\cos\alpha_3) \\
&\quad + 2(\cos\alpha_1 - \cos\alpha_2 - \cos\alpha_3)
\end{aligned}$$

$$
\begin{aligned}
&= 4\left\{\cos^2\frac{\alpha_1}{2} - \left(\sin\frac{\alpha_2}{2} - \sin\frac{\alpha_3}{2}\right)^2\right\} \\
&\quad \times \left\{-\cos^2\frac{\alpha_1}{2} + \left(\sin\frac{\alpha_2}{2} + \sin\frac{\alpha_3}{2}\right)^2\right\},
\end{aligned}
\tag{8.39b}
$$

$$
\begin{aligned}
E_2:\quad m_0^2 &= \sin^2\alpha_1 + \sin^2\alpha_2 + \sin^2\alpha_3 \\
x \quad &\quad + 2(-\cos\alpha_1\cos\alpha_2 + \cos\alpha_1\cos\alpha_3 - \cos\alpha_2\cos\alpha_3) \\
&\quad + 2(-\cos\alpha_1 + \cos\alpha_2 - \cos\alpha_3) \\
&= 4\left\{\cos^2\frac{\alpha_2}{2} - \left(\sin\frac{\alpha_1}{2} - \sin\frac{\alpha_3}{2}\right)^2\right\} \\
&\quad \times \left\{-\cos^2\frac{\alpha_2}{2} + \left(\sin\frac{\alpha_1}{2} + \sin\frac{\alpha_3}{2}\right)^2\right\},
\end{aligned}
\tag{8.39c}
$$

$$
\begin{aligned}
E_3:\quad m_0^2 &= \sin^2\alpha_1 + \sin^2\alpha_2 + \sin^2\alpha_3 \\
&\quad + 2(\cos\alpha_1\cos\alpha_2 - \cos\alpha_1\cos\alpha_3 - \cos\alpha_2\cos\alpha_3) \\
&\quad + 2(-\cos\alpha_1 - \cos\alpha_2 + \cos\alpha_3) \\
&= 4\left\{\cos^2\frac{\alpha_3}{2} - \left(\sin\frac{\alpha_1}{2} - \sin\frac{\alpha_2}{2}\right)^2\right\} \\
&\quad \times \left\{-\cos^2\frac{\alpha_3}{2} + \left(\sin\frac{\alpha_1}{2} + \sin\frac{\alpha_2}{2}\right)^2\right\}.
\end{aligned}
\tag{8.39d}
$$

The ingyrocenter E_0 of any gyrotriangle $A_1A_2A_3$ exists. Hence, its constant m_0^2 must be positive. Hence, by the extreme right-hand side of the fourth equation in (8.39a), (8.39b), (8.39c), (8.39d), we have the inequality

$$
\left|\cos\frac{\alpha_2}{2} - \cos\frac{\alpha_3}{2}\right| < \cos\frac{\alpha_1}{2} < \cos\frac{\alpha_2}{2} + \cos\frac{\alpha_3}{2}
\tag{8.40}
$$

for the gyroangles α_k, $k = 1, 2, 3$, of any gyrotriangle $A_1A_2A_3$ in an Einstein gyrovector space.

It can be shown that in addition to (8.40), gyroangles α_k, $k = 1, 2, 3$, of any gyrotriangle $A_1A_2A_3$ satisfy the inequality

$$
\left|\sin\frac{\alpha_2}{2} - \sin\frac{\alpha_3}{2}\right| < \cos\frac{\alpha_1}{2}.
\tag{8.41}
$$

Inequality (8.41), along with the identities in (8.39a), (8.39b), (8.39c), (8.39d) for the constant m_0^2 of each of the gyrotriangle incenter and excenters E_k, $k = 0, 1, 2, 3$, imply the following existence conditions for E_k:

0. Ingyrocenter E_0 of gyrotriangle $A_1A_2A_3$ always exists. Equivalently, ingyrocenter E_0 of gyrotriangle $A_1A_2A_3$ exists if and only if

$$\cos\frac{\alpha_1}{2} < \cos\frac{\alpha_2}{2} + \cos\frac{\alpha_3}{2}. \tag{8.42}$$

1. Exgyrocenter E_1 of gyrotriangle $A_1A_2A_3$ exists if and only if

$$\cos\frac{\alpha_1}{2} < \sin\frac{\alpha_2}{2} + \sin\frac{\alpha_3}{2}. \tag{8.43}$$

2. Exgyrocenter E_2 of gyrotriangle $A_1A_2A_3$ exists if and only if

$$\cos\frac{\alpha_2}{2} < \sin\frac{\alpha_1}{2} + \sin\frac{\alpha_3}{2}. \tag{8.44}$$

3. Exgyrocenter E_3 of gyrotriangle $A_1A_2A_3$ exists if and only if

$$\cos\frac{\alpha_3}{2} < \sin\frac{\alpha_1}{2} + \sin\frac{\alpha_2}{2}. \tag{8.45}$$

8.8 In-Excenter Trigonometric Barycentric Representations

The gyrotrigonometric gyrobarycentric coordinates (8.37a), (8.37b), (8.37c) (8.37d) remain invariant in form under the Euclidean limit $s \to \infty$, resulting in the following corollary of Theorem 8.5:

Corollary 8.6 (In-Excenters) *Let $S = \{A_1, A_2, A_3\}$ be a pointwise independent set of three points in a Euclidean vector space $\mathbb{R}^n$. The incenter E_0 and the excenters E_k, $k = 1, 2, 3$, of triangle $A_1A_2A_3$ have the following trigonometric barycentric coordinate representations*:

$$E_0 = \frac{\sin\alpha_1 A_1 + \sin\alpha_2 A_2 + \sin\alpha_3 A_3}{\sin\alpha_1 + \sin\alpha_2 + \sin\alpha_3}, \tag{8.46a}$$

$$E_1 = \frac{-\sin\alpha_1 A_1 + \sin\alpha_2 A_2 + \sin\alpha_3 A_3}{-\sin\alpha_1 + \sin\alpha_2 + \sin\alpha_3}, \tag{8.46b}$$

$$E_2 = \frac{\sin\alpha_1 A_1 - \sin\alpha_2 A_2 + \sin\alpha_3 A_3}{\sin\alpha_1 - \sin\alpha_2 + \sin\alpha_3}, \tag{8.46c}$$

$$E_3 = \frac{\sin\alpha_1 A_1 + \sin\alpha_2 A_2 - \sin\alpha_3 A_3}{\sin\alpha_1 + \sin\alpha_2 - \sin\alpha_3}. \tag{8.46d}$$

8.9 Exgyrocircle Points of Tangency, Part I

The exgyrocircle points of tangency where the A_k-excircle of the gyrotriangle meets the opposite side of A_k, Fig. 8.4, are associated with the gyrotriangle gyrocenter called Nagel gyropoint, N_a, shown in Fig. 8.4, and studied in Sect. 8.12

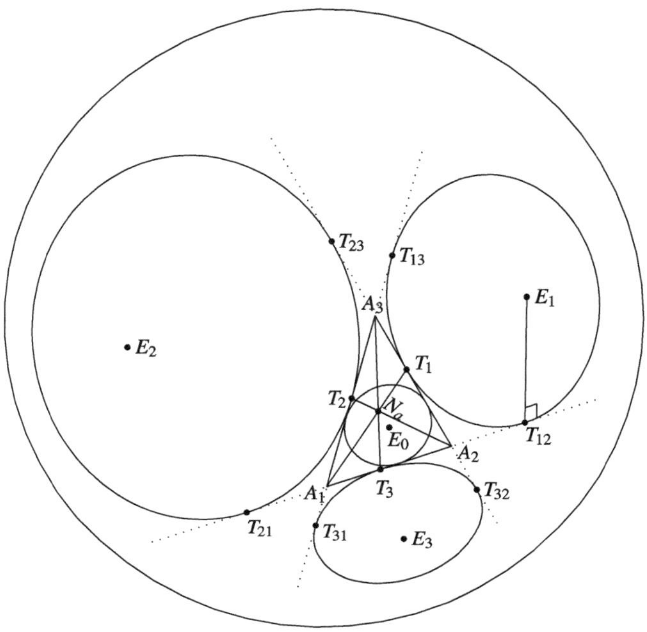

Fig. 8.4 Nagel Gyropoint, N_a, of a gyrotriangle in an Einstein gyrovector space. The point E_k, $k = 1, 2, 3$, is the A_k-exgyrocenter of gyrotriangle $A_1A_2A_3$ opposite to A_k, and the point T_k is the point in which the A_k-excircle of the gyrotriangle meets the gyrotriangle side opposite to A_k. The Nagel gyropoint of gyrotriangle $A_1A_2A_3$ is the point of concurrency of the three gyrolines A_kT_k, determined in (8.69), p. 245. The points of tangency T_1, etc., are given by Theorem 8.7, p. 239, and the points of tangency T_{12}, etc., are given by Theorem 8.12, p. 247

Let us consider the point of tangency T_3 where the A_3 excircle of a gyrotriangle $A_1A_2A_3$ meets the gyrotriangle side opposite to A_3, shown in Fig. 8.4. It is the perpendicular foot of the gyrotriangle exgyrocenter E_3 on the gyroline A_1A_2. Accordingly, T_3 is the gyroaltitude foot of gyrotriangle $A_1A_2E_3$, drawn from E_3, as shown in Fig. 8.4.

Hence, by Theorem 7.16, p. 183, the gyroaltitude foot T_3 possesses the gyrobarycentric coordinate representation

$$T_3 = \frac{m_1\gamma_{A_1}A_1 + m_2\gamma_{A_2}A_2}{m_1\gamma_{A_1} + m_2\gamma_{A_2}} \tag{8.47a}$$

with respect to the set $\{A_1, A_2\}$, with gyrobarycentric coordinates

$$\begin{aligned} m_1 &= \gamma_{\ominus A_1 \oplus A_2}\gamma_{\ominus A_2 \oplus E_3} - \gamma_{\ominus A_1 \oplus E_3} = \gamma_{12}\gamma_{\ominus A_2 \oplus E_3} - \gamma_{\ominus A_1 \oplus E_3}, \\ m_2 &= \gamma_{\ominus A_1 \oplus A_2}\gamma_{\ominus A_1 \oplus E_3} - \gamma_{\ominus A_2 \oplus E_3} = \gamma_{12}\gamma_{\ominus A_1 \oplus E_3} - \gamma_{\ominus A_2 \oplus E_3}. \end{aligned} \tag{8.47b}$$

The gyrobarycentric coordinates m_1 and m_2 in (8.47b) involve the gamma factors $\gamma_{\ominus A_1 \oplus E_3}$ and $\gamma_{\ominus A_2 \oplus E_3}$, which we calculate below.

Being the A_3-exgyrocenter of gyrotriangle $A_1A_2A_3$, E_3 is given by, (8.14c), p. 225,

$$E_3 = \frac{\gamma_{23}a_{23}\gamma_{A_1}A_1 + \gamma_{13}a_{13}\gamma_{A_2}A_2 - \gamma_{12}a_{12}\gamma_{A_3}A_3}{\gamma_{23}a_{23}\gamma_{A_1} + \gamma_{13}a_{13}\gamma_{A_2} - \gamma_{12}a_{12}\gamma_{A_3}}. \tag{8.48}$$

Hence, by Theorem 4.6, p. 90,

$$\gamma_{X\oplus E_3} = \frac{\gamma_{23}a_{23}\gamma_{X\oplus A_1} + \gamma_{13}a_{13}\gamma_{X\oplus A_2} - \gamma_{12}a_{12}\gamma_{X\oplus A_3}}{m_0} \tag{8.49}$$

for all $X \in \mathbb{R}_s^n$, where $m_0 > 0$ is the constant of the gyrobarycentric coordinate representation of E_3 in (8.48). This constant need not be specified as we will see below in the transition from (8.51) to (8.52).

Following (8.49) with $X = \ominus A_k$, $k = 1, 2, 3$, we have, respectively,

$$\begin{aligned}
\gamma_{\ominus A_1 \oplus E_3} &= \frac{\gamma_{23}a_{23} + \gamma_{13}a_{13}\gamma_{12} - \gamma_{12}a_{12}\gamma_{13}}{m_0}, \\
\gamma_{\ominus A_2 \oplus E_3} &= \frac{\gamma_{23}a_{23}\gamma_{12} + \gamma_{13}a_{13} - \gamma_{12}a_{12}\gamma_{23}}{m_0}, \\
\gamma_{\ominus A_3 \oplus E_3} &= \frac{\gamma_{23}a_{23}\gamma_{13} + \gamma_{13}a_{13}\gamma_{23} - \gamma_{12}a_{12}}{m_0}.
\end{aligned} \tag{8.50}$$

Substituting from (8.50) into (8.47b), we have

$$\begin{aligned}
m_1 &= \frac{(\gamma_{12}^2 - 1)\gamma_{23}a_{23} - (\gamma_{12}\gamma_{23} - \gamma_{13})\gamma_{12}a_{12}}{m_0}, \\
m_2 &= \frac{(\gamma_{12}^2 - 1)\gamma_{13}a_{13} - (\gamma_{12}\gamma_{13} - \gamma_{23})\gamma_{12}a_{12}}{m_0}.
\end{aligned} \tag{8.51}$$

Being homogeneous, a common nonzero factor of gyrobarycentric coordinates is irrelevant, so that convenient gyrobarycentric coordinates m_1 and m_2 of the point T_3 in (8.47a) are obtained from (8.51):

$$\begin{aligned}
m_1 &= \left(\gamma_{12}^2 - 1\right)\gamma_{23}a_{23} - (\gamma_{12}\gamma_{23} - \gamma_{13})\gamma_{12}a_{12}, \\
m_2 &= \left(\gamma_{12}^2 - 1\right)\gamma_{13}a_{13} - (\gamma_{12}\gamma_{13} - \gamma_{23})\gamma_{12}a_{12}.
\end{aligned} \tag{8.52}$$

Substituting from (7.143)–(7.147) into (8.52), along with the abbreviation $F = F(\alpha_1, \alpha_2, \alpha_3)$, we have

$$\begin{aligned} m_1 &= \frac{8F^{3/2}}{\sin^2\alpha_1 \sin^2\alpha_2 \sin\alpha_3} \frac{1-\cos\alpha_2}{\sin\alpha_2} = \frac{8F^{3/2}}{\sin^2\alpha_1 \sin^2\alpha_2 \sin\alpha_3} \tan\frac{\alpha_2}{2}, \\ m_2 &= \frac{8F^{3/2}}{\sin^2\alpha_1 \sin^2\alpha_2 \sin\alpha_3} \frac{1-\cos\alpha_1}{\sin\alpha_1} = \frac{8F^{3/2}}{\sin^2\alpha_1 \sin^2\alpha_2 \sin\alpha_3} \tan\frac{\alpha_1}{2}. \end{aligned} \tag{8.53}$$

Since gyrobarycentric coordinates are homogeneous, a nonzero common factor of a system of gyrobarycentric coordinates is irrelevant. Hence, it follows from (8.53) that convenient gyrobarycentric coordinates for the point T_3 in (8.47a) are

$$\begin{aligned} m_1 &= \tan\frac{\alpha_2}{2}, \\ m_2 &= \tan\frac{\alpha_1}{2} \end{aligned} \tag{8.54}$$

so that, by (8.47a), we have

$$T_3 = \frac{\tan\frac{\alpha_2}{2}\gamma_{A_1} A_1 + \tan\frac{\alpha_1}{2}\gamma_{A_2} A_2}{\tan\frac{\alpha_2}{2}\gamma_{A_1} + \tan\frac{\alpha_1}{2}\gamma_{A_2}}. \tag{8.55}$$

We have thus obtained the following theorem:

Theorem 8.7 *Let $A_1A_2A_3$ be a gyrotriangle in an Einstein gyrovector space $\mathbb{R}_s^n$ and let T_k, $k = 1, 2, 3$, be the point where the A_k-exgyrocircle of the gyrotriangle meets the opposite side of A_k, Fig.* 8.4. *Gyrotrigonometric gyrobarycentric coordinate representations of the points T_k are*:

$$\begin{aligned} T_1 &= \frac{\tan\frac{\alpha_3}{2}\gamma_{A_2} A_2 + \tan\frac{\alpha_2}{2}\gamma_{A_3} A_3}{\tan\frac{\alpha_3}{2}\gamma_{A_2} + \tan\frac{\alpha_2}{2}\gamma_{A_3}}, \\ T_2 &= \frac{\tan\frac{\alpha_3}{2}\gamma_{A_1} A_1 + \tan\frac{\alpha_1}{2}\gamma_{A_3} A_3}{\tan\frac{\alpha_3}{2}\gamma_{A_1} + \tan\frac{\alpha_1}{2}\gamma_{A_3}}, \\ T_3 &= \frac{\tan\frac{\alpha_2}{2}\gamma_{A_1} A_1 + \tan\frac{\alpha_1}{2}\gamma_{A_2} A_2}{\tan\frac{\alpha_2}{2}\gamma_{A_1} + \tan\frac{\alpha_1}{2}\gamma_{A_2}}. \end{aligned} \tag{8.56}$$

Proof The proof of the third equation in (8.56) is given by (8.47a)–(8.55). The proof of the first and the second equation in (8.56) is obtained from the first by vertex permutations. □

The three points T_k, $k = 1, 2, 3$, of Theorem 8.7 are shown in Fig. 8.4, where they are determined by the results of the Theorem. As Fig. 8.4 indicates, the three gyrolines A_1T_1, A_2T_2 and A_3T_3 are concurrent. We show in Sect. 8.12 that this is indeed the case, giving rise to the Nagel gyropoint.

8.10 Excircle Points of Tangency, Part I

The gyrotrigonometric gyrobarycentric coordinates in (8.56) remain invariant in form under the Euclidean limit $s \to \infty$, resulting in the following corollary of Theorem 8.7:

Corollary 8.8 *Let $A_1A_2A_3$ be a triangle in a Euclidean vector space $\mathbb{R}^n$ and let T_k, $k=1,2,3$, be the point where the A_k-excircle of the triangle meets the opposite side of A_k. Trigonometric barycentric coordinate representations of the points T_k are*:

$$\begin{aligned} T_1 &= \frac{\tan\frac{\alpha_3}{2}A_2 + \tan\frac{\alpha_2}{2}A_3}{\tan\frac{\alpha_3}{2} + \tan\frac{\alpha_2}{2}}, \\ T_2 &= \frac{\tan\frac{\alpha_3}{2}A_1 + \tan\frac{\alpha_1}{2}A_3}{\tan\frac{\alpha_3}{2} + \tan\frac{\alpha_1}{2}}, \\ T_3 &= \frac{\tan\frac{\alpha_2}{2}A_1 + \tan\frac{\alpha_1}{2}A_2}{\tan\frac{\alpha_2}{2} + \tan\frac{\alpha_1}{2}}. \end{aligned} \tag{8.57}$$

8.11 Left Gyrotranslated Exgyrocircles

Left gyrotranslating gyrotriangle $A_1A_2A_3$ by $\ominus A_1$, the gyrotriangle becomes gyrotriangle $O(\ominus A_1 \oplus A_2)(\ominus A_1 \oplus A_3)$, where $O = \ominus A_1 \oplus A_1$ is the arbitrarily selected origin of the Einstein gyrovector space $(\mathbb{R}^n_s, \oplus, \otimes)$, $O = (0, 0, \ldots, 0)$. The gyrotriangle exgyrocenters E_k and tangent points T_k, $k = 1, 2, 3$, are left gyrotranslated as well, becoming, respectively, $\ominus A_1 \oplus E_k$ and $\ominus A_1 \oplus T_k$.

Applying to (8.20), p. 229, the Gyrobarycentric Coordinate Representation Gyrocovariance Theorem 4.6, p. 90, we have from Identity (4.29a), p. 91, with $X = \ominus A_1$, using the standard gyrotriangle index notation, shown in Fig. 8.4, in Fig. 6.1, p. 128, and in (6.1), p. 127:

$$\begin{aligned} &\ominus A_1 \oplus E_1 \\ &= \ominus A_1 \oplus \frac{-\gamma_{23}a_{23}\gamma_{A_1}A_1 + \gamma_{13}a_{13}\gamma_{A_2}A_2 + \gamma_{12}a_{12}\gamma_{A_3}A_3}{-\gamma_{23}a_{23}\gamma_{A_1} + \gamma_{13}a_{13}\gamma_{A_2} + \gamma_{12}a_{12}\gamma_{A_3}} \\ &= \frac{-\gamma_{23}a_{23}\gamma_{\ominus A_1 \oplus A_1}(\ominus A_1 \oplus A_1) + \gamma_{13}a_{13}\gamma_{\ominus A_1 \oplus A_2}(\ominus A_1 \oplus A_2) + \gamma_{12}a_{12}\gamma_{\ominus A_1 \oplus A_3}(\ominus A_1 \oplus A_3)}{-\gamma_{23}a_{23}\gamma_{\ominus A_1 \oplus A_1} + \gamma_{13}a_{13}\gamma_{\ominus A_1 \oplus A_2} + \gamma_{12}a_{12}\gamma_{\ominus A_1 \oplus A_3}} \end{aligned}$$

$$= \frac{\gamma_{13}a_{13}\gamma_{12}\mathbf{a}_{12} + \gamma_{12}a_{12}\gamma_{13}\mathbf{a}_{13}}{-\gamma_{23}a_{23} + \gamma_{13}a_{13}\gamma_{12} + \gamma_{12}a_{12}\gamma_{13}}$$

$$= \frac{\sin\alpha_2\gamma_{12}\mathbf{a}_{12} + \sin\alpha_3\gamma_{13}\mathbf{a}_{13}}{-\sin\alpha_1 + \sin\alpha_2\gamma_{12} + \sin\alpha_3\gamma_{13}}. \tag{8.58a}$$

The last equation in (8.58a) is obtained by substitutions from (7.143).

Similarly,

$$\ominus A_1 \oplus E_2 = \ominus A_1 \oplus \frac{\gamma_{23}a_{23}\gamma_{A_1}A_1 - \gamma_{13}a_{13}\gamma_{A_2}A_2 + \gamma_{12}a_{12}\gamma_{A_3}A_3}{\gamma_{23}a_{23}\gamma_{A_1} - \gamma_{13}a_{13}\gamma_{A_2} + \gamma_{12}a_{12}\gamma_{A_3}}$$

$$= \frac{-\gamma_{13}a_{13}\gamma_{12}\mathbf{a}_{12} + \gamma_{12}a_{12}\gamma_{13}\mathbf{a}_{13}}{\gamma_{23}a_{23} - \gamma_{13}a_{13}\gamma_{12} + \gamma_{12}a_{12}\gamma_{13}}$$

$$= \frac{-\sin\alpha_2\gamma_{12}\mathbf{a}_{12} + \sin\alpha_3\gamma_{13}\mathbf{a}_{13}}{\sin\alpha_1 - \sin\alpha_2\gamma_{12} + \sin\alpha_3\gamma_{13}} \tag{8.58b}$$

and

$$\ominus A_1 \oplus E_3 = \ominus A_1 \oplus \frac{\gamma_{23}a_{23}\gamma_{A_1}A_1 + \gamma_{13}a_{13}\gamma_{A_2}A_2 - \gamma_{12}a_{12}\gamma_{A_3}A_3}{\gamma_{23}a_{23}\gamma_{A_1} + \gamma_{13}a_{13}\gamma_{A_2} - \gamma_{12}a_{12}\gamma_{A_3}}$$

$$= \frac{\gamma_{13}a_{13}\gamma_{12}\mathbf{a}_{12} - \gamma_{12}a_{12}\gamma_{13}\mathbf{a}_{13}}{\gamma_{23}a_{23} + \gamma_{13}a_{13}\gamma_{12} - \gamma_{12}a_{12}\gamma_{13}}$$

$$= \frac{\sin\alpha_2\gamma_{12}\mathbf{a}_{12} - \sin\alpha_3\gamma_{13}\mathbf{a}_{13}}{\sin\alpha_1 + \sin\alpha_2\gamma_{12} - \sin\alpha_3\gamma_{13}}. \tag{8.58c}$$

Note that, by Definition 4.5, p. 89, the set of points $S = \{A_1, A_2, A_3\}$ is pointwise independent in the Einstein gyrovector space $(\mathbb{R}^n_s, \oplus, \otimes)$. Hence, the two gyrovectors $\mathbf{a}_{12} = \ominus A_1 \oplus A_2$ and $\mathbf{a}_{13} = \ominus A_1 \oplus A_3$ in $\mathbb{R}^n_s \subset \mathbb{R}^n$ in (8.58a), (8.58b), (8.58c) are linearly independent in $\mathbb{R}^n$.

Similarly, under a left gyrotranslation by $\ominus A_1$, the tangent points T_k, $k = 1, 2, 3$, in Fig. 8.4 and in (8.56) become:

$$\ominus A_1 \oplus T_1 = \ominus A_1 \oplus \frac{\tan\frac{\alpha_3}{2}\gamma_{A_2}A_2 + \tan\frac{\alpha_2}{2}\gamma_{A_3}A_3}{\tan\frac{\alpha_3}{2}\gamma_{A_2} + \tan\frac{\alpha_2}{2}\gamma_{A_3}}$$

$$= \frac{\tan\frac{\alpha_3}{2}\gamma_{12}\mathbf{a}_{12} + \tan\frac{\alpha_2}{2}\gamma_{13}\mathbf{a}_{13}}{\tan\frac{\alpha_3}{2}\gamma_{12} + \tan\frac{\alpha_2}{2}\gamma_{13}}, \tag{8.59a}$$

$$\ominus A_1 \oplus T_2 = \ominus A_1 \oplus \frac{\tan\frac{\alpha_3}{2}\gamma_{A_1} A_1 + \tan\frac{\alpha_1}{2}\gamma_{A_3} A_3}{\tan\frac{\alpha_3}{2}\gamma_{A_1} + \tan\frac{\alpha_1}{2}\gamma_{A_3}}$$

$$= \frac{\tan\frac{\alpha_1}{2}\gamma_{13}\mathbf{a}_{13}}{\tan\frac{\alpha_3}{2} + \tan\frac{\alpha_1}{2}\gamma_{13}} \tag{8.59b}$$

and

$$\ominus A_1 \oplus T_3 = \ominus A_1 \oplus \frac{\tan\frac{\alpha_2}{2}\gamma_{A_1} A_1 + \tan\frac{\alpha_1}{2}\gamma_{A_2} A_2}{\tan\frac{\alpha_2}{2}\gamma_{A_1} + \tan\frac{\alpha_1}{2}\gamma_{A_2}}$$

$$= \frac{\tan\frac{\alpha_1}{2}\gamma_{12}\mathbf{a}_{12}}{\tan\frac{\alpha_2}{2} + \tan\frac{\alpha_1}{2}\gamma_{12}}. \tag{8.59c}$$

8.12 Nagel Gyropoint

Definition 8.9 (Nagel Gyropoint) Let $A_1A_2A_3$ be a gyrotriangle in an Einstein gyrovector space $(\mathbb{R}^n_s, \oplus, \otimes)$ and let T_k, $k = 1, 2, 3$, be the tangent point in which the A_k-excircle of the gyrotriangle meets the gyrotriangle side opposite to A_k, Fig. 8.4, p. 237. The gyrotriangle vertices A_k and the gyrotriangle points of tangency T_k form the three gyrolines A_kT_k that are concurrent. Owing to analogies with Euclidean geometry, this point of concurrency is called the Nagel gyropoint, N_a of the gyrotriangle.

Let the Nagel gyropoint N of gyrotriangle $A_1A_2A_3$ in an Einstein gyrovector space $(\mathbb{R}^n_s, \oplus, \otimes)$, Fig. 8.4, be given by its gyrobarycentric coordinate representation with respect to the set $S = \{A_1, A_2, A_3\}$ of the gyrotriangle,

$$N = \frac{m_1\gamma_{A_1} A_1 + m_2\gamma_{A_2} A_2 + m_3\gamma_{A_3} A_3}{m_1\gamma_{A_1} + m_2\gamma_{A_2} + m_3\gamma_{A_3}}, \tag{8.60}$$

where the gyrobarycentric coordinates $(m_1 : m_2 : m_3)$ of N in (8.60) are to be determined.

Left gyrotranslating gyrotriangle $A_1A_2A_3$ by $\ominus A_1$, the gyrotriangle becomes gyrotriangle $O(\ominus A_1 \oplus A_2)(\ominus A_1 \oplus A_3)$, where $O = \ominus A_1 \oplus A_1$ is the arbitrarily selected origin of the Einstein gyrovector space $(\mathbb{R}^n_s, \oplus, \otimes)$, $O = (0, 0, \ldots, 0)$. The gyrotriangle tangent points T_k, $k = 1, 2, 3$, are left gyrotranslated as well, becoming, respectively, $\ominus A_1 \oplus T_k$, which are given by (8.59a), (8.59b), (8.59c).

Similarly, the gyrotriangle Nagel gyropoint of the left gyrotranslated gyrotriangle becomes $P = \ominus A_1 \oplus N$, given by

$$
\begin{aligned}
P &= \ominus A_1 \oplus N \\
&= \frac{m_2 \gamma_{\ominus A_1 \oplus A_2} (\ominus A_1 \oplus A_2) + m_3 \gamma_{\ominus A_1 \oplus A_3} (\ominus A_1 \oplus A_3)}{m_1 + m_2 \gamma_{\ominus A_1 \oplus A_2} + m_3 \gamma_{\ominus A_1 \oplus A_3}} \\
&= \frac{m_2 \gamma_{12} \mathbf{a}_{12} + m_3 \gamma_{13} \mathbf{a}_{13}}{m_1 + m_2 \gamma_{12} + m_3 \gamma_{13}}.
\end{aligned}
\tag{8.61}
$$

1. The tangent point $\ominus A_1 \oplus T_1$ and the vertex $O = \ominus A_1 \oplus A_1 = (0, 0, \ldots, 0)$ of the left gyrotranslated gyrotriangle $O(\ominus A_1 \oplus A_2)(\ominus A_1 \oplus A_3)$ form the Euclidean line

$$
L_1 = (\ominus A_1 \oplus T_1) t_1 = \frac{\tan \frac{\alpha_3}{2} \gamma_{12} \mathbf{a}_{12} + \tan \frac{\alpha_2}{2} \gamma_{13} \mathbf{a}_{13}}{\tan \frac{\alpha_3}{2} \gamma_{12} + \tan \frac{\alpha_2}{2} \gamma_{13}} t_1 \tag{8.62a}
$$

   as we see from (8.59a), where $t_1 \in \mathbb{R}$ is the line parameter.

2. The tangent point $\ominus A_1 \oplus T_2$ and the vertex $\mathbf{a}_{12} = \ominus A_1 \oplus A_2$ of the left gyrotranslated gyrotriangle $O(\ominus A_1 \oplus A_2)(\ominus A_1 \oplus A_3)$ form the Euclidean line

$$
\begin{aligned}
L_2 &= (\ominus A_1 \oplus A_2) + \bigl(-(\ominus A_1 \oplus A_2) + (\ominus A_1 \oplus T_2)\bigr) t_2 \\
&= \mathbf{a}_{12} + \left(-\mathbf{a}_{12} + \frac{\tan \frac{\alpha_1}{2} \gamma_{13} \mathbf{a}_{13}}{\tan \frac{\alpha_3}{2} + \tan \frac{\alpha_1}{2} \gamma_{13}}\right) t_2
\end{aligned}
\tag{8.62b}
$$

   as we see from (8.59b), where $t_2 \in \mathbb{R}$ is the line parameter.

3. The tangent point $\ominus A_1 \oplus T_3$ and the vertex $\mathbf{a}_{13} = \ominus A_1 \oplus A_3$ of the left gyrotranslated gyrotriangle $O(\ominus A_1 \oplus A_2)(\ominus A_1 \oplus A_3)$ form the Euclidean line

$$
\begin{aligned}
L_3 &= (\ominus A_1 \oplus A_3) + \bigl(-(\ominus A_1 \oplus A_3) + (\ominus A_1 \oplus T_3)\bigr) t_3 \\
&= \mathbf{a}_{13} + \left(-\mathbf{a}_{13} + \frac{\tan \frac{\alpha_1}{2} \gamma_{12} \mathbf{a}_{12}}{\tan \frac{\alpha_2}{2} + \tan \frac{\alpha_1}{2} \gamma_{12}}\right) t_3
\end{aligned}
\tag{8.62c}
$$

   as we see from (8.59c), where $t_3 \in \mathbb{R}$ is the line parameter.

Since the point P lies on each of the three lines L_k, $k = 1, 2, 3$, there exist values $t_{k,0}$ of the line parameters t_k, $k = 1, 2, 3$, respectively, such that

$$
\begin{aligned}
P - \frac{\tan\frac{\alpha_3}{2}\gamma_{12}\mathbf{a}_{12} + \tan\frac{\alpha_2}{2}\gamma_{13}\mathbf{a}_{13}}{\tan\frac{\alpha_3}{2}\gamma_{12} + \tan\frac{\alpha_2}{2}\gamma_{13}} t_{1,0} &= 0, \\
P - \mathbf{a}_{12} - \left(-\mathbf{a}_{12} + \frac{\tan\frac{\alpha_1}{2}\gamma_{13}\mathbf{a}_{13}}{\tan\frac{\alpha_3}{2} + \tan\frac{\alpha_1}{2}\gamma_{13}}\right) t_{2,0} &= 0, \\
P - \mathbf{a}_{13} - \left(-\mathbf{a}_{13} + \frac{\tan\frac{\alpha_1}{2}\gamma_{12}\mathbf{a}_{12}}{\tan\frac{\alpha_2}{2} + \tan\frac{\alpha_1}{2}\gamma_{12}}\right) t_{3,0} &= 0,
\end{aligned}
\tag{8.63}
$$

where P is given by (8.61).

The system of equations (8.63) was obtained by methods of gyroalgebra, and will be solved below by a common method of linear algebra.

Substituting P from (8.61) into (8.63), and rewriting each equation in (8.63) as a linear combination of $\mathbf{a}_{12}$ and $\mathbf{a}_{13}$ equals zero, we obtain the following homogeneous linear system of three gyrovector equations

$$
\begin{aligned}
c_{11}\mathbf{a}_{12} + c_{12}\mathbf{a}_{13} &= \mathbf{0}, \\
c_{21}\mathbf{a}_{12} + c_{22}\mathbf{a}_{13} &= \mathbf{0}, \\
c_{31}\mathbf{a}_{12} + c_{32}\mathbf{a}_{13} &= \mathbf{0},
\end{aligned}
\tag{8.64}
$$

where each coefficient c_{ij}, $i = 1, 2, 3$, $j = 1, 2$, is a function of the gyrotriangle parameters $\gamma_{12}, \gamma_{13}, \gamma_{23}$ and α_k, and the six unknowns $t_{k,0}$ and m_k, $k = 1, 2, 3$.

Since the set $S = \{A_1, A_2, A_3\}$ is pointwise independent, the two gyrovectors $\mathbf{a}_{12} = \ominus A_1 \oplus A_2$ and $\mathbf{a}_{13} = \ominus A_1 \oplus A_3$ in $\mathbb{R}^n_s$, considered as vectors in $\mathbb{R}^n$, are linearly independent in $\mathbb{R}^n$. Hence, each coefficient c_{ij} in (8.64) equals zero. Accordingly, the three gyrovector equations in (8.64) are equivalent to the following six scalar equations,

$$
c_{11} = c_{12} = c_{21} = c_{22} = c_{31} = c_{32} = 0 \tag{8.65}
$$

for the six unknowns $t_{k,0}$ and m_k, $k = 1, 2, 3$.

An explicit presentation of the resulting system (8.65) reveals that it is slightly nonlinear. Like the system (7.221), p. 208, however, it is linear in the unknowns $t_{1,0}, t_{2,0}, t_{3,0}$. Solving three equations of the system for $t_{1,0}, t_{2,0}, t_{3,0}$, and substituting these into the remaining equations of the system we obtain a system that determines the ratios m_1/m_3 and m_2/m_3 uniquely, from which convenient (homogeneous) gyrobarycentric coordinates $(m_1 : m_2 : m_3)$ are obtained.

The unique determination of m_1/m_3 and m_2/m_3 turns out to be

$$\begin{aligned}\frac{m_1}{m_3} &= \frac{\tan\frac{\alpha_3}{2}}{\tan\frac{\alpha_1}{2}},\\ \frac{m_2}{m_3} &= \frac{\tan\frac{\alpha_3}{2}}{\tan\frac{\alpha_2}{2}}\end{aligned} \tag{8.66}$$

from which two convenient gyrobarycentric coordinates result. These are:

$$(m_1 : m_2 : m_3) = \left(\tan\frac{\alpha_2}{2}\tan\frac{\alpha_3}{2} : \tan\frac{\alpha_1}{2}\tan\frac{\alpha_3}{2} : \tan\frac{\alpha_1}{2}\tan\frac{\alpha_2}{2}\right) \tag{8.67}$$

and, equivalently,

$$(m_1 : m_2 : m_3) = \left(\cot\frac{\alpha_1}{2} : \cot\frac{\alpha_2}{2} : \cot\frac{\alpha_3}{2}\right). \tag{8.68}$$

Formalizing the main result of this section, we have the following theorem:

Theorem 8.10 (Nagel Gyropoint) *Let $A_1A_2A_3$ be a gyrotriangle in an Einstein gyrovector space $(\mathbb{R}^n_s, \oplus, \otimes)$. A gyrotrigonometric gyrobarycentric coordinate representation of the gyrotriangle Nagel gyropoint N_a, Fig.* 8.4, *p.* 237, *is given by*

$$N_a = \frac{\cot\frac{\alpha_1}{2}\gamma_{A_1}A_1 + \cot\frac{\alpha_2}{2}\gamma_{A_2}A_2 + \cot\frac{\alpha_3}{2}\gamma_{A_3}A_3}{\cot\frac{\alpha_1}{2}\gamma_{A_1} + \cot\frac{\alpha_2}{2}\gamma_{A_2} + \cot\frac{\alpha_3}{2}\gamma_{A_3}}. \tag{8.69}$$

Proof The proof follows immediately from (8.60) and (8.68). □

The gyrotrigonometric gyrobarycentric coordinates of Nagel gyropoint in (8.69) remain invariant in form under the Euclidean limit $s \to \infty$, resulting in the following corollary of Theorem 8.10:

Corollary 8.11 (Nagel Point) *Let $A_1A_2A_3$ be a triangle in a Euclidean vector space $\mathbb{R}^n$. A trigonometric barycentric coordinate representation of the triangle Nagel point N_a is given by*

$$N_a = \frac{\cot\frac{\alpha_1}{2}A_1 + \cot\frac{\alpha_2}{2}A_2 + \cot\frac{\alpha_3}{2}A_3}{\cot\frac{\alpha_1}{2} + \cot\frac{\alpha_2}{2} + \cot\frac{\alpha_3}{2}}. \tag{8.70}$$

The remarkable similarity in form between the trigonometric barycentric coordinate representation (8.70) of Nagel point and the gyrotrigonometric gyrobarycentric coordinate representation (8.69) of Nagel gyropoint demonstrates the way gyrotrigonometric gyrobarycentric coordinate representations capture analogies that Euclidean and hyperbolic geometry share.

8.13 Exgyrocircle Points of Tangency, Part II

Let us consider the point of tangency T_{12} where the A_1 excircle of a gyrotriangle $A_1A_2A_3$ meets the gyrotriangle side A_1A_2, shown in Fig. 8.4. It is the perpendicular foot of the gyrotriangle exgyrocenter E_1 on the gyroline A_1A_2. Accordingly, T_{12} is the gyroaltitude foot of gyrotriangle $A_1A_2E_1$, drawn from E_1, as shown in Fig. 8.4.

Hence, by Theorem 7.16, p. 183, the gyroaltitude foot T_{12} possesses the gyrobarycentric coordinate representation

$$T_{12} = \frac{m_1\gamma_{A_1}A_1 + m_2\gamma_{A_2}A_2}{m_1\gamma_{A_1} + m_2\gamma_{A_2}} \tag{8.71a}$$

with respect to the set $\{A_1, A_2\}$, with gyrobarycentric coordinates

$$\begin{aligned} m_1 &= \gamma_{\ominus A_1\oplus A_2}\gamma_{\ominus A_2\oplus E_1} - \gamma_{\ominus A_1\oplus E_1} = \gamma_{12}\gamma_{\ominus A_2\oplus E_1} - \gamma_{\ominus A_1\oplus E_1}, \\ m_2 &= \gamma_{\ominus A_1\oplus A_2}\gamma_{\ominus A_1\oplus E_1} - \gamma_{\ominus A_2\oplus E_1} = \gamma_{12}\gamma_{\ominus A_1\oplus E_1} - \gamma_{\ominus A_2\oplus E_1}. \end{aligned} \tag{8.71b}$$

The gyrobarycentric coordinates m_1 and m_2 in (8.47b) involve the gamma factors $\gamma_{\ominus A_1\oplus E_1}$ and $\gamma_{\ominus A_2\oplus E_1}$, which we calculate below.

Being the A_1-exgyrocenter of gyrotriangle $A_1A_2A_3$, E_1 is given by, (8.14a), p. 225,

$$E_1 = \frac{-\gamma_{23}a_{23}\gamma_{A_1}A_1 + \gamma_{13}a_{13}\gamma_{A_2}A_2 + \gamma_{12}a_{12}\gamma_{A_3}A_3}{-\gamma_{23}a_{23}\gamma_{A_1} + \gamma_{13}a_{13}\gamma_{A_2} + \gamma_{12}a_{12}\gamma_{A_3}}. \tag{8.72}$$

Hence, by Theorem 4.6, p. 90,

$$\gamma_{X\oplus E_1} = \frac{-\gamma_{23}a_{23}\gamma_{X\oplus A_1} + \gamma_{13}a_{13}\gamma_{X\oplus A_2} + \gamma_{12}a_{12}\gamma_{X\oplus A_3}}{m_0} \tag{8.73}$$

for all $X \in \mathbb{R}^n_s$, where $m_0 > 0$ is the constant of the gyrobarycentric coordinate representation of E_1 in (8.72). This constant need not be specified as we will see below in the transition from (8.75) to (8.76).

Following (8.73) with $X = \ominus A_k$, $k = 1, 2, 3$, we have, respectively,

$$\begin{aligned} \gamma_{\ominus A_1\oplus E_1} &= \frac{-\gamma_{23}a_{23} + \gamma_{13}a_{13}\gamma_{12} + \gamma_{12}a_{12}\gamma_{13}}{m_0}, \\ \gamma_{\ominus A_2\oplus E_1} &= \frac{-\gamma_{23}a_{23}\gamma_{12} + \gamma_{13}a_{13} + \gamma_{12}a_{12}\gamma_{23}}{m_0}, \\ \gamma_{\ominus A_3\oplus E_1} &= \frac{-\gamma_{23}a_{23}\gamma_{13} + \gamma_{13}a_{13}\gamma_{23} + \gamma_{12}a_{12}}{m_0}. \end{aligned} \tag{8.74}$$

Substituting from (8.74) into (8.71b), we have

$$
\begin{aligned}
m_1 &= \frac{-(\gamma_{12}^2-1)\gamma_{23}a_{23}+(\gamma_{12}\gamma_{23}-\gamma_{13})\gamma_{12}a_{12}}{m_0}, \\
m_2 &= \frac{(\gamma_{12}^2-1)\gamma_{13}a_{13}+(\gamma_{12}\gamma_{13}-\gamma_{23})\gamma_{12}a_{12}}{m_0}.
\end{aligned}
\tag{8.75}
$$

Being homogeneous, a common nonzero factor of gyrobarycentric coordinates is irrelevant, so that convenient gyrobarycentric coordinates m_1 and m_2 of the point T_3 in (8.71a) are obtained from (8.75):

$$
\begin{aligned}
m_1 &= -(\gamma_{12}^2-1)\gamma_{23}a_{23}+(\gamma_{12}\gamma_{23}-\gamma_{13})\gamma_{12}a_{12}, \\
m_2 &= (\gamma_{12}^2-1)\gamma_{13}a_{13}+(\gamma_{12}\gamma_{13}-\gamma_{23})\gamma_{12}a_{12}.
\end{aligned}
\tag{8.76}
$$

Substituting from (7.143)–(7.147), p. 187, into (8.76), along with the abbreviation $F=F(\alpha_1,\alpha_2,\alpha_3)$, we have

$$
\begin{aligned}
m_1 &= \frac{8F^{3/2}}{\sin^2\alpha_1\sin^2\alpha_2\sin\alpha_3}\frac{\cos\alpha_2-1}{\sin\alpha_2} = -\frac{8F^{3/2}}{\sin^2\alpha_1\sin^2\alpha_2\sin\alpha_3}\tan\frac{\alpha_2}{2}, \\
m_2 &= \frac{8F^{3/2}}{\sin^2\alpha_1\sin^2\alpha_2\sin\alpha_3}\frac{\cos\alpha_1+1}{\sin\alpha_1} = \frac{8F^{3/2}}{\sin^2\alpha_1\sin^2\alpha_2\sin\alpha_3}\cot\frac{\alpha_1}{2}.
\end{aligned}
\tag{8.77}
$$

Since gyrobarycentric coordinates are homogeneous, a nonzero common factor of a system of gyrobarycentric coordinates is irrelevant. Hence, it follows from (8.77) that convenient gyrobarycentric coordinates for the point T_{12} in (8.71a) are

$$
\begin{aligned}
m_1 &= \tan\frac{\alpha_2}{2}, \\
m_2 &= -\cot\frac{\alpha_1}{2}
\end{aligned}
\tag{8.78}
$$

so that, by (8.71a), we have

$$
T_{12} = \frac{\tan\frac{\alpha_2}{2}\gamma_{A_1}A_1 - \cot\frac{\alpha_1}{2}\gamma_{A_2}A_2}{\tan\frac{\alpha_2}{2}\gamma_{A_1} - \cot\frac{\alpha_1}{2}\gamma_{A_2}}.
\tag{8.79}
$$

We have thus obtained the following theorem:

Theorem 8.12 (The Exgyrocircle Points of Tangency Theorem) *Let $A_1A_2A_3$ be a gyrotriangle in an Einstein gyrovector space $(\mathbb{R}_s^n,\oplus,\otimes)$, let T_{ij} be the point of tangency where the A_i-exgyrocircle meets the extension of the gyrotriangle side A_iA_j, and let T_i be the point where the A_i-exgyrocircle of the gyrotriangle meets the opposite side of A_i, Fig. 8.4, p. 237.*

Then, the gyrotrigonometric gyrobarycentric coordinate representations of the points of tangency on the A_k-exgyrocircle, $k=1,2,3$, with respect to the set $S=\{A_1,A_2,A_3\}$ are given by the equations listed below.

For $k=1$,

$$T_{12}=\frac{\tan\frac{\alpha_2}{2}\gamma_{A_1}A_1-\cot\frac{\alpha_1}{2}\gamma_{A_2}A_2}{\tan\frac{\alpha_2}{2}\gamma_{A_1}-\cot\frac{\alpha_1}{2}\gamma_{A_2}},$$

$$T_{13}=\frac{\tan\frac{\alpha_3}{2}\gamma_{A_1}A_1-\cot\frac{\alpha_1}{2}\gamma_{A_3}A_3}{\tan\frac{\alpha_3}{2}\gamma_{A_1}-\cot\frac{\alpha_1}{2}\gamma_{A_3}}, \tag{8.80a}$$

$$T_1=\frac{\tan\frac{\alpha_3}{2}\gamma_{A_2}A_2+\tan\frac{\alpha_2}{2}\gamma_{A_3}A_3}{\tan\frac{\alpha_3}{2}\gamma_{A_2}+\tan\frac{\alpha_2}{2}\gamma_{A_3}}.$$

For $k=2$,

$$T_{21}=\frac{-\cot\frac{\alpha_2}{2}\gamma_{A_1}A_1+\tan\frac{\alpha_1}{2}\gamma_{A_2}A_2}{-\cot\frac{\alpha_2}{2}\gamma_{A_1}+\tan\frac{\alpha_1}{2}\gamma_{A_2}},$$

$$T_{23}=\frac{\tan\frac{\alpha_3}{2}\gamma_{A_2}A_2-\cot\frac{\alpha_2}{2}\gamma_{A_3}A_3}{\tan\frac{\alpha_3}{2}\gamma_{A_2}-\cot\frac{\alpha_2}{2}\gamma_{A_3}}, \tag{8.80b}$$

$$T_2=\frac{\tan\frac{\alpha_3}{2}\gamma_{A_1}A_1+\tan\frac{\alpha_1}{2}\gamma_{A_3}A_3}{\tan\frac{\alpha_3}{2}\gamma_{A_1}+\tan\frac{\alpha_1}{2}\gamma_{A_3}}.$$

And for $k=3$,

$$T_{31}=\frac{-\cot\frac{\alpha_3}{2}\gamma_{A_1}A_1+\tan\frac{\alpha_1}{2}\gamma_{A_3}A_3}{-\cot\frac{\alpha_3}{2}\gamma_{A_1}+\tan\frac{\alpha_1}{2}\gamma_{A_3}},$$

$$T_{32}=\frac{-\cot\frac{\alpha_3}{2}\gamma_{A_2}A_2+\tan\frac{\alpha_2}{2}\gamma_{A_3}A_3}{-\cot\frac{\alpha_3}{2}\gamma_{A_2}+\tan\frac{\alpha_2}{2}\gamma_{A_3}}, \tag{8.80c}$$

$$T_3=\frac{\tan\frac{\alpha_2}{2}\gamma_{A_1}A_1+\tan\frac{\alpha_1}{2}\gamma_{A_2}A_2}{\tan\frac{\alpha_2}{2}\gamma_{A_1}+\tan\frac{\alpha_1}{2}\gamma_{A_2}}.$$

Proof The proof of the first equation in (8.80a) is established in (8.79). The proof of the second equation in (8.80a) is obtained from the first by interchanging vertices A_1 and A_2. The third equation in (8.80a) is the result (8.56) of Theorem 8.7, p. 239. Finally, (8.80b) and (8.80c) are obtained from (8.80a) by cyclic permutations of the gyrotriangle vertices. □

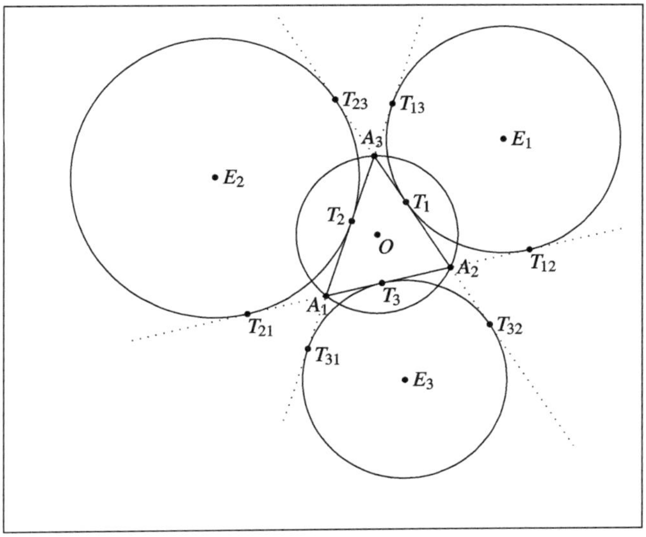

Fig. 8.5 Excircle points of tangency. The points of tangency T_1, etc., are given by Corollary 8.8, p. 240, of Theorem 8.7, and the points of tangency T_{12}, etc., are given by Corollary 8.13, p. 249, of Theorem 8.12. The triangle circumcenter is O, so that its circumradius is $R = \| - A_k + O\|$, $k = 1, 2, 3$

8.14 Excircle Points of Tangency, Part II

The way gyrotrigonometric gyrobarycentric coordinate representations in Einstein gyrovector spaces capture analogies between Euclidean and hyperbolic geometry, shown in Fig. 8.4, p. 237, and in Fig. 8.5, is now demonstrated.

The gyrotrigonometric gyrobarycentric coordinates of tangency points in Theorem 8.12 remain invariant in form under the Euclidean limit $s \to \infty$, resulting in the following corollary of Theorem 8.12:

Corollary 8.13 (The Excircle Points of Tangency Theorem) *Let $A_1A_2A_3$ be a triangle in a Euclidean vector space $\mathbb{R}^n$, let T_{ij} be the point of tangency where the A_i-excircle meets the extension of the triangle side A_iA_j, and let T_i be the point where the A_i-excircle of the triangle meets the opposite side of A_i.* 8.4, *p.* 237, *as shown in Fig.* 8.5.

Then, trigonometric barycentric coordinate representations of points of tangency on the A_k-excircle, $k = 1, 2, 3$, are as listed below.

For $k = 1$,

$$\begin{aligned} T_{12} &= \frac{\tan\frac{\alpha_2}{2}A_1 - \cot\frac{\alpha_1}{2}A_2}{\tan\frac{\alpha_2}{2} - \cot\frac{\alpha_1}{2}}, \\ T_{13} &= \frac{\tan\frac{\alpha_3}{2}A_1 - \cot\frac{\alpha_1}{2}A_3}{\tan\frac{\alpha_3}{2} - \cot\frac{\alpha_1}{2}}, \\ T_1 &= \frac{\tan\frac{\alpha_3}{2}A_2 + \tan\frac{\alpha_2}{2}A_3}{\tan\frac{\alpha_3}{2} + \tan\frac{\alpha_2}{2}}. \end{aligned} \tag{8.81a}$$

For $k = 2$,

$$\begin{aligned} T_{21} &= \frac{-\cot\frac{\alpha_2}{2}A_1 + \tan\frac{\alpha_1}{2}A_2}{-\cot\frac{\alpha_2}{2} + \tan\frac{\alpha_1}{2}}, \\ T_{23} &= \frac{\tan\frac{\alpha_3}{2}A_2 - \cot\frac{\alpha_2}{2}A_3}{\tan\frac{\alpha_3}{2} - \cot\frac{\alpha_2}{2}}, \\ T_2 &= \frac{\tan\frac{\alpha_3}{2}A_1 + \tan\frac{\alpha_1}{2}A_3}{\tan\frac{\alpha_3}{2} + \tan\frac{\alpha_1}{2}}. \end{aligned} \tag{8.81b}$$

And for $k = 3$,

$$\begin{aligned} T_{31} &= \frac{-\cot\frac{\alpha_3}{2}A_1 + \tan\frac{\alpha_1}{2}A_3}{-\cot\frac{\alpha_3}{2} + \tan\frac{\alpha_1}{2}}, \\ T_{32} &= \frac{-\cot\frac{\alpha_3}{2}A_2 + \tan\frac{\alpha_2}{2}A_3}{-\cot\frac{\alpha_3}{2} + \tan\frac{\alpha_2}{2}}, \\ T_3 &= \frac{\tan\frac{\alpha_2}{2}A_1 + \tan\frac{\alpha_1}{2}A_2}{\tan\frac{\alpha_2}{2} + \tan\frac{\alpha_1}{2}}. \end{aligned} \tag{8.81c}$$

8.15 Gyrodistance Between Gyrotriangle Tangency Points

In subsections of this section, we determine the gyrodistances between various tangency points of exgyrocircles. These, in turn, will be used to calculate the measures of some gyroangles that exgyrocircles generate.

8.15.1 The Gyrodistance Between T_{12} and T_{13}

Let $A_1A_2A_3$ be a gyrotriangle in an Einstein gyrovector space $(\mathbb{R}_s^n, \oplus, \otimes)$, and let T_{12} and T_{13} be the tangency points where the gyrotriangle A_1-exgyrocircle

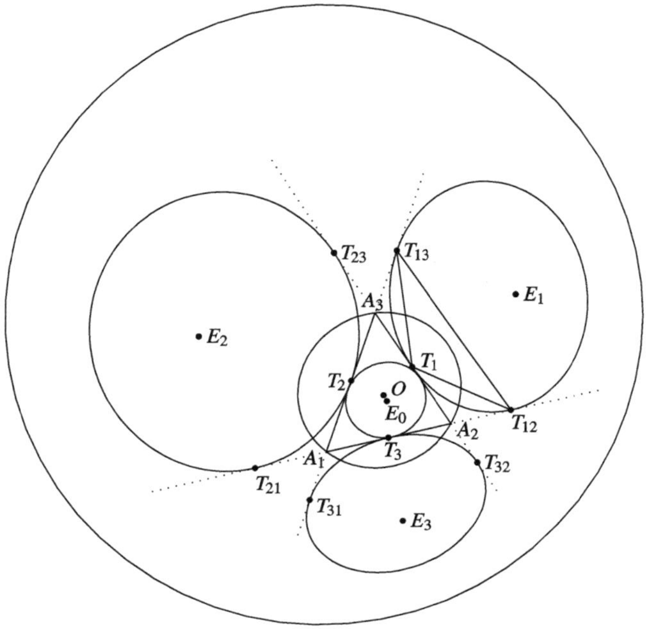

Fig. 8.6 Exgyrocircle points of tangency. The points of tangency T_1, etc., are given by Theorem 8.7, p. 239, and the points of tangency T_{12}, etc., are given by Theorem 8.12, p. 247. The gyroangle $\beta_1 = \angle T_{12}T_1T_{13}$ that the tangency points of the A_1-exgyrocircle generates is determined in terms of the gyroangles α_k, $k = 1, 2, 3$, of the reference gyrotriangle $A_1A_2A_3$ in Sect. 8.16, p. 258

meets the extensions of sides A_1A_2 and A_1A_2, as shown in Fig. 8.6. Their gyrobarycentric coordinate representations with respect to the pointwise independent set $S = \{A_1, A_2, A_3\}$, as given by Theorem 8.12, are:

$$T_{12} = \frac{m'_1\gamma_{A_1}A_1 + m'_2\gamma_{A_2}A_2}{m'_1\gamma_{A_1} + m'_2\gamma_{A_2}}, \tag{8.82a}$$

where the gyrobarycentric coordinates m'_k, $k = 1, 2$, are given by

$$\begin{aligned} m'_1 &= \tan\frac{\alpha_2}{2}, \\ m'_2 &= -\cot\frac{\alpha_1}{2}, \end{aligned} \tag{8.82b}$$

and

$$T_{13} = \frac{m_1\gamma_{A_1}A_1 + m_3\gamma_{A_3}A_3}{m_1\gamma_{A_1} + m_3\gamma_{A_3}}, \tag{8.83a}$$

where the gyrobarycentric coordinates m_k, $k = 1, 3$, are given by

$$\begin{aligned} m_1 &= \tan\frac{\alpha_3}{2}, \\ m_3 &= -\cot\frac{\alpha_1}{2}. \end{aligned} \tag{8.83b}$$

Hence, by (4.121), p. 113,

$$\gamma_{\ominus T_{12}\oplus T_{13}} = \frac{1}{m_0 m_0'}\left\{m_1 m_2' \gamma_{12} + m_1' m_3 \gamma_{13} + m_2' m_3 \gamma_{23} + m_1 m_1'\right\}, \tag{8.84}$$

where, by (4.118b) and (4.119b), p. 112, $m_0 > 0$ and $m_0' > 0$ are given by

$$\begin{aligned} m_0^2 &= m_1^2 + m_3^2 + 2m_1 m_3 \gamma_{13}, \\ (m_0')^2 &= (m_1')^2 + (m_2')^2 + 2m_1' m_2' \gamma_{12}. \end{aligned} \tag{8.85}$$

Substituting (8.82b) and (8.83b) into (8.84) (and squaring, but subsequently taking a square root), we obtain the gamma factor $\gamma_{\ominus T_{12}\oplus T_{13}}$ expressed in terms of the gyrotriangle gyroangles α_k, $k = 1, 2, 3$,

$$\begin{aligned} \gamma_{\ominus T_{12}\oplus T_{13}} = \frac{1}{16}\frac{1}{\sin^2\frac{\alpha_2}{2}\sin^2\frac{\alpha_3}{2}} &\left\{2\cos^2\alpha_1 + 2(\cos\alpha_2 + \cos\alpha_3 - 1)^2\right. \\ &+ \cos(\alpha_1 + \alpha_2 + \alpha_3) + \cos(-\alpha_1 + \alpha_2 + \alpha_3) \\ &\left.+ \cos(\alpha_1 - \alpha_2 + \alpha_3) + \cos(\alpha_1 + \alpha_2 - \alpha_3)\right\}. \end{aligned} \tag{8.86}$$

Substituting (8.86) into the identity, (1.9), p. 5,

$$\|\ominus T_{12}\oplus T_{13}\|^2 = s^2 \frac{\gamma^2_{\ominus T_{12}\oplus T_{13}} - 1}{\gamma^2_{\ominus T_{12}\oplus T_{13}}}, \tag{8.87}$$

we obtain the desired gyrodistance,

$$\begin{aligned} \|\ominus T_{12}\oplus T_{13}\|^2 = &\frac{8s^2 \cos\frac{\alpha_1+\alpha_2+\alpha_3}{2}\cos\frac{-\alpha_1+\alpha_2+\alpha_3}{2}\cos\frac{\alpha_1-\alpha_2+\alpha_3}{2}\cos\frac{\alpha_1+\alpha_2-\alpha_3}{2}}{\{2(1-\cos\alpha_2-\cos\alpha_3)^2 + 2\cos^2\alpha_1 + E(\alpha_1,\alpha_2,\alpha_3)\}^2} \\ &\times \left\{2 + 4(1-\cos\alpha_2-\cos\alpha_3)^2 + 2\left(\cos^2\alpha_1\right.\right. \\ &\left.\left.- \cos^2\alpha_2 - \cos^2\alpha_3\right) + E(\alpha_1,\alpha_2,\alpha_3)\right\}, \end{aligned} \tag{8.88}$$

where

$$\begin{aligned} E(\alpha_1,\alpha_2,\alpha_3) = &\cos(\alpha_1 + \alpha_2 + \alpha_3) + \cos(-\alpha_1 + \alpha_2 + \alpha_3) \\ &+ \cos(\alpha_1 - \alpha_2 + \alpha_3) + \cos(\alpha_1 + \alpha_2 - \alpha_3). \end{aligned} \tag{8.89}$$

Eliminating the factor $s^2 \cos \frac{\alpha_1+\alpha_2+\alpha_3}{2}$ between (8.88) and (7.35), p. 163, and taking a square root of both sides of the resulting equation, we obtain the following equation:

$$\begin{aligned}
&\|\ominus T_{12}\oplus T_{13}\| \\
&\quad = \frac{2\sqrt{2}R\cos\frac{-\alpha_1+\alpha_2+\alpha_3}{2}\cos\frac{\alpha_1-\alpha_2+\alpha_3}{2}\cos\frac{\alpha_1+\alpha_2-\alpha_3}{2}}{2(\cos\alpha_2+\cos\alpha_3-1)^2+2\cos^2\alpha_1+E(\alpha_1,\alpha_2,\alpha_3)} \\
&\qquad \times\sqrt{2+4(1-\cos\alpha_2-\cos\alpha_3)^2+2(\cos^2\alpha_1-\cos^2\alpha_2-\cos^2\alpha_3)+E(\alpha_1,\alpha_2,\alpha_3)},
\end{aligned} \tag{8.90}$$

where R is the circumgyroradius of the reference gyrotriangle $A_1A_2A_3$.

8.15.2 The Gyrodistance Between T_1 and T_{12}, T_{13}

As in Sect. 8.15.1, we calculate here the gyrodistance between gyrotriangle tangency points.

Following (8.80a), p. 248, we have

$$T_{12} = \frac{m'_1\gamma_{A_1}A_1 + m'_2\gamma_{A_2}A_2}{m'_1\gamma_{A_1} + m'_2\gamma_{A_2}}, \tag{8.91a}$$

where the gyrobarycentric coordinates m'_k, $k = 1, 2$, are given by

$$\begin{aligned}
m'_1 &= \tan\frac{\alpha_2}{2}, \\
m'_2 &= -\cot\frac{\alpha_1}{2},
\end{aligned} \tag{8.91b}$$

and

$$T_1 = \frac{m_2\gamma_{A_2}A_2 + m_3\gamma_{A_3}A_3}{m_2\gamma_{A_2} + m_3\gamma_{A_3}}, \tag{8.92a}$$

where the gyrobarycentric coordinates m_k, $k = 2, 3$, are given by

$$\begin{aligned}
m_2 &= \tan\frac{\alpha_3}{2}, \\
m_3 &= \tan\frac{\alpha_2}{2}.
\end{aligned} \tag{8.92b}$$

Hence, by (4.121), p. 113,

$$\gamma_{\ominus T_1\oplus T_{12}} = \frac{1}{m_0m'_0}\{m'_1m_2\gamma_{12} + m'_1m_3\gamma_{13} + m'_2m_3\gamma_{23} + m_2m'_2\}, \tag{8.93}$$

where, by (4.118b) and (4.119b), p. 112, $m_0 > 0$ and $m'_0 > 0$ are given by

$$\begin{aligned} m_0^2 &= m_2^2 + m_3^2 + 2m_1 m_3 \gamma_{23}, \\ (m'_0)^2 &= (m'_1)^2 + (m'_2)^2 + 2m'_1 m'_2 \gamma_{12}. \end{aligned} \tag{8.94}$$

Substituting (8.91b) and (8.92b) into (8.93), we obtain the gamma factor $\gamma_{\ominus T_1 \oplus T_{12}}$ expressed in terms of the gyrotriangle gyroangles α_k, $k = 1, 2, 3$,

$$\begin{aligned} \gamma_{\ominus T_1 \oplus T_{12}} &= \frac{1}{16} \frac{1}{\cos^2 \frac{\alpha_1}{2} \sin^2 \frac{\alpha_3}{2}} \{2\cos^2 \alpha_2 + 2(1 + \cos\alpha_1 - \cos\alpha_3)^2 \\ &\quad + \cos(\alpha_1 + \alpha_2 + \alpha_3) + \cos(-\alpha_1 + \alpha_2 + \alpha_3) \\ &\quad + \cos(\alpha_1 - \alpha_2 + \alpha_3) + \cos(\alpha_1 + \alpha_2 - \alpha_3)\}. \end{aligned} \tag{8.95}$$

Substituting (8.95) into the identity, (1.9), p. 5,

$$\|\ominus T_1 \oplus T_{12}\|^2 = s^2 \frac{\gamma^2_{\ominus T_{12} \oplus T_{13}} - 1}{\gamma^2_{\ominus T_{12} \oplus T_{13}}}, \tag{8.96}$$

we obtain the desired gyrodistance,

$$\begin{aligned} \|\ominus T_1 \oplus T_{12}\|^2 &= \frac{8s^2 \cos\frac{\alpha_1+\alpha_2+\alpha_3}{2} \cos\frac{-\alpha_1+\alpha_2+\alpha_3}{2} \cos\frac{\alpha_1-\alpha_2+\alpha_3}{2} \cos\frac{\alpha_1+\alpha_2-\alpha_3}{2}}{\{2(1+\cos\alpha_1 - \cos\alpha_3)^2 + 2\cos^2\alpha_2 + E(\alpha_1, \alpha_2, \alpha_3)\}^2} \\ &\quad \times \{2 + 4(1+\cos\alpha_1 - \cos\alpha_3)^2 + 2(\cos^2\alpha_1 - \cos^2\alpha_2 + \cos^2\alpha_3) \\ &\quad + E(\alpha_1, \alpha_2, \alpha_3)\}, \end{aligned} \tag{8.97}$$

where $E(\alpha_1, \alpha_2, \alpha_3)$ is given by (8.89), p. 252.

Eliminating the factor $s^2 \cos\frac{\alpha_1+\alpha_2+\alpha_3}{2}$ between (8.97) and (7.35), p. 163, and taking a square root of both sides of the resulting equation, we obtain the following equation:

$$\begin{aligned} &\|\ominus T_1 \oplus T_{12}\| \\ &\quad = \frac{2\sqrt{2}R \cos\frac{-\alpha_1+\alpha_2+\alpha_3}{2} \cos\frac{\alpha_1-\alpha_2+\alpha_3}{2} \cos\frac{\alpha_1+\alpha_2-\alpha_3}{2}}{2(1+\cos\alpha_1 - \cos\alpha_3)^2 + 2\cos^2\alpha_2 + E(\alpha_1, \alpha_2, \alpha_3)} \\ &\qquad \times \sqrt{2 + 4(1+\cos\alpha_1 - \cos\alpha_3)^2 + 2(-\cos^2\alpha_1 + \cos^2\alpha_2 - \cos^2\alpha_3) + E(\alpha_1, \alpha_2, \alpha_3)}, \end{aligned} \tag{8.98}$$

where R is the circumgyroradius of the reference gyrotriangle $A_1A_2A_3$.

8.15.3 Resulting Gyrodistances Between Tangency Points

Formalizing the results of Sects. 8.15.1–8.15.2, we obtain the following theorem in which gyrodistances between various tangency points are related to the gyroangles of the reference gyrotriangle and its circumgyroradius.

Theorem 8.14 (Gyrodistances Between Exgyrocircle Points of Tangency) *Let $A_1A_2A_3$ be a gyrotriangle in an Einstein gyrovector space $(\mathbb{R}^n_s, \oplus, \otimes)$, let T_{ij} be the point of tangency where the A_i-exgyrocircle meets the extension of the gyrotriangle side A_iA_j, and let T_i be the point where the A_i-exgyrocircle of the gyrotriangle meets the opposite side of A_i, Fig.* 8.6, *p.* 251. *Furthermore, let α_k, $k = 1, 2, 3$, and R be the gyrotriangle gyroangles and its circumgyroradius.*

Then, the gyrodistances between tangency points of the A_1-exgyrocircle are:

$$\begin{aligned}
&\|\ominus T_{12}\oplus T_{13}\| \\
&\quad= \frac{2\sqrt{2}R\cos\frac{-\alpha_1+\alpha_2+\alpha_3}{2}\cos\frac{\alpha_1-\alpha_2+\alpha_3}{2}\cos\frac{\alpha_1+\alpha_2-\alpha_3}{2}}{2\cos^2\alpha_1+2(1-\cos\alpha_2-\cos\alpha_3)^2+E(\alpha_1,\alpha_2,\alpha_3)} \\
&\qquad\times\sqrt{2+2(\cos^2\alpha_1-\cos^2\alpha_2-\cos^2\alpha_3)+4(1-\cos\alpha_2-\cos\alpha_3)^2+E(\alpha_1,\alpha_2,\alpha_3)},
\end{aligned} \tag{8.99a}$$

$$\begin{aligned}
&\|\ominus T_{1}\oplus T_{12}\| \\
&\quad= \frac{2\sqrt{2}R\cos\frac{-\alpha_1+\alpha_2+\alpha_3}{2}\cos\frac{\alpha_1-\alpha_2+\alpha_3}{2}\cos\frac{\alpha_1+\alpha_2-\alpha_3}{2}}{2\cos^2\alpha_2+2(1+\cos\alpha_1-\cos\alpha_3)^2+E(\alpha_1,\alpha_2,\alpha_3)} \\
&\qquad\times\sqrt{2+2(-\cos^2\alpha_1+\cos^2\alpha_2-\cos^2\alpha_3)+4(1+\cos\alpha_1-\cos\alpha_3)^2+E(\alpha_1,\alpha_2,\alpha_3)},
\end{aligned} \tag{8.99b}$$

$$\begin{aligned}
&\|\ominus T_{1}\oplus T_{13}\| \\
&\quad= \frac{2\sqrt{2}R\cos\frac{-\alpha_1+\alpha_2+\alpha_3}{2}\cos\frac{\alpha_1-\alpha_2+\alpha_3}{2}\cos\frac{\alpha_1+\alpha_2-\alpha_3}{2}}{2\cos^2\alpha_3+2(1+\cos\alpha_1-\cos\alpha_2)^2+E(\alpha_1,\alpha_2,\alpha_3)} \\
&\qquad\times\sqrt{2+2(-\cos^2\alpha_1-\cos^2\alpha_2+\cos^2\alpha_3)+4(1+\cos\alpha_1-\cos\alpha_2)^2+E(\alpha_1,\alpha_2,\alpha_3)},
\end{aligned} \tag{8.99c}$$

where $E(\alpha_1, \alpha_2, \alpha_3)$ is given by (8.89), *i.e.,*

$$\begin{aligned}
E(\alpha_1,\alpha_2,\alpha_3) &= \cos(\alpha_1+\alpha_2+\alpha_3)+\cos(-\alpha_1+\alpha_2+\alpha_3) \\
&\quad+\cos(\alpha_1-\alpha_2+\alpha_3)+\cos(\alpha_1+\alpha_2-\alpha_3),
\end{aligned} \tag{8.99d}$$

and where R is the circumgyroradius of the reference gyrotriangle $A_1A_2A_3$.

Furthermore, the gamma factors of these gyrodistances are:

$$\gamma_{\ominus T_{12}\oplus T_{13}} = \frac{1}{16}\frac{1}{\sin^2\frac{\alpha_2}{2}\sin^2\frac{\alpha_3}{2}}\bigl\{2\cos^2\alpha_1+2(1-\cos\alpha_2-\cos\alpha_3)^2+E(\alpha_1,\alpha_2,\alpha_3)\bigr\}, \tag{8.100a}$$

$$\gamma_{\ominus T_1 \oplus T_{12}} = \frac{1}{16} \frac{1}{\cos^2 \frac{\alpha_1}{2} \sin^2 \frac{\alpha_3}{2}} \{2\cos^2 \alpha_2 + 2(1+\cos\alpha_1 - \cos\alpha_3)^2 + E(\alpha_1, \alpha_2, \alpha_3)\}, \tag{8.100b}$$

$$\gamma_{\ominus T_1 \oplus T_{13}} = \frac{1}{16} \frac{1}{\cos^2 \frac{\alpha_1}{2} \sin^2 \frac{\alpha_2}{2}} \{2\cos^2 \alpha_3 + 2(1+\cos\alpha_1 - \cos\alpha_2)^2 + E(\alpha_1, \alpha_2, \alpha_3)\}. \tag{8.100c}$$

Proof Equations (8.99a) and (8.100a) are established in (8.90) and (8.86), respectively. Equations (8.99b) and (8.100b) are established in (8.98) and (8.95), respectively. Finally, (8.99c) and (8.100c) are obtained, respectively, from (8.99b) and (8.100b) by interchanging the vertices A_2 and A_3 of the reference gyrotriangle $A_1A_2A_3$. □

Interestingly, (8.99a), (8.99b), (8.99c), (8.99d) remain invariant in form under the Euclidean limit $s \to \infty$, so that the equations are valid in Euclidean geometry as well. However, for application in Euclidean geometry (8.99a), (8.99b), (8.99c), (8.99d) can be simplified owing to the fact that triangle angle sum in π.

Indeed, under the condition

$$\alpha_1 + \alpha_2 + \alpha_3 = \pi, \tag{8.101a}$$

we have the trigonometric identities similar to (7.22b), p. 159,

$$\begin{aligned} \cos\frac{-\alpha_1 + \alpha_2 + \alpha_3}{2} &= \sin\alpha_1, \\ \cos\frac{\alpha_1 - \alpha_2 + \alpha_3}{2} &= \sin\alpha_2, \\ \cos\frac{\alpha_1 + \alpha_2 - \alpha_3}{2} &= \sin\alpha_3, \end{aligned} \tag{8.101b}$$

and

$$\begin{aligned} \cos(\alpha_1 + \alpha_2 + \alpha_3) &= -1, \\ \cos(-\alpha_1 + \alpha_2 + \alpha_3) &= -\cos 2\alpha_1, \\ \cos(\alpha_1 - \alpha_2 + \alpha_3) &= -\cos 2\alpha_2, \\ \cos(\alpha_1 + \alpha_2 - \alpha_3) &= -\cos 2\alpha_3. \end{aligned} \tag{8.101c}$$

Hence, we obtain the following corollary of Theorem 8.14:

Corollary 8.15 (Distances Between Excircle Points of Tangency, I) *Let $A_1A_2A_3$ be a triangle in a Euclidean vector space $\mathbb{R}^n$, let T_{ij} be the point of tangency where the A_i-excircle meets the extension of the triangle side A_iA_j, and let T_i be the point* *where the A_i-excircle of the triangle meets the opposite side of A_i, Fig. 8.5, p. 249. Furthermore, let α_k, $k = 1, 2, 3$, and R be the triangle angles and its circumradius.*

Then, the distances between tangency points of the A_1-excircle are:

$$\begin{aligned}
&\|-T_{12}+T_{13}\| \\
&\quad = \frac{2\sqrt{2}R\sin\alpha_1\sin\alpha_2\sin\alpha_3}{2\cos^2\alpha_1+2(1-\cos\alpha_2-\cos\alpha_3)^2+E(\alpha_1,\alpha_2,\alpha_3)} \\
&\qquad \times\sqrt{2+2(\cos^2\alpha_1-\cos^2\alpha_2-\cos^2\alpha_3)+4(1-\cos\alpha_2-\cos\alpha_3)^2+E(\alpha_1,\alpha_2,\alpha_3)},
\end{aligned} \tag{8.102a}$$

$$\begin{aligned}
&\|-T_{1}+T_{12}\| \\
&\quad = \frac{2\sqrt{2}R\sin\alpha_1\sin\alpha_2\sin\alpha_3}{2\cos^2\alpha_2+2(1+\cos\alpha_1-\cos\alpha_3)^2+E(\alpha_1,\alpha_2,\alpha_3)} \\
&\qquad \times\sqrt{2+2(-\cos^2\alpha_1+\cos^2\alpha_2-\cos^2\alpha_3)+4(1+\cos\alpha_1-\cos\alpha_3)^2+E(\alpha_1,\alpha_2,\alpha_3)},
\end{aligned} \tag{8.102b}$$

and

$$\begin{aligned}
&\|-T_{1}+T_{13}\| \\
&\quad = \frac{2\sqrt{2}R\sin\alpha_1\sin\alpha_2\sin\alpha_3}{2\cos^2\alpha_3+2(1+\cos\alpha_1-\cos\alpha_2)^2+E(\alpha_1,\alpha_2,\alpha_3)} \\
&\qquad \times\sqrt{2+2(-\cos^2\alpha_1-\cos^2\alpha_2+\cos^2\alpha_3)+4(1+\cos\alpha_1-\cos\alpha_2)^2+E(\alpha_1,\alpha_2,\alpha_3)},
\end{aligned} \tag{8.102c}$$

where $E(\alpha_1, \alpha_2, \alpha_3)$ is given by

$$E(\alpha_1,\alpha_2,\alpha_3) = -(1+\cos 2\alpha_1+\cos 2\alpha_2+\cos 2\alpha_3) \tag{8.102d}$$

and where R is the circumradius of the reference triangle $A_1A_2A_3$.

Equations (8.99a), (8.99b), (8.99c), (8.99d) of Theorem 8.14 about gyrodistances in hyperbolic geometry remain valid in Euclidean geometry as well, where they form equations about corresponding distances in Euclidean geometry, in which Einstein addition becomes vector addition. However, when Theorem 8.14 is considered in Euclidean geometry, its results can be simplified owing to Condition (8.101a) that triangle angles obey. The trigonometric simplifications that Condition (8.101a) offers, (8.101b)–(8.101c), thus give rise to Corollary 8.15.

Interestingly, Corollary 8.15 can be further simplified by employing rather involved trigonometric simplifications that Theorem 8.14 uncovers, as follows:

Theorem 8.14 remains invariant in form under the Euclidean limit, $s \to \infty$. Hence, it survives unimpaired in the transition from hyperbolic to Euclidean geometry. But in that Euclidean limit, gamma factors tend to 1. Hence, in particular, in the application of Theorem 8.14 in Euclidean geometry the gamma factors in (8.100a), (8.100b), (8.100c) are 1, resulting in the following trigonometric simplifications that become available under Condition (8.101a):

$$\begin{aligned}
16\sin^2\frac{\alpha_2}{2}\sin^2\frac{\alpha_3}{2} &= 2\cos^2\alpha_1 + 2(1-\cos\alpha_2-\cos\alpha_3)^2 + E(\alpha_1,\alpha_2,\alpha_3),\\
16\cos^2\frac{\alpha_1}{2}\sin^2\frac{\alpha_3}{2} &= 2\cos^2\alpha_2 + 2(1+\cos\alpha_1-\cos\alpha_3)^2 + E(\alpha_1,\alpha_2,\alpha_3),\\
16\cos^2\frac{\alpha_1}{2}\sin^2\frac{\alpha_2}{2} &= 2\cos^2\alpha_3 + 2(1+\cos\alpha_1-\cos\alpha_2)^2 + E(\alpha_1,\alpha_2,\alpha_3),\\
\alpha_1+\alpha_2+\alpha_3 &= \pi.
\end{aligned} \tag{8.103}$$

Owing to Identities (8.103) that triangle angles possess, Corollary 8.15 can be simplified as follows:

Corollary 8.16 (Distances Between Excircle Points of Tangency, II) *Let $A_1A_2A_3$ be a triangle in a Euclidean vector space $\mathbb{R}^n$, let T_{ij} be the point of tangency where the A_i-excircle meets the extension of the triangle side A_iA_j, and let T_i be the point where the A_i-excircle of the triangle meets the opposite side of A_i, Fig.* 8.5, *p.* 249. *Then, the distances between tangency points of the A_1-excircle are*:

$$\begin{aligned}
\|-T_{12}+T_{13}\| &= 4R\sin\alpha_1\cos\frac{\alpha_2}{2}\cos\frac{\alpha_3}{2},\\
\|-T_1+T_{12}\| &= 4R\sin\frac{\alpha_1}{2}\sin\alpha_2\cos\frac{\alpha_3}{2},\\
\|-T_1+T_{13}\| &= 4R\sin\frac{\alpha_1}{2}\cos\frac{\alpha_2}{2}\sin\alpha_3,
\end{aligned} \tag{8.104}$$

where R is the circumradius of the reference triangle $A_1A_2A_3$.

Distances between tangency points of the A_2- and the A_3-excircle of triangle $A_1A_2A_3$ can be obtained from (8.104) by cyclic vertex permutations.

8.16 Exgyrocircle Gyroangles

Let $A_1A_2A_3$ be a gyrotriangle in an Einstein gyrovector space $(\mathbb{R}^n_s, \oplus, \otimes)$ that possesses an A_1-exgyrocircle. The gyroangle

$$\beta_1 = \angle T_{12}T_1T_{13} \tag{8.105}$$

generated by the tangency points T_1, T_{12} and T_{13} of the A_1-exgyrocircle of the gyrotriangle, as shown in Fig. 8.7, is called the A_1-exgyrocircle gyroangle.

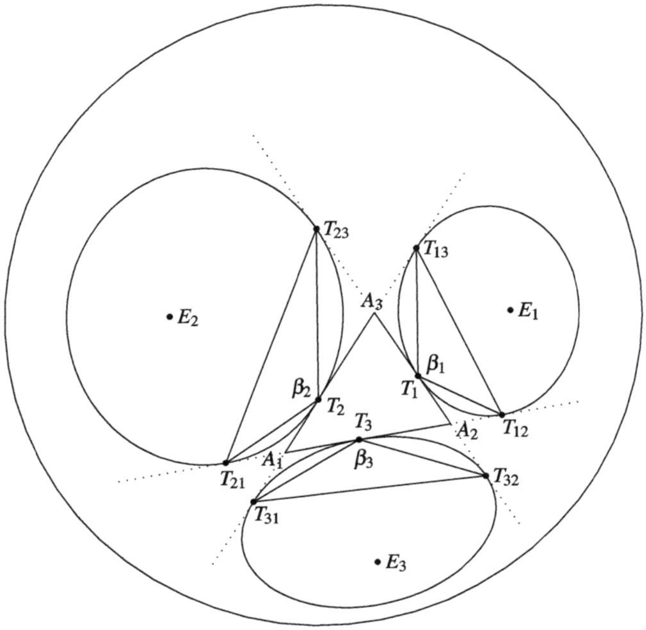

Fig. 8.7 Gyroangles β_k, $k = 1, 2, 3$, generated by exgyrocircle points of tangency

In this section, we face the task of determining exgyrocircle gyroangles β_k, $k = 1, 2, 3$, in terms of the gyroangles of their reference gyrotriangle $A_1A_2A_3$.

Following the *SSS* to *AAA* conversion law in Theorem 6.2, (6.22), p. 135, and (6.20), p. 134, the gyrosine and gyrocosine of gyroangle β_1 are given by the equations

$$\sin^2\beta_1 = \frac{1 + 2\gamma_{\ominus T_1 \oplus T_{12}}\gamma_{\ominus T_1 \oplus T_{13}}\gamma_{\ominus T_{12} \oplus T_{13}} - \gamma^2_{\ominus T_1 \oplus T_{12}} - \gamma^2_{\ominus T_1 \oplus T_{13}} - \gamma^2_{\ominus T_{12} \oplus T_{13}}}{(\gamma^2_{\ominus T_1 \oplus T_{12}} - 1)(\gamma^2_{\ominus T_1 \oplus T_{13}} - 1)},$$

$$\cos^2\beta_1 = \frac{(\gamma_{\ominus T_1 \oplus T_{12}}\gamma_{\ominus T_1 \oplus T_{13}} - \gamma_{\ominus T_{12} \oplus T_{13}})^2}{(\gamma^2_{\ominus T_1 \oplus T_{12}} - 1)(\gamma^2_{\ominus T_1 \oplus T_{13}} - 1)}. \tag{8.106a}$$

Substituting the gamma factors from (8.100a), p. 255, into (8.106a) and simplifying the results gyrotrigonometrically, we obtain the equations

$$\sin\beta_1 = \frac{f_s}{\sqrt{f_s^2 + f_c^2}},$$
$$\cos\beta_1 = \frac{f_c}{\sqrt{f_s^2 + f_c^2}}, \tag{8.106b}$$

where

$$
\begin{aligned}
f_s &= 32\cos^3\tfrac{\alpha_1}{2}\sin\tfrac{\alpha_2}{2}\sin\tfrac{\alpha_3}{2} > 0,\\
f_c &= 2\bigl(\sin^2\alpha_1+\cos^2\alpha_2+\cos^2\alpha_3\bigr) - 8\cos^2\tfrac{\alpha_1}{2}(\cos\alpha_2+\cos\alpha_3)\\
&\quad + E(\alpha_1,\alpha_2,\alpha_3) < 0,
\end{aligned}
\tag{8.106c}
$$

where, as in (8.99d), p. 255,

$$
\begin{aligned}
E(\alpha_1,\alpha_2,\alpha_3) &= \cos(\alpha_1+\alpha_2+\alpha_3)+\cos(-\alpha_1+\alpha_2+\alpha_3)\\
&\quad +\cos(\alpha_1-\alpha_2+\alpha_3)+\cos(\alpha_1+\alpha_2-\alpha_3).
\end{aligned}
\tag{8.106d}
$$

Interestingly, the sum $f_s^2+f_c^2$ of squares can be written as a product,

$$
f_s^2+f_c^2 = f_{123}f_{132},
\tag{8.107a}
$$

where

$$
\begin{aligned}
f_{123} &= 2\bigl(\cos^2\alpha_1+\cos^2\alpha_2+\cos^2\alpha_3\bigr)+32\cos^2\frac{\alpha_1}{2}\sin^2\frac{\alpha_2}{2}+E(\alpha_1,\alpha_2,\alpha_3)-2,\\
f_{132} &= 2\bigl(\cos^2\alpha_1+\cos^2\alpha_2+\cos^2\alpha_3\bigr)+32\cos^2\frac{\alpha_1}{2}\sin^2\frac{\alpha_3}{2}+E(\alpha_1,\alpha_2,\alpha_3)-2.
\end{aligned}
\tag{8.107b}
$$

Formalizing the main result of this section, we obtain the following theorem:

Theorem 8.17 (Gyroangles Generated by Exgyrocircle Points of Tangency) *Let $A_1A_2A_3$ be a gyrotriangle in an Einstein gyrovector space $(\mathbb{R}_s^n,\oplus,\otimes)$ with gyroangles α_k, $k=1,2,3$, and let $E(\alpha_1,\alpha_2,\alpha_3)$ be given by the equation*

$$
\begin{aligned}
E(\alpha_1,\alpha_2,\alpha_3) &= \cos(\alpha_1+\alpha_2+\alpha_3)+\cos(-\alpha_1+\alpha_2+\alpha_3)\\
&\quad +\cos(\alpha_1-\alpha_2+\alpha_3)+\cos(\alpha_1+\alpha_2-\alpha_3).
\end{aligned}
\tag{8.108}
$$

Furthermore, let T_{ij} be the point of tangency where the A_i-exgyrocircle meets the extension of the gyrotriangle side A_iA_j, and let T_i be the point where the A_i-exgyrocircle of the gyrotriangle meets the opposite side of A_i, Fig. 8.7. *Then, the measures of gyroangles*

$$
\begin{aligned}
\beta_1 &= \angle T_{12}T_1T_{12},\\
\beta_2 &= \angle T_{21}T_2T_{23},\\
\beta_3 &= \angle T_{31}T_2T_{32}
\end{aligned}
\tag{8.109}
$$

that the exgyrocircle tangency points generate are determined by the following equations:

$$
\sin\beta_1 = \frac{f_{s1}}{\sqrt{f_{s1}^2+f_{c1}^2}}, \qquad \cos\beta_1 = \frac{f_{c1}}{\sqrt{f_{s1}^2+f_{c1}^2}},
\tag{8.110a}
$$

where

$$\begin{aligned} f_{s1} &= 32\cos^3\frac{\alpha_1}{2}\sin\frac{\alpha_2}{2}\sin\frac{\alpha_3}{2} > 0, \\ f_{c1} &= 2\left(\sin^2\alpha_1 + \cos^2\alpha_2 + \cos^2\alpha_3\right) \\ &\quad - 8\cos^2\frac{\alpha_1}{2}(\cos\alpha_2 + \cos\alpha_3) + E(\alpha_1, \alpha_2, \alpha_3) \\ &< 0, \end{aligned} \tag{8.110b}$$

$$\sin\beta_2 = \frac{f_{s2}}{\sqrt{f_{s2}^2 + f_{c2}^2}}, \qquad \cos\beta_2 = \frac{f_{c2}}{\sqrt{f_{s2}^2 + f_{c2}^2}}, \tag{8.111a}$$

where

$$\begin{aligned} f_{s2} &= 32\sin\frac{\alpha_1}{2}\cos^3\frac{\alpha_2}{2}\sin\frac{\alpha_3}{2} > 0, \\ f_{c2} &= 2\left(\cos^2\alpha_1 + \sin^2\alpha_2 + \cos^2\alpha_3\right) \\ &\quad - 8\cos^2\frac{\alpha_2}{2}(\cos\alpha_1 + \cos\alpha_3) + E(\alpha_1, \alpha_2, \alpha_3) \\ &< 0, \end{aligned} \tag{8.111b}$$

$$\sin\beta_3 = \frac{f_{s3}}{\sqrt{f_{s2}^2 + f_{c3}^2}}, \qquad \cos\beta_3 = \frac{f_{c3}}{\sqrt{f_{s3}^2 + f_{c3}^2}}, \tag{8.112a}$$

where

$$\begin{aligned} f_{s3} &= 32\sin\frac{\alpha_1}{2}\sin\frac{\alpha_2}{2}\cos^3\frac{\alpha_3}{2} > 0, \\ f_{c3} &= 2\left(\cos^2\alpha_1 + \cos^2\alpha_2 + \sin^2\alpha_3\right) \\ &\quad - 8\cos^2\frac{\alpha_3}{2}(\cos\alpha_1 + \cos\alpha_2) + E(\alpha_1, \alpha_2, \alpha_3) \\ &< 0. \end{aligned} \tag{8.112b}$$

Proof The determination of gyroangle β_1 in (8.110a), (8.110b) is established in (8.106a), (8.106b), (8.106c), (8.106d). The determination of gyroangles β_2 and β_3 follows from that of β_1 in (8.106a), (8.106b), (8.106c), (8.106d) by cyclic permutations of the vertices of the reference gyrotriangle $A_1A_2A_3$. □

8.17 Exgyrocircle Gyroangle Sum

Let $A_1A_2A_3$ be a gyrotriangle in an Einstein gyrovector space $(\mathbb{R}_s^n, \oplus, \otimes)$ that possesses the A_k-exgyrocircle, $k = 1, 2, 3$. The gyrotriangle exgyrocircles, in turn, generate the exgyrocircle gyroangles β_k, $k = 1, 2, 3$, the sum of which, $\beta_1 + \beta_2 + \beta_3$, is the gyrotriangle exgyrocircle gyroangle sum.

In this section, we face the task of determining the gyrotriangle exgyrocircle gyroangle sum in terms of the gyroangles of the reference gyrotriangle.

Let $A_1A_2A_3$ be a gyrotriangle in the Einstein gyrovector space that possesses three exgyrocircles and let β_k, $k = 1, 2, 3$, be the resulting gyrotriangle exgyrocircle gyroangles. The gyrosine and gyrocosine of the gyrotriangle exgyrocircle gyroangle sum are

$$\begin{aligned} \sin(\beta_1 + \beta_2 + \beta_3) &= -\sin\beta_1 \sin\beta_2 \sin\beta_3 + \sin\beta_1 \cos\beta_2 \cos\beta_3 \\ &\quad + \cos\beta_1 \sin\beta_2 \cos\beta_3 + \cos\beta_1 \cos\beta_2 \sin\beta_3, \\ \cos(\beta_1 + \beta_2 + \beta_3) &= \cos\beta_1 \cos\beta_2 \cos\beta_3 - \cos\beta_1 \sin\beta_2 \sin\beta_3 \\ &\quad - \sin\beta_1 \cos\beta_2 \sin\beta_3 - \sin\beta_1 \sin\beta_2 \cos\beta_3. \end{aligned} \tag{8.113}$$

Substituting $\sin\beta_k$ and $\cos\beta_k$, $k = 1, 2, 3$, from (8.110a), (8.110b), (8.111a), (8.111b), (8.112a), (8.112b) into (8.113), we obtain the equations

$$\begin{aligned} \sin(\beta_1 + \beta_2 + \beta_3) &= \cos\frac{\alpha_1 + \alpha_2 + \alpha_3}{2} f_1(\alpha_1, \alpha_2, \alpha_3), \\ \cos(\beta_1 + \beta_2 + \beta_3) &= f_2(\alpha_1, \alpha_2, \alpha_3), \end{aligned} \tag{8.114}$$

where the functions f_1 and f_2 of α_k, $k = 1, 2, 3$, are too involved and hence are not presented here explicitly.

The factor $\cos\frac{\alpha_1+\alpha_2+\alpha_3}{2}$ on the right-hand side of the first equation in (8.114) enables an interesting result of Euclidean geometry to be uncovered. In the Euclidean limit, $s \to \infty$, a gyrotriangle gyroangle sum tends to a corresponding triangle angle sum, which is π. Hence, in that limit, the factor tends to $\cos\frac{\pi}{2} = 0$, implying that $\sin(\beta_1 + \beta_2 + \beta_3) = 0$ in Euclidean geometry. The latter, in turn, implies that in Euclidean geometry the excircle angle sum is $\beta_1 + \beta_2 + \beta_3 = 2\pi$.

8.18 Exgyrocenter-Point-of-Tangency Gyrocenter

Definition 8.18 (Exgyrocenter-Point-of-Tangency Gyrocenter, P) Let $A_1A_2A_3$ be a gyrotriangle in an Einstein gyrovector space $(\mathbb{R}^n_s, \oplus, \otimes)$ along with its exgyrocenters E_k and points of tangency T_k, $k = 1, 2, 3$, Fig. 8.8. The points E_k and T_k form the three gyrolines E_kT_k that are concurrent. This point of concurrency is called the exgyrocenter-point-of-tangency Gyrocenter, P, Fig. 8.8, of the gyrotriangle.

Let the gyrocenter P of gyrotriangle $A_1A_2A_3$ in an Einstein gyrovector space $(\mathbb{R}^n_s, \oplus, \otimes)$ be given by its gyrobarycentric coordinate representation with respect to the set $S = \{A_1, A_2, A_3\}$ of the gyrotriangle vertices,

$$P = \frac{m_1\gamma_{A_1} A_1 + m_2\gamma_{A_2} A_2 + m_3\gamma_{A_3} A_3}{m_1\gamma_{A_1} + m_2\gamma_{A_2} + m_3\gamma_{A_3}}, \tag{8.115}$$

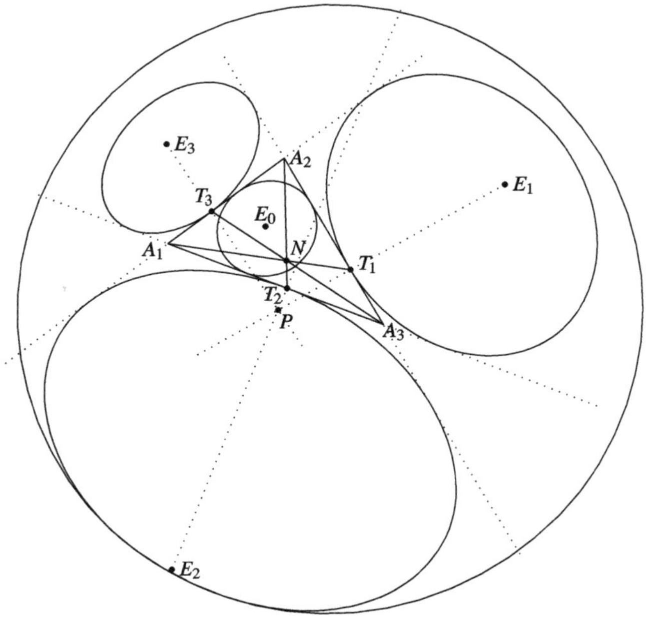

Fig. 8.8 The Exgyrocenter-Point-of-Tangency Gyrocenter, P, of a gyrotriangle $A_1A_2A_3$. This gyrotriangle gyrocenter, P, is the point of concurrency of the three gyrolines E_kT_k, $k = 1, 2, 3$, where E_k are the gyrotriangle exgyrocenters and T_k are the gyrotriangle points of tangency with its exgyrocircles. Gyrotrigonometric gyrobarycentric coordinates of the gyrotriangle gyrocenter P are given in Theorem 8.19. Note that the three points A_k, E_0 and E_k for each k, $k = 1, 2, 3$, are gyrocollinear

where the gyrobarycentric coordinates $(m_1 : m_2 : m_3)$ of P in (8.115) are to be determined.

Similarly to (8.58a), (8.58b), (8.58c), (8.59a), (8.59b) and (8.59c), under the left gyrotranslation by $\ominus A_1$ the gyrotriangle gyrocenter P in (8.115) becomes

$$\ominus A_1 \oplus P = \frac{m_2 \gamma_{\ominus A_1 \oplus A_2} (\ominus A_1 \oplus A_2) + m_3 \gamma_{\ominus A_1 \oplus A_3} (\ominus A_1 \oplus A_3)}{m_1 + m_2 \gamma_{\ominus A_1 \oplus A_2} + m_3 \gamma_{\ominus A_1 \oplus A_3}}$$

$$= \frac{m_2 \gamma_{12} \mathbf{a}_{12} + m_3 \gamma_{13} \mathbf{a}_{13}}{m_1 + m_2 \gamma_{12} + m_3 \gamma_{13}}. \tag{8.116}$$

1. The exgyrocenter $\ominus A_1 \oplus E_1$ and the tangent point $\ominus A_1 \oplus T_1$ of the left gyrotranslated gyrotriangle $O(\ominus A_1 \oplus A_2)(\ominus A_1 \oplus A_3)$, as calculated in (8.58a) and (8.59a),

are:

$$\ominus A_1 \oplus E_1 = \frac{\sin\alpha_2\gamma_{12}\mathbf{a}_{12} + \sin\alpha_3\gamma_{13}\mathbf{a}_{13}}{-\sin\alpha_1 + \sin\alpha_2\gamma_{12} + \sin\alpha_3\gamma_{13}},$$
$$\ominus A_1 \oplus T_1 = \frac{\tan\frac{\alpha_3}{2}\gamma_{12}\mathbf{a}_{12} + \tan\frac{\alpha_2}{2}\gamma_{13}\mathbf{a}_{13}}{\tan\frac{\alpha_3}{2}\gamma_{12} + \tan\frac{\alpha_2}{2}\gamma_{13}}. \tag{8.117a}$$

These points are contained in the Euclidean line

$$L_1 = (\ominus A_1 \oplus T_1) + \big(-(\ominus A_1 \oplus T_1) + (\ominus A_1 \oplus E_1)\big)t_1, \tag{8.117b}$$

where $t_1 \in \mathbb{R}$ is the line parameter. This line passes through the point $(\ominus A_1 \oplus T_1) \in \mathbb{R}_s^n \subset \mathbb{R}^n$ when $t_1 = 0$, and it passes through the point $(\ominus A_1 \oplus E_1) \in \mathbb{R}_s^n \subset \mathbb{R}^n$ when $t_1 = 1$.

2. Similarly to (8.117a), (8.117b), The exgyrocenter $\ominus A_1 \oplus E_2$ and the tangent point $\ominus A_1 \oplus T_2$ of the left gyrotranslated gyrotriangle $O(\ominus A_1 \oplus A_2)(\ominus A_1 \oplus A_3)$, as calculated in (8.58b) and (8.59b), are:

$$\ominus A_1 \oplus E_2 = \frac{-\sin\alpha_2\gamma_{12}\mathbf{a}_{12} + \sin\alpha_3\gamma_{13}\mathbf{a}_{13}}{\sin\alpha_1 - \sin\alpha_2\gamma_{12} + \sin\alpha_3\gamma_{13}},$$
$$\ominus A_1 \oplus T_2 = \frac{\tan\frac{\alpha_1}{2}\gamma_{13}\mathbf{a}_{13}}{\tan\frac{\alpha_3}{2} + \tan\frac{\alpha_1}{2}\gamma_{13}}. \tag{8.118a}$$

These points are contained in the Euclidean line

$$L_2 = (\ominus A_1 \oplus T_2) + \big(-(\ominus A_1 \oplus T_2) + (\ominus A_1 \oplus E_2)\big)t_2 \tag{8.118b}$$

where $t_2 \in \mathbb{R}$ is the line parameter. This line passes through the point $(\ominus A_1 \oplus T_2) \in \mathbb{R}_s^n \subset \mathbb{R}^n$ when $t_2 = 0$, and it passes through the point $(\ominus A_1 \oplus E_2) \in \mathbb{R}_s^n \subset \mathbb{R}^n$ when $t_2 = 1$.

3. Similarly to (8.117a), (8.117b), (8.118a) and (8.118b), The exgyrocenter $\ominus A_1 \oplus E_3$ and the tangent point $\ominus A_1 \oplus T_3$ of the left gyrotranslated gyrotriangle $O(\ominus A_1 \oplus A_2)(\ominus A_1 \oplus A_3)$, as calculated in (8.58c) and (8.59c), are:

$$\ominus A_1 \oplus E_3 = \frac{\sin\alpha_2\gamma_{12}\mathbf{a}_{12} - \sin\alpha_3\gamma_{13}\mathbf{a}_{13}}{\sin\alpha_1 + \sin\alpha_2\gamma_{12} - \sin\alpha_3\gamma_{13}},$$
$$\ominus A_1 \oplus T_3 = \frac{\tan\frac{\alpha_1}{2}\gamma_{12}\mathbf{a}_{12}}{\tan\frac{\alpha_2}{2} + \tan\frac{\alpha_1}{2}\gamma_{12}}. \tag{8.119a}$$

These points are contained in the Euclidean line

$$L_3 = (\ominus A_1 \oplus T_3) + \big(-(\ominus A_1 \oplus T_3) + (\ominus A_1 \oplus E_3)\big)t_3, \tag{8.119b}$$

where $t_3 \in \mathbb{R}$ is the line parameter. This line passes through the point $(\ominus A_1 \oplus T_3) \in \mathbb{R}^n_s \subset \mathbb{R}^n$ when $t_3 = 0$, and it passes through the point $(\ominus A_1 \oplus E_3) \in \mathbb{R}^n_s \subset \mathbb{R}^n$ when $t_3 = 1$.

If the Euclidean lines L_k, $k = 1, 2, 3$, in (8.117a), (8.117b), (8.118a), (8.118b), (8.119a), (8.119b) are concurrent, then their concurrency point is the gyrocenter P of the gyrotranslated gyrotriangle $O(\ominus A_1 \oplus A_2)(\ominus A_1 \oplus A_3)$, given by (8.116).

As in (7.219), p. 207, and in (8.64), p. 244, the condition that gyrocenter $\ominus A_1 \oplus P$ lies on each of the three Euclidean lines L_k, $k = 1, 2, 3$, along with the linear independence of the gyrovectors $\mathbf{a}_{12}$ and $\mathbf{a}_{13}$ in $\mathbb{R}^n_s$, when considered as vectors in $\mathbb{R}^n$, gives rise to a system of six homogeneous linear equations for the six unknowns t_k and m_k, $k = 1, 2, 3$. Here the three unknowns t_k are the line parameters that determine the gyrocenter P on each of the three lines L_k, and the three unknowns m_k are the gyrobarycentric coordinates of the gyrocenter P of gyrotriangle $A_1A_2A_3$ with respect to the set $S = \{A_1, A_2, A_3\}$ of the gyrotriangle vertices, in (8.115) and (8.116).

The resulting six homogeneous linear equations for the six unknowns t_1, t_2, t_3 and m_1, m_2, m_3 are not linearly independent. Indeed, they determine uniquely the five unknowns t_1, t_2, t_3 and $m_1/m_3, m_2/m_3$, resulting in

$$\begin{aligned} \frac{m_1}{m_3} &= \frac{\sin\alpha_1(1 + \cos\alpha_1 - \cos\alpha_2 - \cos\alpha_3)}{\sin\alpha_3(1 - \cos\alpha_1 - \cos\alpha_2 + \cos\alpha_3)}, \\ \frac{m_2}{m_3} &= \frac{\sin\alpha_1(1 - \cos\alpha_1 + \cos\alpha_2 - \cos\alpha_3)}{\sin\alpha_3(1 - \cos\alpha_1 - \cos\alpha_2 + \cos\alpha_3)}. \end{aligned} \tag{8.120}$$

Gyrobarycentric coordinates are homogeneous, so that a common nonzero factor is irrelevant. Hence, (8.120) gives rise to the following theorem:

Theorem 8.19 (The Exgyrocenter-Point-of-Tangency Gyrocenter Theorem) *Let $A_1A_2A_3$ be a gyrotriangle in an Einstein gyrovector space $(\mathbb{R}^n_s, \oplus, \otimes)$, and let P be the exgyrocenter-point-of-tangency gyrocenter of the gyrotriangle, Fig.* 8.8.

Then, in the standard gyrotriangle index notation, the gyrotriangle gyrocenter P is given by its gyrobarycentric coordinate representation,

$$P = \frac{m_1\gamma_{A_1}A_1 + m_2\gamma_{A_2}A_2 + m_3\gamma_{A_3}A_3}{m_1\gamma_{A_1} + m_2\gamma_{A_2} + m_3\gamma_{A_3}}, \tag{8.121}$$

where gyrotrigonometric gyrobarycentric coordinates $(m_1 : m_2 : m_3)$ of P in (8.121) *are*

$$\begin{aligned} m_1 &= \sin\alpha_1(1 + \cos\alpha_1 - \cos\alpha_2 - \cos\alpha_3), \\ m_2 &= \sin\alpha_2(1 - \cos\alpha_1 + \cos\alpha_2 - \cos\alpha_3), \\ m_3 &= \sin\alpha_3(1 - \cos\alpha_1 - \cos\alpha_2 + \cos\alpha_3). \end{aligned} \tag{8.122}$$

8.19 Problems

Problem 8.1 The Gyrotriangle Exgyroradius:
Derive the system (8.7), p. 223, of three equations by substituting successively (8.4a), (8.4b), (8.4c) and (8.2) into (8.5).

Problem 8.2 The Constant of the Gyrotrigonometric Gyrobarycentric Coordinate Representation of the Gyrotriangle Excenters and Incenter:
Derive the constants m_0^2 in (8.39a), (8.39b), (8.39c), (8.39d), p. 235, of each of the gyrotrigonometric gyrobarycentric coordinate representation of E_k, $k = 1, 2, 3, 0$, from (8.15), p. 226, and (8.37a), (8.37b), (8.37c), (8.37d), (8.38a), (8.38b), (8.38c), (8.38d), p. 234.

Problem 8.3 A Gyrotriangle Gyroangle Inequality:
Prove Inequality (8.41), p. 235, for the gyroangles α_k, $k = 1, 2, 3$ of any gyrotriangle $A_1A_2A_3$ in an Einstein gyrovector space.

Problem 8.4 A Substitution:
Derive (8.53), p. 239, by substitutions from (7.143) and (7.147) into (8.52).

Problem 8.5 Gyrotriangle Gyrotrigonometric Substitutions:
Complete the proof of Theorem 8.3, p. 230, in detail by substituting gyrotriangle gyrotrigonometric identities from Sect. 7.12, p. 187, into (8.21)–(8.22) and employing the result of Theorem 7.4, p. 163.

Problem 8.6 In-Exgyroradii Relation:
Derive (8.26) and (8.27) from (8.25), p. 230, by using the software *Mathematica* for computer algebra to manipulate trigonometric functions by commands like `TrigToExp`, `ExpToTrig`, `TrigReduce` and `TrigFactor`.

Problem 8.7 Gyrotrigonometric Substitutions:
Derive the gamma factor (8.86), p. 252, by substituting (8.82b) and (8.83b) into (8.84) (and squaring, but subsequently taking a square root).

Problem 8.8 Gyrotrigonometric Substitutions:
Derive (8.88), p. 252, by substituting (8.86) in Identity (8.87), which is a special case of Identity (1.9), p. 5.

Problem 8.9 Gyrotrigonometric Substitutions:
Derive the gamma factor (8.95), p. 254, by substituting (8.91b) and (8.92b) into (8.93).

Problem 8.10 Gyrotrigonometric Substitutions:
Derive (8.97), p. 254, by substituting (8.95) into Identity (8.96), which is a special case of Identity (1.9), p. 5.

Problem 8.11 Trigonometric Substitutions:
Simplify the results of Corollary 8.15, p. 256, into those of Corollary 8.16, p. 258, in Euclidean geometry by substituting Identities (8.103) that triangle angles possess and employing well-known, elementary trigonometric identities.

Problem 8.12 The *SSS* to *AAA* conversion law:
Apply the *SSS* to *AAA* conversion law, Theorem 6.2, p. 134, to obtain the equations in (8.106a), p. 259.

Problem 8.13 Gyrotrigonometric Substitutions:
Substitute the gamma factors from (8.100a), p. 255, into (8.106a), p. 259, and simplify the results gyrotrigonometrically (= trigonometrically) to obtain (8.106b).

Problem 8.14 A Gyrotrigonometric/Trigonometric Identity:
Prove the gyrotrigonometric/trigonometric Identity (8.107a), (8.107b), p. 260.

Problem 8.15 Gyrotrigonometric Substitutions:
Substitute the $\sin\beta_k$ and $\cos\beta_k$, $k = 1,2,3$, from (8.110a), (8.110b), (8.111a), (8.111b), (8.112a), (8.112b) into (8.113) to obtain (8.114), p. 262, explicitly, and deduce that

$$\lim_{s\to\infty} \sin(\beta_1 + \beta_2 + \beta_3) = 0. \tag{8.123}$$

Furthermore, show that (8.123) implies that in Euclidean geometry, the excircle angle sum is $\beta_1 + \beta_2 + \beta_3 = 2\pi$.

Problem 8.16 Gyrocollinearity:
Explain why the three points A_k, E_0 and E_k for each k, $k = 1,2,3$, in Fig. 8.8, p. 1263, are gyrocollinear.

Chapter 9
Gyrotriangle Gyrocevians

Abstract Triangle Cevians play an important role in the study of Euclidean triangles. Accordingly, hyperbolic triangle Cevians, called gyrotriangle gyrocevians, play an important role in the study of hyperbolic triangles as well. The determination of the gyrocevian gyrolength and the measure of gyroangles that gyrocevians generate in gyrotriangles is presented. As an application, a special gyrocevian that generates special ingyrocircles is studied. Furthermore, gyrocevian concurrency conditions are uncovered and the hyperbolic version of the Theorem of Ceva is presented along with the related hyperbolic Brocard points.

9.1 Gyrocevians and the Hyperbolic Theorem of Ceva

Definition 9.1 (Gyrocevians) A gyrocevian is a gyrosegment that joins a vertex of a gyrotriangle in an Einstein gyrovector space with a point on the opposite side.

Theorem 9.2 (Gyrocevians Point of Concurrency) *Let A_1A_{23} and A_2A_{31} be two gyrocevians of a gyrotriangle $A_1A_2A_3$ in an Einstein gyrovector space $(\mathbb{R}^n_s, \oplus, \otimes)$, Fig. 9.1, and let the gyrocevian feet be given by their gyrobarycentric coordinate representations*

$$
\begin{aligned}
A_{31} &= \frac{m_{21}\gamma_{A_1} A_1 + m_{23}\gamma_{A_3} A_3}{m_{21}\gamma_{A_1} + m_{23}\gamma_{A_3}}, \\
A_{23} &= \frac{m_{12}\gamma_{A_2} A_2 + m_{13}\gamma_{A_3} A_3}{m_{12}\gamma_{A_2} + m_{13}\gamma_{A_3}}
\end{aligned}
\qquad (9.1)
$$

with respect to the set $S = \{A_1, A_2, A_3\}$.

A.A. Ungar, *Hyperbolic Triangle Centers*,
Fundamental Theories of Physics 166,
DOI 10.1007/978-90-481-8637-2_9,

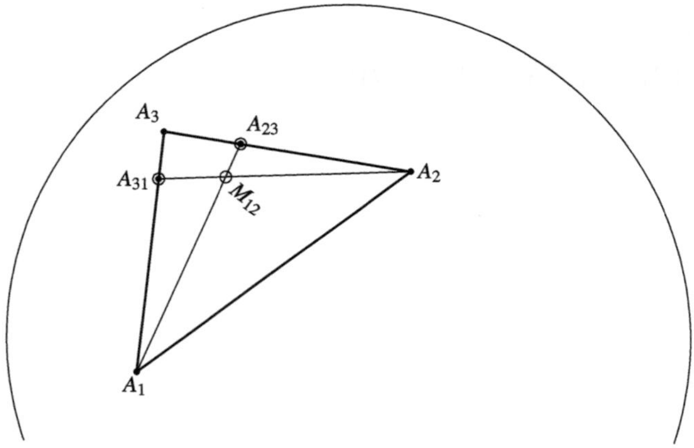

Fig. 9.1 Gyrocevians Point of Concurrency. The point of concurrency, M_{12}, of two gyrocevians A_1A_{23} and A_2A_{31} of a gyrotriangle $A_1A_2A_3$ in an Einstein gyrovector space is determined in (9.2)

Then the two gyrocevians intersect at a point M_{12} on the interior of the gyrotriangle, given by the gyrobarycentric coordinate representation

$$M_{12} = \frac{m_{13}m_{21}\gamma_{A_1} A_1 + m_{12}m_{23}\gamma_{A_2} A_2 + m_{13}m_{23}\gamma_{A_3} A_3}{m_{13}m_{21}\gamma_{A_1} + m_{12}m_{23}\gamma_{A_2} + m_{13}m_{23}\gamma_{A_3}} \tag{9.2}$$

with respect to the set $S = \{A_1, A_2, A_3\}$.

Proof Let the point of intersection, M_{12}, of the gyrocevians A_1A_{23} and A_2A_{31}, Fig. 9.1, be given by its its gyrobarycentric coordinate representation

$$M_{12} = \frac{m_1\gamma_{A_1} A_1 + m_2\gamma_{A_2} A_2 + m_3\gamma_{A_3} A_3}{m_1\gamma_{A_1} + m_2\gamma_{A_2} + m_3\gamma_{A_3}} \tag{9.3}$$

with respect to the set $S = \{A_1, A_2, A_3\}$, where the gyrobarycentric coordinates are to be determined in (9.10).

Let us left gyrotranslate by $\ominus A_1$ the two gyrocevians A_1A_{23} and A_2A_{31} along with their point of intersection, M_{12}, obtaining the following left gyrotranslated gyrocevians $(\ominus A_1 \oplus A_1)(\ominus A_1 \oplus A_{23}) = O(\ominus A_1 \oplus A_{23})$ and $(\ominus A_1 \oplus A_1)(\ominus A_1 \oplus A_{31})$, along with their point of intersection $\ominus A_1 \oplus M_{12}$. Here $\ominus A_1 \oplus A_1 = O$ is the origin of $\mathbb{R}_s^n$, given by $O = \mathbf{0} = (0, \ldots, 0)$ with respect to its Cartesian coordinates.

By the Gyrobarycentric Coordinate Representation Gyrocovariance Theorem 4.6, p. 90, the left gyrotranslations of the points A_{31}, A_{23} and M_{12} by $\ominus A_1$ are

$$\ominus A_1 \oplus A_{31} = \ominus A_1 \oplus \frac{m_{21}\gamma_{A_1} A_1 + m_{23}\gamma_{A_3} A_3}{m_{21}\gamma_{A_1} + m_{23}\gamma_{A_3}}$$

$$= \frac{m_{21}\gamma_{\ominus A_1 \oplus A_1}(\ominus A_1 \oplus A_1) + m_{23}\gamma_{\ominus A_1 \oplus A_3}(\ominus A_1 \oplus A_3)}{m_{21}\gamma_{\ominus A_1 \oplus A_1} + m_{23}\gamma_{\ominus A_1 \oplus A_3}}$$

$$= \frac{m_{23}\gamma_{13}\mathbf{a}_{13}}{m_{21} + m_{23}\gamma_{13}}, \tag{9.4a}$$

$$\ominus A_1 \oplus A_{23} = \ominus A_1 \oplus \frac{m_{12}\gamma_{A_2}A_2 + m_{13}\gamma_{A_3}A_3}{m_{12}\gamma_{A_2} + m_{13}\gamma_{A_3}}$$

$$= \frac{m_{12}\gamma_{\ominus A_1 \oplus A_2}(\ominus A_1 \oplus A_2) + m_{13}\gamma_{\ominus A_1 \oplus A_3}(\ominus A_1 \oplus A_3)}{m_{12}\gamma_{\ominus A_1 \oplus A_2} + m_{13}\gamma_{\ominus A_1 \oplus A_3}}$$

$$= \frac{m_{12}\gamma_{12}\mathbf{a}_{12} + m_{13}\gamma_{13}\mathbf{a}_{13}}{m_{12}\gamma_{12} + m_{13}\gamma_{13}}, \tag{9.4b}$$

$$\ominus A_1 \oplus M_{12}$$

$$= \ominus A_1 \oplus \frac{m_1\gamma_{A_1}A_1 + m_2\gamma_{A_2}A_2 + m_3\gamma_{A_3}A_3}{m_1\gamma_{A_1} + m_2\gamma_{A_2} + m_3\gamma_{A_3}}$$

$$= \frac{m_1\gamma_{\ominus A_1 \oplus A_1}(\ominus A_1 \oplus A_1) + m_2\gamma_{\ominus A_1 \oplus A_2}(\ominus A_1 \oplus A_2) + m_3\gamma_{\ominus A_1 \oplus A_3}(\ominus A_1 \oplus A_3)}{m_1\gamma_{\ominus A_1 \oplus A_1} + m_2\gamma_{\ominus A_1 \oplus A_2} + m_3\gamma_{\ominus A_1 \oplus A_3}}$$

$$= \frac{m_2\gamma_{12}\mathbf{a}_{12} + m_3\gamma_{13}\mathbf{a}_{13}}{m_1 + m_2\gamma_{12} + m_3\gamma_{13}}. \tag{9.4c}$$

1. The first left gyrotranslated gyrocevian, $(\ominus A_1 \oplus A_1)(\ominus A_1 \oplus A_{23})$, is a gyrosegment that joins its foot $\ominus A_1 \oplus A_{23}$ with vertex $\ominus A_1 \oplus A_1 = O = \mathbf{0}$ of the left gyrotranslated gyrotriangle. It lies on the Euclidean line L_1 in $\mathbb{R}^n$ that passes through the points $O = \mathbf{0}$ and $\ominus A_1 \oplus A_{23}$. Hence, by (9.4b),

$$L_1 = (\ominus A_1 \oplus A_1) + \big(-(\ominus A_1 \oplus A_1) + (\ominus A_1 \oplus A_{23})\big)t_1$$

$$= \frac{m_{12}\gamma_{12}\mathbf{a}_{12} + m_{13}\gamma_{13}\mathbf{a}_{13}}{m_{12}\gamma_{12} + m_{13}\gamma_{13}}t_1, \tag{9.5a}$$

   where $t_1 \in \mathbb{R}$ is the line parameter.
2. The second left gyrotranslated gyrocevian, $(\ominus A_1 \oplus A_2)(\ominus A_1 \oplus A_{31})$, is a gyrosegment that joins its foot $\ominus A_1 \oplus A_{31}$ with vertex $\ominus A_1 \oplus A_2 = \mathbf{a}_{12}$ of the left gyrotranslated gyrotriangle. It lies on the Euclidean line L_2 in $\mathbb{R}^n$ that passes

through the points $\ominus A_1 \oplus A_2 = \mathbf{a}_{12}$ and $\ominus A_1 \oplus A_{31}$. Hence, by (9.4a),

$$\begin{aligned} L_2 &= (\ominus A_1 \oplus A_2) + \bigl(-(\ominus A_1 \oplus A_2) + (\ominus A_1 \oplus A_{31})\bigr) t_1 \\ &= \mathbf{a}_{12} + \left(-\mathbf{a}_{12} + \frac{m_{23}\gamma_{13}\mathbf{a}_{13}}{m_{21} + m_{23}\gamma_{13}}\right) t_2, \end{aligned} \tag{9.5b}$$

where $t_2 \in \mathbb{R}$ is the line parameter.

The point $\ominus A_1 \oplus M_{12}$ lies on each of the two lines L_k, $k = 1, 2$. Hence, there exist values $t_{k,0}$ of the line parameters t_k, $k = 1, 2$, respectively, such that

$$\begin{aligned} (\ominus A_1 \oplus M_{12}) - \frac{m_{12}\gamma_{12}\mathbf{a}_{12} + m_{13}\gamma_{13}\mathbf{a}_{13}}{m_{12}\gamma_{12} + m_{13}\gamma_{13}} t_{1,0} &= 0, \\ (\ominus A_1 \oplus M_{12}) - \mathbf{a}_{12} - \left(-\mathbf{a}_{12} + \frac{m_{23}\gamma_{13}\mathbf{a}_{13}}{m_{21} + m_{23}\gamma_{13}}\right) t_{2,0} &= 0. \end{aligned} \tag{9.6}$$

Substituting $\ominus A_1 \oplus M_{12}$ from (9.4c) into (9.6) and rewriting each equation in (9.6) as a linear combination of $\mathbf{a}_{12}$ and $\mathbf{a}_{13}$ equals zero, we obtain the following homogeneous linear system of two gyrovector equations

$$\begin{aligned} c_{11}\mathbf{a}_{12} + c_{12}\mathbf{a}_{13} &= \mathbf{0}, \\ c_{21}\mathbf{a}_{12} + c_{22}\mathbf{a}_{13} &= \mathbf{0}, \end{aligned} \tag{9.7}$$

where the coefficients c_{ij}, $i, j = 1, 2$, are functions of the unknown values $t_{1,0}, t_{2,0}$ of the line parameters t_1, t_2 and of the gyrobarycentric coordinates $(m_1 : m_2 : m_3)$ of the gyrocevians point of intersection M_{12} in (9.3).

Since the set $S = \{A_1, A_2, A_3\}$ is pointwise independent, the two gyrovectors $\mathbf{a}_{12} = \ominus A_1 \oplus A_2$ and $\mathbf{a}_{13} = \ominus A_1 \oplus A_3$ in $\mathbb{R}^n_s$, considered as vectors in $\mathbb{R}^n$, are linearly independent in $\mathbb{R}^n$. Hence, each coefficient c_{ij} in (9.7) equals zero. Accordingly, the two gyrovector equations in (9.7) are equivalent to the following four scalar equations,

$$c_{11} = c_{12} = c_{21} = c_{22} = 0. \tag{9.8}$$

The resulting four equations are linear in the unknowns $t_{1,0}, t_{2,0}$. Solving the equation $c_{k1} = 0$ for the unknown $t_{k,0}$, and substituting the resulting $t_{k,0}$ into the equation $c_{k2} = 0$, $k = 1, 2$, we obtain the following two equations for gyrobarycentric coordinates:

$$\frac{m_1}{m_3} = \frac{m_{21}}{m_{23}}, \qquad \frac{m_2}{m_3} = \frac{m_{12}}{m_{13}}. \tag{9.9}$$

Since gyrobarycentric coordinates are homogeneous, one may select $m_3 = m_{13}m_{23}$, obtaining

$$(m_1 : m_2 : m_3) = (m_{13}m_{21} : m_{12}m_{23} : m_{13}m_{23}), \tag{9.10}$$

as desired. □

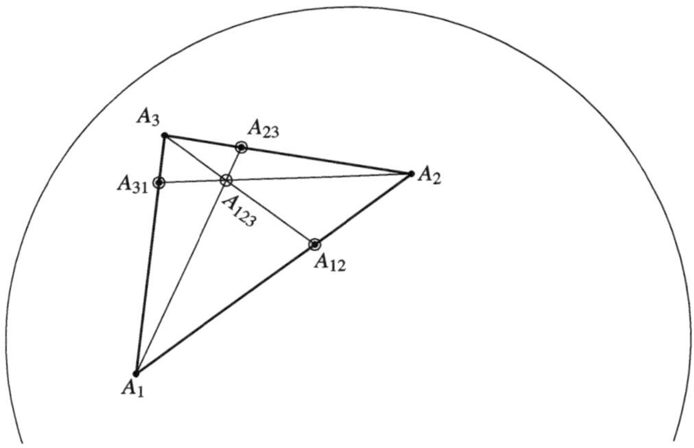

Fig. 9.2 The Hyperbolic Theorem of Ceva. The three gyrocevians A_1A_{23}, A_2A_{31} and A_3A_{12} are concurrent if and only if (9.11) holds

Theorem 9.3 (The Hyperbolic Theorem of Ceva) *Let $A_1A_2A_3$ be a gyrotriangle in an Einstein gyrovector space $(\mathbb{R}^n_s, \oplus, \otimes)$ and let A_{ij} be a point on side A_iA_j of the gyrotriangle, $1 \le i \ne j \le 3$. The three gyrocevians A_1A_{12}, A_1A_{31} and A_1A_{23} are concurrent, Fig.* 9.2, *if and only if*

$$\frac{\gamma_{\ominus A_1 \oplus A_{12}} \|\ominus A_1 \oplus A_{12}\| \gamma_{\ominus A_3 \oplus A_{31}} \|\ominus A_3 \oplus A_{31}\| \gamma_{\ominus A_2 \oplus A_{23}} \|\ominus A_2 \oplus A_{23}\|}{\gamma_{\ominus A_2 \oplus A_{12}} \|\ominus A_2 \oplus A_{12}\| \gamma_{\ominus A_1 \oplus A_{31}} \|\ominus A_1 \oplus A_{31}\| \gamma_{\ominus A_3 \oplus A_{23}} \|\ominus A_3 \oplus A_{23}\|} = 1. \tag{9.11}$$

Proof The proof is presented in two parts.

Part I Let the feet, A_{12}, A_{31} and A_{23}, Fig. 9.2, of the three Cevians in the theorem be given by their gyrobarycentric coordinate representation with respect to the set $S = \{A_1, A_2, A_3\}$ of vertices of the gyrotriangle $A_1A_2A_3$,

$$\begin{aligned} A_{12} &= \frac{m_{31}\gamma_{A_1} A_1 + m_{32}\gamma_{A_2} A_2}{m_{31}\gamma_{A_1} + m_{32}\gamma_{A_2}}, \\ A_{31} &= \frac{m_{21}\gamma_{A_1} A_1 + m_{23}\gamma_{A_3} A_3}{m_{21}\gamma_{A_1} + m_{23}\gamma_{A_3}}, \\ A_{23} &= \frac{m_{12}\gamma_{A_2} A_2 + m_{13}\gamma_{A_3} A_3}{m_{12}\gamma_{A_2} + m_{13}\gamma_{A_3}}. \end{aligned} \tag{9.12}$$

Applying the hyperbolic lever law relation (4.79), p. 102, to each of the gyrobarycentric coordinate representations in (9.12), we obtain the ratios

$$\begin{aligned}\frac{m_{31}}{m_{32}} &= \frac{\gamma_{\ominus A_2 \oplus a_{12}} \|\ominus A_2 \oplus A_{12}\|}{\gamma_{\ominus A_1 \oplus a_{12}} \|\ominus A_1 \oplus A_{12}\|}, \\ \frac{m_{21}}{m_{23}} &= \frac{\gamma_{\ominus A_3 \oplus a_{13}} \|\ominus A_3 \oplus A_{31}\|}{\gamma_{\ominus A_1 \oplus a_{13}} \|\ominus A_1 \oplus A_{31}\|}, \\ \frac{m_{12}}{m_{13}} &= \frac{\gamma_{\ominus A_3 \oplus a_{23}} \|\ominus A_3 \oplus A_{23}\|}{\gamma_{\ominus A_2 \oplus a_{23}} \|\ominus A_2 \oplus A_{23}\|}.\end{aligned} \tag{9.13}$$

By the Gyrobarycentric Coordinate Representation Gyrocovariance Theorem 4.6, p. 90, the left gyrotranslations of the Cevian feet in (9.12) by $\ominus A_1$ are

$$\begin{aligned}\ominus A_1 \oplus A_{12} &= \frac{m_{31}\gamma_{\ominus A_1 \oplus A_1}(\ominus A_1 \oplus A_1) + m_{32}\gamma_{\ominus A_1 \oplus A_2}(\ominus A_1 \oplus A_2)}{m_{31}\gamma_{\ominus A_1 \oplus A_1} + m_{32}\gamma_{\ominus A_1 \oplus A_2}} \\ &= \frac{m_{32}\gamma_{12}\mathbf{a}_{12}}{m_{31} + m_{32}\gamma_{12}}, \\ \ominus A_1 \oplus A_{31} &= \frac{m_{23}\gamma_{13}\mathbf{a}_{13}}{m_{21} + m_{23}\gamma_{13}}, \\ \ominus A_1 \oplus A_{23} &= \frac{m_{12}\gamma_{12}\mathbf{a}_{12} + m_{13}\gamma_{13}\mathbf{a}_{13}}{m_{12}\gamma_{12} + m_{13}\gamma_{13}}.\end{aligned} \tag{9.14}$$

The three gyrocevians A_1A_{23}, A_2A_{31} and A_3A_{12} of gyrotriangle $A_1A_2A_3$, Fig. 9.2, are left gyrotranslated by $\ominus A_1$, respectively, into the three gyrocevians $O(\ominus A_1 \oplus A_{23})$, $(\ominus A_1 \oplus A_2)(\ominus A_1 \oplus A_{31})$ and $(\ominus A_1 \oplus A_3)(\ominus A_1 \oplus A_{12})$, where $O = \mathbf{0} = (0, \ldots, 0)$ is the origin of the Cartesian coordinate system of both $\mathbb{R}^n_s$ and $\mathbb{R}^n$. Owing to the gyrocovariance of points of concurrency, the left gyrotranslated gyrocevians of gyrotriangle $A_1A_2A_3$ are gyrocevians of the left gyrotranslated gyrotriangle $((\ominus A_1 \oplus A_1) = \mathbf{0})(\ominus A_1 \oplus A_2)(\ominus A_1 \oplus A_3)$, and the original gyrocevians of gyrotriangle $A_1A_2A_3$ are concurrent if and only if their left gyrotranslated counterparts are concurrent.

1. The first left gyrotranslated gyrocevian is a gyrosegment that joins its foot $\ominus A_1 \oplus A_{23}$ with vertex $\ominus A_1 \oplus A_1 = O = \mathbf{0}$ of the left gyrotranslated gyrotriangle. It lies on the Euclidean line L_1 in $\mathbb{R}^n$ that passes through the points $O = \mathbf{0}$

and $\ominus A_1 \oplus A_{23}$. Hence, by (9.14),

$$L_1 = (\ominus A_1 \oplus A_1) + \big(-(\ominus A_1 \oplus A_1) + (\ominus A_1 \oplus A_{23})\big)t_1$$

$$= \frac{m_{12}\gamma_{12}\mathbf{a}_{12} + m_{13}\gamma_{13}\mathbf{a}_{13}}{m_{12}\gamma_{12} + m_{13}\gamma_{13}} t_1, \tag{9.15a}$$

where $t_1 \in \mathbb{R}$ is the line parameter.

2. The second left gyrotranslated gyrocevian is a gyrosegment that joins its foot $\ominus A_1 \oplus A_{31}$ with vertex $\ominus A_1 \oplus A_2 = \mathbf{a}_{12}$ of the left gyrotranslated gyrotriangle. It lies on the Euclidean line L_2 in $\mathbb{R}^n$ that passes through the points $\mathbf{a}_{12} = \ominus A_1 \oplus A_2$ and $\ominus A_1 \oplus A_{31}$. Hence, by (9.14),

$$L_2 = (\ominus A_1 \oplus A_2) + \big(-(\ominus A_1 \oplus A_2) + (\ominus A_1 \oplus A_{31})\big)t_1$$

$$= \mathbf{a}_{12} + \left(-\mathbf{a}_{12} + \frac{m_{23}\gamma_{13}\mathbf{a}_{13}}{m_{21} + m_{23}\gamma_{13}}\right)t_2, \tag{9.15b}$$

where $t_2 \in \mathbb{R}$ is the line parameter.

3. The third left gyrotranslated gyrocevian is a gyrosegment that joins its foot $\ominus A_1 \oplus A_{12}$ with vertex $\ominus A_1 \oplus A_3 = \mathbf{a}_{13}$ of the left gyrotranslated gyrotriangle. It lies on the Euclidean line L_3 in $\mathbb{R}^n$ that passes through the points $\mathbf{a}_{13} = \ominus A_1 \oplus A_3$ and $\ominus A_1 \oplus A_{12}$. Hence, by (9.14),

$$L_3 = (\ominus A_1 \oplus A_3) + \big(-(\ominus A_1 \oplus A_3) + (\ominus A_1 \oplus A_{12})\big)t_3$$

$$= \mathbf{a}_{13} + \left(-\mathbf{a}_{13} + \frac{m_{32}\gamma_{12}\mathbf{a}_{12}}{m_{31} + m_{32}\gamma_{12}}\right)t_3, \tag{9.15c}$$

where $t_3 \in \mathbb{R}$ is the line parameter.

Let us assume that there exists a point $P = A_{123} \in \mathbb{R}_s^n$ that lies on each of the three gyrocevians of gyrotriangle $A_1A_2A_3$, Fig. 9.2. Then, its left gyrotranslated point, $\ominus A_1 \oplus P$, by $\ominus A_1$, lies on each of the three lines L_k, $k = 1, 2, 3$. Hence, there exist values $t_{k,0}$ of the line parameters t_k, $k = 1, 2, 3$, respectively, such that

$$\begin{aligned}
&(\ominus A_1 \oplus P) - \frac{m_{12}\gamma_{12}\mathbf{a}_{12} + m_{13}\gamma_{13}\mathbf{a}_{13}}{m_{12}\gamma_{12} + m_{13}\gamma_{13}} t_{1,0} = 0, \\
&(\ominus A_1 \oplus P) - \mathbf{a}_{12} - \left(-\mathbf{a}_{12} + \frac{m_{23}\gamma_{13}\mathbf{a}_{13}}{m_{21} + m_{23}\gamma_{13}}\right)t_{2,0} = 0, \\
&(\ominus A_1 \oplus P) - \mathbf{a}_{13} - \left(-\mathbf{a}_{13} + \frac{m_{32}\gamma_{12}\mathbf{a}_{12}}{m_{31} + m_{32}\gamma_{12}}\right)t_{3,0} = 0,
\end{aligned} \tag{9.16}$$

where P is given by its gyrobarycentric coordinate representation

$$P = \frac{m_1\gamma_{A_1} A_1 + m_2\gamma_{A_2} A_2 + m_3\gamma_{A_3} A_3}{m_1\gamma_{A_1} + m_2\gamma_{A_2} + m_3\gamma_{A_3}} \tag{9.17}$$

with respect to the set $S = \{A_1, A_2, A_3\}$, so that

$$\ominus A_1 \oplus P = \frac{m_2\gamma_{12}\mathbf{a}_{12} + m_3\gamma_{13}\mathbf{a}_{13}}{m_1 + m_2\gamma_{12} + m_3\gamma_{13}}. \tag{9.18}$$

Substituting $\ominus A_1 \oplus P$ from (9.18) into (9.16) and rewriting each equation in (9.16) as a linear combination of $\mathbf{a}_{12}$ and $\mathbf{a}_{13}$ equals zero, we obtain the following homogeneous linear system of three gyrovector equations

$$\begin{aligned} c_{11}\mathbf{a}_{12} + c_{12}\mathbf{a}_{13} &= \mathbf{0}, \\ c_{21}\mathbf{a}_{12} + c_{22}\mathbf{a}_{13} &= \mathbf{0}, \\ c_{31}\mathbf{a}_{12} + c_{32}\mathbf{a}_{13} &= \mathbf{0}, \end{aligned} \tag{9.19}$$

where the coefficients c_{ij}, $i = 1, 2, 3$, $j = 1, 2$, are functions of the unknown values $t_{1,0}, t_{2,0}, t_{3,0}$ of the line parameters t_1, t_2, t_3 and of the gyrobarycentric coordinates $(m_1 : m_2 : m_3)$, $(m_{12}; m_{13})$, $(m_{21}; m_{23})$ and $(m_{31}; m_{32})$.

Since the set $S = \{A_1, A_2, A_3\}$ is pointwise independent, the two gyrovectors $\mathbf{a}_{12} = \ominus A_1 \oplus A_2$ and $\mathbf{a}_{13} = \ominus A_1 \oplus A_3$ in $\mathbb{R}^n_s$, considered as vectors in $\mathbb{R}^n$, are linearly independent in $\mathbb{R}^n$. Hence, each coefficient c_{ij} in (9.19) equals zero. Accordingly, the three gyrovector equations in (9.19) are equivalent to the following six scalar equations,

$$c_{11} = c_{12} = c_{21} = c_{22} = c_{31} = c_{32} = 0. \tag{9.20}$$

The resulting six equations are linear in the unknowns $t_{1,0}, t_{2,0}, t_{3,0}$. Solving the equation $c_{k1} = 0$ for the unknown $t_{k,0}$, and substituting the resulting $t_{k,0}$ into the equation $c_{k2} = 0$, $k = 1, 2, 3$, we obtain the following three equations for gyrobarycentric coordinates:

$$\frac{m_1}{m_2} = \frac{m_{31}}{m_{32}}, \qquad \frac{m_2}{m_3} = \frac{m_{12}}{m_{13}}, \qquad \frac{m_3}{m_1} = \frac{m_{23}}{m_{21}}. \tag{9.21}$$

Hence,

$$\frac{m_{31}}{m_{32}}\frac{m_{12}}{m_{13}}\frac{m_{23}}{m_{21}} = \frac{m_1}{m_2}\frac{m_2}{m_3}\frac{m_3}{m_1} = 1. \tag{9.22}$$

Finally, the result, (9.11), of the theorem follows immediately from (9.22) and the hyperbolic lever relations in (9.13)

Part II Conversely, given (9.11), we will show that the gyrocevians A_1A_{23}, A_2A_{31} and A_3A_{12}, Fig. 9.2, are concurrent.

By Theorem 9.2, p. 269, the two gyrocevians A_1A_{23} and A_2A_{31} intersect at some point A'_{123} on the interior of gyrotriangle $A_1A_2A_3$.

Let $A_3A'_{12}$ be the gyrocevian that contains the points A_3 and A'_{123}, with foot A'_{12} that lies on side A_1A_2 of the gyrotriangle. Now we have the three concurrent gyrocevians A_1A_{23} and A_2A_{31} and $A_3A'_{12}$.

Hence, by Part I of the proof, (9.11) with A_{12} replaced by A'_{12} is valid, that is,

$$\frac{\gamma_{\ominus A_1\oplus A'_{12}}\|\ominus A_1\oplus A'_{12}\|\,\gamma_{\ominus A_3\oplus A_{31}}\|\ominus A_3\oplus A_{31}\|\,\gamma_{\ominus A_2\oplus A_{23}}\|\ominus A_2\oplus A_{23}\|}{\gamma_{\ominus A_2\oplus A'_{12}}\|\ominus A_2\oplus A'_{12}\|\,\gamma_{\ominus A_1\oplus A_{31}}\|\ominus A_1\oplus A_{31}\|\,\gamma_{\ominus A_3\oplus A_{23}}\|\ominus A_3\oplus A_{23}\|}=1. \tag{9.23}$$

Following (9.23) and assumption (9.11), we have

$$\frac{m_1}{m_2}:=\frac{\gamma_{\ominus A_1\oplus A_{12}}\|\ominus A_1\oplus A_{12}\|}{\gamma_{\ominus A_2\oplus A_{12}}\|\ominus A_2\oplus A_{12}\|}=\frac{\gamma_{\ominus A_1\oplus A'_{12}}\|\ominus A_1\oplus A'_{12}\|}{\gamma_{\ominus A_2\oplus A'_{12}}\|\ominus A_2\oplus A'_{12}\|}. \tag{9.24}$$

The equality in (9.24) follows from (9.23) and (9.11), and $m_1 > 0$ and $m_2 > 0$ are any two positive numbers the ratio of which is defined in (9.24).

By Theorem 4.12, p. 103, both A_{12} and A'_{12} in (9.24) possess the same gyrobarycentric coordinate representation with respect to the set $S = \{A_1, A_2\}$,

$$A_{12}=A'_{12}=\frac{m_1\gamma_{A_1}A_1+m_2\gamma_{A_2}A_2}{m_1\gamma_{A_1}+m_2\gamma_{A_2}} \tag{9.25}$$

so that, in particular, $A_{12} = A'_{12}$. This equality, in turn, implies that the three gyrocevians A_1A_{23} A_2A_{31} and A_3A_{12} are concurrent, as desired. □

Owing to the importance of the result in (9.22), obtained in the process of proving the Hyperbolic Ceva Theorem 9.3, we elevate it into the following Corollary:

Corollary 9.4 (A Gyrocevian Concurrence Condition) *Let the feet, A_{12}, A_{31} and A_{23}, Fig. 9.2, p. 273, of three Cevians that emanate from the three vertices of a gyrotriangle $A_1A_2A_3$ in an Einstein gyrovector space $(\mathbb{R}^n_s, \oplus, \otimes)$ be given by their gyrobarycentric coordinate representation with respect to the set $S = \{A_1, A_2, A_3\}$ of vertices of the gyrotriangle $A_1A_2A_3$,*

$$\begin{aligned}
A_{12}&=\frac{m_{31}\gamma_{A_1}A_1+m_{32}\gamma_{A_2}A_2}{m_{31}\gamma_{A_1}+m_{32}\gamma_{A_2}},\\
A_{31}&=\frac{m_{21}\gamma_{A_1}A_1+m_{23}\gamma_{A_3}A_3}{m_{21}\gamma_{A_1}+m_{23}\gamma_{A_3}},\\
A_{23}&=\frac{m_{12}\gamma_{A_2}A_2+m_{13}\gamma_{A_3}A_3}{m_{12}\gamma_{A_2}+m_{13}\gamma_{A_3}}.
\end{aligned} \tag{9.26}$$

The three gyrocevians A_1A_{12}, A_1A_{31} and A_1A_{23} are concurrent, Fig. 9.2, *if and only if*

$$\frac{m_{31}}{m_{32}}\frac{m_{12}}{m_{13}}\frac{m_{23}}{m_{21}} = 1. \tag{9.27}$$

Proof The result (9.27) of the corollary is established in the result (9.22) of the proof of Theorem 9.3, p. 273. □

9.2 The Gyrocevian Gyroangles Theorem

An A_3-gyrocevian of a gyrotriangle $A_1A_2A_3$ with foot P_3 is shown in Fig. 9.3. It generates two pairs of gyroangles on the interior of the gyrotriangle. These gyroangles, shown in Fig. 9.3, are the pair of gyroangles ε_1 and ε_2, satisfying $\varepsilon_1 + \varepsilon_1 = \pi$, and the pair of gyroangles $\alpha_{3,1}$ and $\alpha_{3,2}$, satisfying $\alpha_{3,1} + \alpha_{3,2} = \alpha_3$. The location of the gyrocevian foot, P_3, on the interior of the gyrotriangle side A_1A_2 is determined by its gyrobarycentric coordinates $m_1 > 0$ and $m_2 > 0$ with respect to the set of points $S = \{A_1, A_2\}$ according to the equation

$$P_3 = \frac{m_1\gamma_{A_1} A_1 + m_2\gamma_{A_2} A_2}{m_1\gamma_{A_1} + m_2\gamma_{A_2}}. \tag{9.28}$$

The following theorem determines the two pairs $(\varepsilon_1, \varepsilon_2)$ and $(\alpha_{3,1}, \alpha_{3,2})$ of gyroangles in terms of the gyrotriangle gyroangles $\alpha_1, \alpha_2, \alpha_3$ and the gyrobarycentric coordinates $(m_1 : m_2)$ of the gyrocevian foot P_3.

Theorem 9.5 (The Gyrocevian Gyroangles Theorem I)

1. *Let $A_1A_2A_3$ be a gyrotriangle in an Einstein gyrovector space* $(\mathbb{R}^n_s, \oplus, \otimes)$.
2. *Let P_3 be the foot of an A_3-gyrocevian, lying on the interior of side A_1A_2, generating the gyroangles $(\varepsilon_1, \varepsilon_2)$ and $(\alpha_{3,1}, \alpha_{3,2})$, as shown in Fig.* 9.3.
3. *Let $(m_1 : m_2)$, $m_1, m_2 > 0$, be gyrobarycentric coordinates of P_3 with respect to the set of points $S = \{A_1, A_2\}$, so that P_3 is given by its gyrobarycentric coordinate representation* (9.28).
4. *Furthermore, let $m_0 > 0$ be the constant of the gyrobarycentric coordinate representation* (9.28), *given by, Definition* 4.5, *p.* 89, (4.27), *p.* 90, (7.146), *p.* 188,

$$\begin{aligned} m_0^2 &= m_1^2 + m_2^2 + 2m_1m_2\gamma_{12} \\ &= (m_1 \pm m_2)^2 + 2m_1m_2(\gamma_{12} \mp 1) \\ &= (m_1 \pm m_2)^2 + 2m_1m_2\frac{\cos\alpha_3 + \cos(\alpha_1 \mp \alpha_2)}{\sin\alpha_1 \sin\alpha_2}. \end{aligned} \tag{9.29}$$

5. *And, finally, let $D > 0$ be given by the equation*

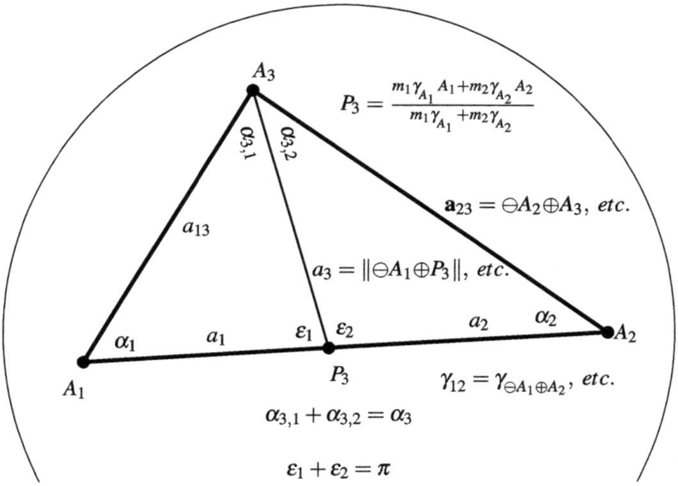

Fig. 9.3 A gyrotriangle gyrocevian A_3P_3 and its foot P_3 of a gyrotriangle $A_1A_2A_3$ in an Einstein gyrovector space $(\mathbb{R}^n_s, \oplus, \otimes)$ is shown for $n = 2$. It forms the two pairs of gyroangles $(\alpha_{3,1}, \alpha_{3,2})$ and $(\varepsilon_1, \varepsilon_2)$ that are determined in (9.31)–(9.32) in terms of the gyrotriangle gyroangles α_k, $k = 1, 2, 3$, and the gyrobarycentric coordinates of the foot P_3

$$D^2 = (m_1 \sin\alpha_2 - m_2 \sin\alpha_1)^2 + 2m_1m_2 \sin\alpha_1 \sin\alpha_2(1 + \cos\alpha_3). \tag{9.30}$$

Then,

$$\begin{aligned}
\sin\varepsilon_1 &= \frac{1}{D} m_0 \sin\alpha_1 \sin\alpha_2, \\
\cos\varepsilon_1 &= \frac{1}{D}(-m_1 \cos\alpha_1 \sin\alpha_2 + m_2 \cos\alpha_2 \sin\alpha_1), \\
\sin\varepsilon_2 &= \frac{1}{D} m_0 \sin\alpha_1 \sin\alpha_2, \\
\cos\varepsilon_2 &= \frac{1}{D}(m_1 \cos\alpha_1 \sin\alpha_2 - m_2 \cos\alpha_2 \sin\alpha_1),
\end{aligned} \tag{9.31}$$

$$\begin{aligned}
\sin\alpha_{3,1} &= \frac{1}{D} m_2 \sin\alpha_1 \sin\alpha_3, \\
\cos\alpha_{3,1} &= \frac{1}{D}(m_1 \sin\alpha_2 + m_2 \sin\alpha_1 \cos\alpha_3), \\
\sin\alpha_{3,2} &= \frac{1}{D} m_1 \sin\alpha_2 \sin\alpha_3, \\
\cos\alpha_{3,2} &= \frac{1}{D}(m_2 \sin\alpha_1 + m_1 \sin\alpha_2 \cos\alpha_3),
\end{aligned} \tag{9.32}$$

and

$$\frac{m_1 \sin\alpha_{3,1}}{m_2 \sin\alpha_{3,2}} = \frac{\sin\alpha_1}{\sin\alpha_2}. \tag{9.33}$$

Proof The proof is divided into four parts for the sake of clarity.

Part I As in the theorem, let $A_1A_2A_3$ be a gyrotriangle in an Einstein gyrovector space $(\mathbb{R}^n_s, \oplus, \otimes)$, and let A_3P_3 be any A_3-gyrocevian of the gyrotriangle, joining the vertex A_3 with its foot P_3, a point on the interior of side A_1A_2 opposite to A_3, as shown in Fig. 9.3. Furthermore, let the gyrocevian foot P_3 be given by its gyrobarycentric coordinates $m_1 > 0$ and $m_2 > 0$ with respect to the set $\{A_1, A_2\}$, (9.28),

$$P_3 = \frac{m_1\gamma_{A_1}A_1 + m_2\gamma_{A_2}A_2}{m_1\gamma_{A_1} + m_2\gamma_{A_2}}. \tag{9.34}$$

Following the Gyrobarycentric Coordinate Representation Gyrocovariance Theorem 4.6, p. 90, the gyrobarycentric coordinate representation of the point P_3 in (9.34) gives rise to the identities in (9.35)–(9.37) below:

$$\ominus X \oplus P_3 = \frac{m_1\gamma_{\ominus X\oplus A_1}(\ominus X\oplus A_1) + m_2\gamma_{\ominus X\oplus A_2}(\ominus X\oplus A_2)}{m_1\gamma_{\ominus X\oplus A_1} + m_2\gamma_{\ominus X\oplus A_2}} \tag{9.35}$$

and

$$\begin{aligned} \gamma_{\ominus X\oplus P_3} &= \frac{m_1\gamma_{\ominus X\oplus A_1} + m_2\gamma_{\ominus X\oplus A_2}}{m_0}, \\ \gamma_{\ominus X\oplus P_3}(\ominus X\oplus P_3) &= \frac{m_1\gamma_{\ominus X\oplus A_1}(\ominus X\oplus A_1) + m_2\gamma_{\ominus X\oplus A_2}(\ominus X\oplus A_2)}{m_0} \end{aligned} \tag{9.36}$$

for any $X \in \mathbb{R}^n_s$, where, in the notation of Fig. 9.3 for the gamma factor γ_{12}, the constant $m_0 > 0$ in (4.28d), p. 91, specializes to

$$m_0^2 = (m_1 + m_2)^2 + 2m_1m_2(\gamma_{12} - 1) \tag{9.37}$$

in (9.36).

Using the notation in Fig. 9.3, it follows from (9.35) with $X = A_1$ that

$$\mathbf{a}_1 := \ominus A_1 \oplus P_3 = \frac{m_2\gamma_{\ominus A_1\oplus A_2}(\ominus A_1\oplus A_2)}{m_1 + m_2\gamma_{\ominus A_1\oplus A_2}} = \frac{m_2\gamma_{12}\mathbf{a}_{12}}{m_1 + m_2\gamma_{12}} \tag{9.38}$$

and, similarly, with $X = A_2$,

$$\mathbf{a}_2 := \ominus A_2 \oplus P_3 = \frac{m_1\gamma_{\ominus A_2\oplus A_1}(\ominus A_2\oplus A_1)}{m_1\gamma_{\ominus A_2\oplus A_1} + m_2} = \frac{m_1\gamma_{21}\mathbf{a}_{21}}{m_1\gamma_{21} + m_2} \tag{9.39}$$

and, similarly, with $X = A_3$,

$$\mathbf{a}_3 := \ominus A_3 \oplus P_3 = \frac{m_1 \gamma_{\ominus A_3 \oplus A_1} (\ominus A_3 \oplus A_1) + m_2 \gamma_{\ominus A_3 \oplus A_2} (\ominus A_3 \oplus A_2)}{m_1 \gamma_{\ominus A_3 \oplus A_1} + m_2 \gamma_{\ominus A_3 \oplus A_2}}$$

$$= \frac{m_1 \gamma_{13} \mathbf{a}_{31} + m_2 \gamma_{23} \mathbf{a}_{32}}{m_1 \gamma_{13} + m_2 \gamma_{23}}. \tag{9.40}$$

Hence, by (9.38)–(9.39), in the notation of Fig. 9.3,

$$\begin{aligned} a_1 &:= \|\mathbf{a}_1\| = \frac{m_2 \gamma_{12} a_{12}}{m_1 + m_2 \gamma_{12}}, \\ a_2 &:= \|\mathbf{a}_2\| = \frac{m_1 \gamma_{12} a_{12}}{m_1 \gamma_{12} + m_2}, \end{aligned} \tag{9.41}$$

noting that while, in general, $\mathbf{a}_{21} = \ominus A_2 \oplus A_1 \neq \ominus A_1 \oplus A_2 = \mathbf{a}_{12}$, we have $a_{21} = \|\ominus A_2 \oplus A_1\| = \|\ominus A_1 \oplus A_2\| = a_{12}$ and, hence, $\gamma_{21} = \gamma_{12}$.

Similarly, it follows from the first identity in (9.36) with $X = A_1$, $X = A_2$ and $X = A_3$, respectively,

$$\begin{aligned} \gamma_{a_1} &= \gamma_{\ominus A_1 \oplus P_3} = \frac{m_1 + m_2 \gamma_{\ominus A_1 \oplus A_2}}{m_0} = \frac{m_1 + m_2 \gamma_{12}}{m_0}, \\ \gamma_{a_2} &= \gamma_{\ominus A_2 \oplus P_3} = \frac{m_1 \gamma_{\ominus A_2 \oplus A_1} + m_2}{m_0} = \frac{m_1 \gamma_{12} + m_2}{m_0}, \\ \gamma_{a_3} &= \gamma_{\ominus A_3 \oplus P_3} = \frac{m_1 \gamma_{\ominus A_3 \oplus A_1} + m_2 \gamma_{\ominus A_3 \oplus A_2}}{m_0} = \frac{m_1 \gamma_{13} + m_2 \gamma_{23}}{m_0}. \end{aligned} \tag{9.42}$$

It follows from (9.41) and (9.42) or, equivalently, from (9.34) and the second identity in (9.36) with $X = A_1$ and $X = A_2$, respectively, that

$$\begin{aligned} \gamma_{a_1} a_1 &= \frac{m_1 + m_2 \gamma_{12}}{m_0} \frac{m_2 \gamma_{12} a_{12}}{m_1 + m_2 \gamma_{12}} = \frac{m_2}{m_0} \gamma_{12} a_{12}, \\ \gamma_{a_2} a_2 &= \frac{m_1 \gamma_{12} + m_2}{m_0} \frac{m_1 \gamma_{12} a_{12}}{m_1 \gamma_{12} + m_2} = \frac{m_1}{m_0} \gamma_{12} a_{12}, \end{aligned} \tag{9.43}$$

implying

$$\frac{\gamma_{a_1} a_1}{\gamma_{a_2} a_2} = \frac{m_2}{m_1}. \tag{9.44}$$

Following the second identity in (9.36) with $X = A_3$, we have

$$\begin{aligned}\gamma_{a_3} a_3 &= \gamma_{\ominus A_3 \oplus P_3}(\ominus A_3 \oplus P_3) \\ &= \frac{m_1 \gamma_{\ominus A_3 \oplus A_1}(\ominus A_3 \oplus A_1) + m_2 \gamma_{\ominus A_3 \oplus A_2}(\ominus A_3 \oplus A_2)}{m_0} \\ &= \frac{m_1 \gamma_{13} \mathbf{a}_{31} + m_2 \gamma_{23} \mathbf{a}_{32}}{m_0}. \end{aligned} \tag{9.45}$$

Noting that the gyrovectors $\mathbf{a}_{31}$ and $\mathbf{a}_{32}$ include the gyroangle α_3, as shown in Fig. 9.3, we have

$$\mathbf{a}_{31} \cdot \mathbf{a}_{32} = a_{13} a_{23} \cos \alpha_3. \tag{9.46}$$

Hence, by (9.45),

$$\begin{aligned}\gamma_{a_3}^2 a_3^2 &= \frac{1}{m_0^2}(m_1 \gamma_{13} \mathbf{a}_{31} + m_2 \gamma_{23} \mathbf{a}_{32})^2 \\ &= \frac{1}{m_0^2}\big(m_1^2 \gamma_{13}^2 a_{13}^2 + m_2^2 \gamma_{23}^2 a_{23}^2 + 2 m_1 m_2 \gamma_{13} \gamma_{23} a_{13} a_{23} \cos \alpha_3\big). \end{aligned} \tag{9.47}$$

Applying the law of gyrosines (6.44), p. 140, to gyrotriangle $A_1 P_3 A_3$ in Fig. 9.3, we have

$$\frac{\sin \varepsilon_1}{\gamma_{13} a_{13}} = \frac{\sin \alpha_1}{\gamma_{a_3} a_3} \tag{9.48}$$

so that, by (9.48) and (9.47),

$$\sin^2 \varepsilon_1 = \frac{\gamma_{13}^2 a_{13}^2}{\gamma_{a_3}^2 a_3^2} \sin^2 \alpha_1 = \frac{1}{D^2} m_0^2 \sin^2 \alpha_1 \sin^2 \alpha_2, \tag{9.49}$$

where $D > 0$ is given by (9.30).

Since a gyrosine of a (proper) gyrotriangle is always positive, (9.49) implies

$$\sin \varepsilon_1 = \frac{1}{D} m_0 \sin \alpha_1 \sin \alpha_2. \tag{9.50}$$

Interchanging the gyrotriangle vertices A_1 and A_2, (9.50) gives

$$\sin \varepsilon_2 = \frac{1}{D} m_0 \sin \alpha_1 \sin \alpha_2 \tag{9.51}$$

so that $\sin \varepsilon_1 = \sin \varepsilon_2$, as expected from the condition $\varepsilon_1 + \varepsilon_2 = \pi$ shown in Fig. 9.3.

Equations (9.50) and (9.51) verify the first and the third equations in (9.31).

Part II By the law of gyrocosines (6.20), p. 134, gyroangle ε_1 of gyrotriangle $A_1P_3A_3$ in Fig. 9.3 satisfies the equation

$$\cos^2\varepsilon_1 = \frac{(-\gamma_{13} + \gamma_{a_1}\gamma_{a_3})^2}{(\gamma_{a_1}^2 - 1)(\gamma_{a_3}^2 - 1)}. \tag{9.52}$$

The substitution of γ_{a_1} and γ_{a_3} from (9.42) into (9.52) gives

$$\cos^2\varepsilon_1 = \frac{\{-m_1(\gamma_{12}\gamma_{13} - \gamma_{23}) + m_2(\gamma_{12}\gamma_{23} - \gamma_{13})\}^2}{(\gamma_{12}^2 - 1)\{m_1^2(\gamma_{13}^2 - 1) + m_2^2(\gamma_{23}^2 - 1) + 2m_1m_2(\gamma_{13}\gamma_{23} - \gamma_{12})\}}. \tag{9.53}$$

Substitutions from the useful gyrotriangle gyrotrigonometric identities in Sect. 7.12, p. 187, into (9.53) give

$$\cos\varepsilon_1 = \frac{1}{D}(-m_1\cos\alpha_1\sin\alpha_2 + m_2\cos\alpha_2\sin\alpha_1), \tag{9.54}$$

where the ambiguous sign that results from the transition from $\cos^2\varepsilon_1$ in (9.53) into $\cos\varepsilon_1$ in (9.54) is determined by the condition, clearly seen in Fig. 9.3, that ε_1 tends to α_2 as m_1 approaches 0.

Interchanging the gyrotriangle vertices A_1 and A_2, (9.54) gives

$$\cos\varepsilon_2 = \frac{1}{D}(m_1\cos\alpha_1\sin\alpha_2 - m_2\cos\alpha_2\sin\alpha_1) \tag{9.55}$$

so that $\cos\varepsilon_1 = -\cos\varepsilon_2$, as expected from the condition $\varepsilon_1 + \varepsilon_2 = \pi$ shown in Fig. 9.3.

Equations (9.54) and (9.55) verify the second and the fourth equations in (9.31).

Part III Applying the law of gyrosines (6.44), p. 140, to gyrotriangle $A_1P_3A_3$ in Fig. 9.3, we have

$$\frac{\sin\alpha_{3,1}}{\gamma_{a_1}a_1} = \frac{\sin\varepsilon_1}{\gamma_{13}a_{13}} \tag{9.56}$$

so that, by (9.56), (9.47) and (9.50),

$$\begin{aligned}\sin^2\alpha_{3,1} &= \frac{\gamma_{a_1}^2a_1^2}{\gamma_{13}^2a_{13}^2}\sin^2\varepsilon_1 = \frac{m_2^2\gamma_{12}^2a_{12}^2}{m_0^2\gamma_{13}^2a_{13}^2}\sin^2\varepsilon_1 = \frac{m_2^2\sin^2\alpha_3}{m_0^2\sin^2\alpha_2}\sin^2\varepsilon_1\\ &= \frac{1}{D^2}m_2^2\sin^2\alpha_1\sin^2\alpha_3.\end{aligned} \tag{9.57}$$

Since a gyrosine of a (proper) gyrotriangle is always positive, (9.57) implies

$$\sin\alpha_{3,1} = \frac{1}{D}m_2\sin\alpha_1\sin\alpha_3. \tag{9.58}$$

Interchanging the gyrotriangle vertices A_1 and A_2, (9.58) gives

$$\sin\alpha_{3,2} = \frac{1}{D} m_1 \sin\alpha_2 \sin\alpha_3. \tag{9.59}$$

Equations (9.58) and (9.59) verify the first and the third equations in (9.32).

Part IV By the law of gyrocosines (6.20), p. 134, gyroangle $\alpha_{3,1}$ of gyrotriangle $A_1P_3A_3$ in Fig. 9.3 satisfies the equation

$$\cos^2\alpha_{3,1} = \frac{(-\gamma_{a_1} + \gamma_{a_3}\gamma_{13})^2}{(\gamma_{a_3}^2 - 1)(\gamma_{13}^2 - 1)}. \tag{9.60}$$

The substitution of γ_{a_1} and γ_{a_3} from (9.42) into (9.52) gives

$$\cos^2\alpha_{3,1} = \frac{\{m_1(\gamma_{13}^2 - 1) + m_2(\gamma_{13}\gamma_{23} - \gamma_{12})\}^2}{(\gamma_{12}^2 - 1)\{m_1^2(\gamma_{13}^2 - 1) + m_2^2(\gamma_{23}^2 - 1) + 2m_1m_2(\gamma_{13}\gamma_{23} - \gamma_{12})\}}. \tag{9.61}$$

Substitutions from the useful gyrotriangle gyrotrigonometric identities in Sect. 7.12, p. 187, into (9.61) give

$$\cos\alpha_{3,1} = \frac{1}{D}(m_1 \sin\alpha_2 + m_2 \sin\alpha_1 \cos\alpha_3), \tag{9.62}$$

where the ambiguous sign that results from the transition from $\cos^2\alpha_{3,1}$ in (9.61) into $\cos\alpha_{3,1}$ in (9.62) is determined by the condition, clearly seen in Fig. 9.3, that $\alpha_{3,1}$ tends to α_3 as m_1 approaches 0.

Interchanging the gyrotriangle vertices A_1 and A_2, (9.62) gives

$$\cos\alpha_{3,2} = \frac{1}{D}(m_2 \sin\alpha_1 + m_1 \sin\alpha_2 \cos\alpha_3). \tag{9.63}$$

Equations (9.62) and (9.63) verify the second and the fourth equations in (9.32). The proof of Theorem 9.5 is thus complete. □

9.3 Gyrocevian Gyrolength

Let $A_1A_2A_3$ be a gyrotriangle in an Einstein gyrovector space $(\mathbb{R}_s^n, \oplus, \otimes)$ and let A_3P_3 be its A_3-gyrocevian with foot P_3 lying on the interior of side A_1A_2, as shown in Fig. 9.3, p. 279. Furthermore, let P_3 be given by its gyrobarycentric coordinate representation with respect to the set $\{A_1, A_2\}$, (9.28),

$$P_3 = \frac{m_1\gamma_{A_1} A_1 + m_2\gamma_{A_2} A_2}{m_1\gamma_{A_1} + m_2\gamma_{A_2}}. \tag{9.64}$$

We wish to express the gyrolength $a_3 = \|\ominus A_3 \oplus P_3\|$ of the gyrocevian A_3P_3, Fig. 9.3, in terms of its gyrotriangle parameters and its gyrobarycentric coordinates $m_1 > 0$ and $m_2 > 0$.

By an application of the AAA to SSS conversion law, (6.33), p. 137, to gyrotriangle $A_1P_3A_3$ in Fig. 9.3, we obtain the equation

$$\gamma_{a_3} = \frac{\cos\alpha_1 + \cos\varepsilon_1 \cos\alpha_{3,1}}{\sin\varepsilon_1 \sin\alpha_{3,1}} \tag{9.65}$$

noting that, by (1.9), p. 5,

$$\frac{a_3^2}{s^2} = \frac{\gamma_{a_3}^2 - 1}{\gamma_{a_3}^2}. \tag{9.66}$$

By successive substitutions into (9.66) from (9.65) and from (9.31)–(9.32), p. 279, we obtain

$$a_3^2 = \frac{4s^2F\{(m_1\sin\alpha_2 - m_2\sin\alpha_1)^2 + 2m_1m_2\sin\alpha_1\sin\alpha_2(1+\cos\alpha_3)\}}{\{m_1\sin\alpha_2(\cos\alpha_2 + \cos\alpha_1\cos\alpha_3) + m_2\sin\alpha_1(\cos\alpha_1 + \cos\alpha_2\cos\alpha_3)\}^2}, \tag{9.67}$$

where, as in (7.144),

$$\begin{aligned} F &= F(\alpha_1, \alpha_2, \alpha_3) \\ &= \cos\frac{\alpha_1+\alpha_2+\alpha_3}{2}\cos\frac{\alpha_1-\alpha_2-\alpha_3}{2}\cos\frac{-\alpha_1+\alpha_2-\alpha_3}{2}\cos\frac{-\alpha_1-\alpha_2+\alpha_3}{2}. \end{aligned} \tag{9.68}$$

By (7.35), p. 163,

$$s^2F = R^2\cos^2\frac{\alpha_1-\alpha_2-\alpha_3}{2}\cos^2\frac{-\alpha_1+\alpha_2-\alpha_3}{2}\cos^2\frac{-\alpha_1-\alpha_2+\alpha_3}{2}, \tag{9.69}$$

where R is the circumgyroradius of gyrotriangle $A_1A_2A_3$, studied in Sect. 7.4.

Eliminating s^2F between (9.67) and (9.69), we obtain an interesting relation between a gyrocevian and its gyrotriangle that is formalized in the following theorem:

Theorem 9.6 (The Gyrocevian Gyroangles Theorem II) *Let $A_1A_2A_3$ be a gyrotriangle with circumgyroradius R in an Einstein gyrovector space $(\mathbb{R}_s^n, \oplus, \otimes)$. Then, in the notation of Fig.* 9.3, *p.* 279, *the gyrolength a_3 of the gyrocevian A_3P_3,*

$$a_3 = \|\ominus A_3 \oplus P_3\|, \tag{9.70}$$

is related to R by the equation

$$\begin{aligned} a_3 &= 2R\frac{\cos\frac{\alpha_1-\alpha-\alpha_3}{2}\cos\frac{-\alpha_1+\alpha-\alpha_3}{2}\cos\frac{-\alpha_1-\alpha+\alpha_3}{2}}{m_1\sin\alpha_2(\cos\alpha_2+\cos\alpha_1\cos\alpha_3)+m_2\sin\alpha_1(\cos\alpha_1+\cos\alpha_2\cos\alpha_3)} \\ &\quad\times\sqrt{(m_1\sin\alpha_2-m_2\sin\alpha_1)^2+2m_1m_2\sin\alpha_1\sin\alpha_2(1+\cos\alpha_3)}. \end{aligned} \tag{9.71}$$

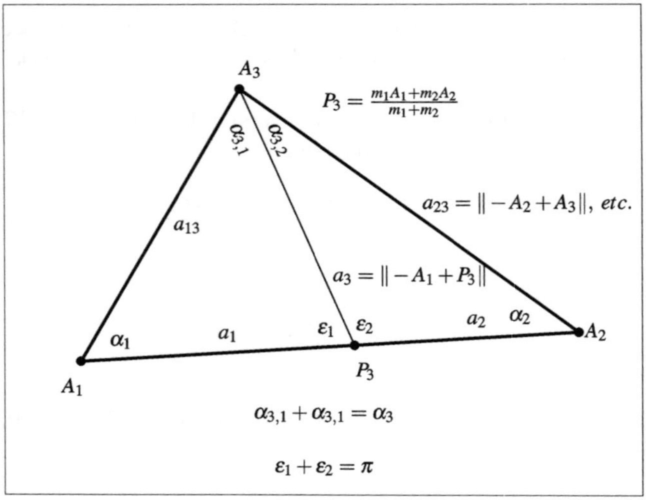

Fig. 9.4 A triangle Cevian A_3P_3 and its foot P_3 of a triangle $A_1A_2A_3$ in a Euclidean space $\mathbb{R}^n$ is shown for $n = 2$. It forms the two pairs of gyroangles $(\alpha_{3,1}, \alpha_{3,2})$ and $(\varepsilon_1, \varepsilon_2)$ that are determined in (9.31)–(9.32) in terms of the triangle angles and the barycentric coordinates of the foot P_3. The equations in (9.31)–(9.32), in turn, can be simplified in Euclidean geometry owing to the condition $\alpha_1 + \alpha_2 + \alpha_3 = \pi$

9.4 Cevian Length

Interestingly, the elegant relation (9.71) remains invariant in form under the Euclidean limit $s \to \infty$, so that it is valid in Euclidean geometry, Fig. 9.4, as well. However, for application in Euclidean geometry the relation (9.71) can be simplified owing to the fact that triangle angle sum in π.

Indeed, under the condition

$$\alpha_1 + \alpha + \alpha_3 = \pi, \qquad \text{(Euclidean Geometry)} \tag{9.72a}$$

we have the following trigonometric identities for triangle angles:

$$\begin{aligned} \cos\frac{\alpha_1 - \alpha - \alpha_3}{2} &= \sin\alpha_1, \\ \cos\frac{-\alpha_1 + \alpha_2 - \alpha_3}{2} &= \sin\alpha_2, \\ \cos\frac{-\alpha_1 - \alpha_2 + \alpha_3}{2} &= \sin\alpha_3, \end{aligned} \tag{9.72b}$$

and

$$\cos\alpha_1 + \cos\alpha_2 \cos\alpha_3 = \sin\alpha_2 \sin\alpha_3,$$
$$\cos\alpha_2 + \cos\alpha_1 \cos\alpha_3 = \sin\alpha_1 \sin\alpha_3, \tag{9.72c}$$
$$\cos\alpha_3 + \cos\alpha_1 \cos\alpha_2 = \sin\alpha_1 \sin\alpha_2.$$

Substituting the simplifications offered by (9.72a), (9.72b), (9.72c) into the result (9.71) of Theorem 9.6, we obtain the following corollary of Theorem 9.6:

Corollary 9.7 *Let $A_1A_2A_3$ be a triangle with angles α_k, $k = 1, 2, 3$, and circumradius R in a Euclidean space $\mathbb{R}^n$. Furthermore, let A_3P_3 be the triangle Cevian that joins the vertex A_3 with a point P_3 that lies on the interior of side A_1A_2, and that possesses the barycentric coordinate representation, Fig.* 9.4,

$$P_3 = \frac{m_1A_1 + m_2A_2}{m_1 + m_2}. \tag{9.73}$$

Then, the length a_3 of the Cevian A_3P_3,

$$a_3 = \| - A_3 + P_3\|, \tag{9.74}$$

is related to R by the Euclidean geometric equation

$$a_3 = \frac{2R}{m_1 + m_2}\sqrt{(m_1 \sin\alpha_2 - m_2 \sin\alpha_1)^2 + 2m_1m_2 \sin\alpha_1 \sin\alpha_2(1 - \cos(\alpha_1 + \alpha_2))}. \tag{9.75}$$

As an example, the special case of Theorem 9.6 when the gyrotriangle is equilateral (that is, it has all three sides congruent) is presented below as an obvious corollary:

Corollary 9.8 *Let $A_1A_2A_3$ be an equilateral gyrotriangle with gyroangles α, gyroaltitudes h and a circumgyroradius R in an Einstein gyrovector space $(\mathbb{R}^n_s, \oplus, \otimes)$. Then*

$$h = \frac{R}{1 - \tan^2 \frac{\alpha}{2}}. \quad \text{(Hyperbolic Geometry)} \tag{9.76}$$

Like the result of Theorem 9.6, the result (9.76) of Corollary (9.8) remains valid in Euclidean geometry as well. In Euclidean geometry, however, the angle α of an equilateral triangle is unique, $\alpha = \frac{\pi}{3}$. Hence, in Euclidean geometry, (9.76) can be simplified,

$$h = \frac{3R}{2}. \quad \text{(Euclidean Geometry)} \tag{9.77}$$

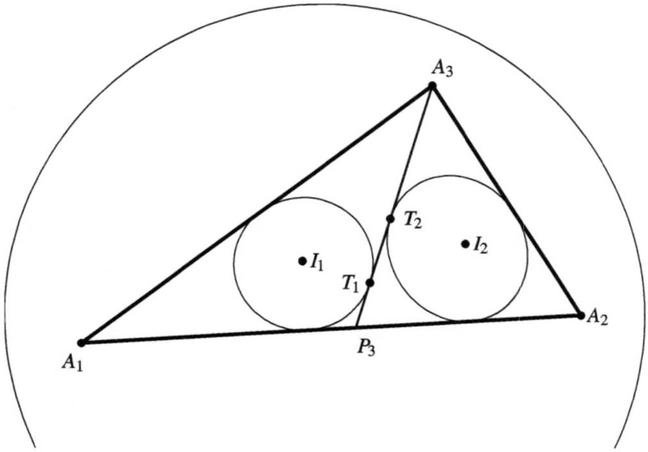

Fig. 9.5 A gyrotriangle split by a gyrocevian into two gyrotriangles, along with the resulting two ingyrocircles. The two ingyrocircles with gyrocenters I_1 and I_2 touch their generating gyrocevian at the points of tangency T_1 and T_2

9.5 A Special Gyrocevian

Let $A_1A_2A_3$ be a gyrotriangle in an Einstein gyrovector space $(\mathbb{R}_s^n, \oplus, \otimes)$ and let P_3 be a point on the interior of its side A_1A_2. The gyrocevian A_3P_3 splits the gyrotriangle into the two gyrotriangles $A_1P_3A_3$ and $A_2P_3A_3$. The ingyrocircles of these gyrotriangles touch the gyrocevian A_3P_3 at the tangency points T_1 and T_2, respectively, as shown in Figs. 9.5 and 9.6.

We wish to determine the unique gyrocevian for which the two points of tangency, T_1 and T_2, are equal,

$$T_1 = T_2 \tag{9.78}$$

as shown in Fig. 9.6.

Let the gyrocevian foot P_3 be given by its gyrobarycentric coordinate representation, (9.64),

$$P_3 = \frac{m_1\gamma_{A_1} A_1 + m_2\gamma_{A_2} A_2}{m_1\gamma_{A_1} + m_2\gamma_{A_2}} \tag{9.79}$$

with respect to the set $S = \{A_1, A_2\}$, where the gyrobarycentric coordinates $m_1 > 0$ and $m_2 > 0$ are to be determined uniquely by Condition (9.78).

Furthermore, let I_1 and I_2 be the ingyrocircle gyrocenters of gyrotriangles $A_1P_3A_3$ and $A_2P_3A_3$, respectively, so that T_1 and T_2 are the points of tangency where the I_1-ingyrocircle and the I_2-ingyrocircle meet, respectively, the gyrocevian A_3P_3, as shown in Figs. 9.5 and 9.6.

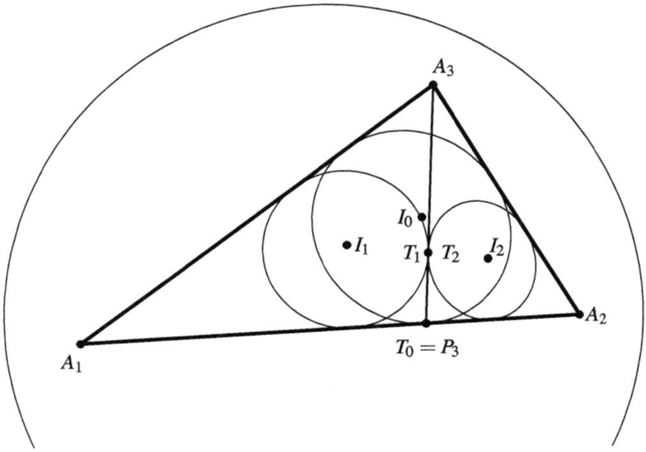

Fig. 9.6 Here the gyrocevian A_3P_3 of gyrotriangle $A_1A_2A_3$ is special. It generates two gyrotriangles that possess the two ingyrocircles with gyrocenters I_1 and I_2 that touch the gyrocevian at the same point of tangency, $T_1 = T_2$. This special gyrocevian A_3P_3 has the foot P_3 that turns out to be the point of tangency T_0 where the ingyrocircle of gyrotriangle $A_1A_2A_3$ touches the side A_1A_2

Then, by (7.174), p. 194,

$$\begin{aligned} T_1 &= \frac{\cot\frac{\alpha_{3,1}}{2}\gamma_{P_3}P_3 + \cot\frac{\varepsilon_1}{2}\gamma_{A_3}A_3}{\cot\frac{\alpha_{3,1}}{2}\gamma_{P_3} + \cot\frac{\varepsilon_1}{2}\gamma_{A_3}}, \\ T_2 &= \frac{\cot\frac{\alpha_{3,2}}{2}\gamma_{P_3}P_3 + \cot\frac{\varepsilon_2}{2}\gamma_{A_3}A_3}{\cot\frac{\alpha_{3,2}}{2}\gamma_{P_3} + \cot\frac{\varepsilon_2}{2}\gamma_{A_3}}, \end{aligned} \tag{9.80}$$

where

$$\varepsilon_1 = \angle A_1P_3A_3, \qquad \varepsilon_2 = \angle A_2P_3A_3, \qquad \varepsilon_1 + \varepsilon_2 = \pi, \tag{9.81}$$

and

$$\alpha_{3,1} = \angle A_1A_3P_3, \qquad \alpha_{3,2} = \angle A_2A_3P_3, \qquad \alpha_{3,1} + \alpha_{3,2} = \alpha_3 \tag{9.82}$$

as shown in Fig. 9.3, p. 279.

Applying Theorem 4.6, p. 90, to the gyrobarycentric coordinate representation (9.80) of T_1 and T_2 with $X = \ominus A_3$, we obtain the following left gyrotranslations of T_1 and T_2 by $\ominus A_3$:

$$\begin{aligned} \ominus A_3 \oplus T_1 &= \frac{\cot\frac{\alpha_{3,1}}{2}\gamma_{\ominus A_3 \oplus P_3}(\ominus A_1 \oplus P_3)}{\cot\frac{\alpha_{3,1}}{2}\gamma_{\ominus A_3 \oplus P_3} + \cot\frac{\varepsilon_1}{2}}, \\ \ominus A_3 \oplus T_2 &= \frac{\cot\frac{\alpha_{3,2}}{2}\gamma_{\ominus A_3 \oplus P_3}(\ominus A_1 \oplus P_3)}{\cot\frac{\alpha_{3,2}}{2}\gamma_{\ominus A_3 \oplus P_3} + \cot\frac{\varepsilon_2}{2}}, \end{aligned} \tag{9.83}$$

where, following (9.81),

$$\cot\frac{\varepsilon_1}{2} = \tan\frac{\varepsilon_2}{2}. \tag{9.84}$$

Condition (9.78) is equivalent to the condition

$$\ominus A_3 \oplus T_1 = \ominus A_3 \oplus T_2 \tag{9.85}$$

which, by (9.83)–(9.84), is equivalent to the condition

$$\frac{\cot\frac{\alpha_{3,1}}{2}}{\cot\frac{\alpha_{3,2}}{2}} = \frac{\cot\frac{\varepsilon_1}{2}}{\cot\frac{\varepsilon_2}{2}}. \tag{9.86}$$

We now face the task of determining the gyrobarycentric coordinates $m_1 > 0$ and $m_2 > 0$ of P_3 in (9.79) from condition (9.86).

Noting the gyrotrigonometric/trigonometric identity

$$\cot\frac{\alpha}{2} = \frac{1+\cos\alpha}{\sin\alpha}. \tag{9.87}$$

Condition (9.86) is equivalent to the condition

$$\frac{1+\cos\alpha_{3,1}}{1+\cos\alpha_{3,2}}\frac{\sin\alpha_{3,2}}{\sin\alpha_{3,1}} = \frac{1+\cos\varepsilon_1}{1-\cos\varepsilon_1}. \tag{9.88}$$

According to Theorem 9.5, p. 278, the gyrocevian A_3P_3 with arbitrary gyrobarycentric coordinates $m_1 > 0$ and $m_2 > 0$ generates the two pairs of gyroangles $(\varepsilon_1, \varepsilon_2)$ and $(\alpha_{3,1}, \alpha_{3,2})$ given by (9.31)–(9.32), p. 279.

Substituting from (9.31)–(9.32) into Condition (9.88), we obtain the equivalent condition

$$\frac{m_1}{m_2} = \frac{\sin\alpha_1(1+\cos\alpha_2)}{\sin\alpha_2(1+\cos\alpha_1)} = \frac{\cot\frac{\alpha_2}{2}}{\cot\frac{\alpha_1}{2}}. \tag{9.89}$$

Hence, by (9.89) and (9.79), the unique gyrocevian foot P_3 of gyrocevian A_3P_3 of gyrotriangle $A_1A_2A_3$ that satisfies Condition (9.78) possesses the gyrobarycentric coordinate representation

$$P_3 = \frac{\cot\frac{\alpha_2}{2}\gamma_{A_1}A_1 + \cot\frac{\alpha_1}{2}\gamma_{A_2}A_2}{\cot\frac{\alpha_2}{2}\gamma_{A_1} + \cot\frac{\alpha_1}{2}\gamma_{A_2}}. \tag{9.90}$$

The gyrocevian foot P_3 in (9.90), Fig. 9.6, satisfies each of the mutually equivalent conditions (9.78), (9.85), (9.86), (9.88), and (9.89). It is recognized by (9.90) as the point of tangency F_3, (7.174), p. 194, in Fig. 7.14, p. 192, where the ingyrocircle of gyrotriangle $A_1A_2A_3$ meets the side A_1A_2. We have thus obtained the following theorem:

Theorem 9.9 *Let $A_1A_2A_3$ be a gyrotriangle with a gyrocevian A_3P_3 that joins vertex A_3 with a point P_3 on the interior of side A_1A_2 in an Einstein gyrovector space*

$(\mathbb{R}_s^n, \oplus, \otimes)$. *Furthermore, let* I_0, I_1 *and* I_2 *be the ingyrocenters of the ingyrocircles of gyrotriangles* $A_1A_2A_3$, $A_1P_3A_3$ *and* $A_2P_3A_3$, *touching the side* A_1A_2 *and the gyrocevian* A_3P_3 *at the tangency points* T_0, T_1 *and* T_2, *respectively, Fig.* 9.6.

Then,

$$T_1 = T_2 \tag{9.91}$$

if and only if

$$T_0 = P_3. \tag{9.92}$$

Clearly, Theorem 9.9 is invariant in form under the Euclidean limit $s \to \infty$, so that it remains valid in Euclidean geometry as well, thus recovering a known result in Euclidean geometry [26, Exer. 1.2, p. 13].

Theorem 9.9 admits the generalization stated the following theorem:

Theorem 9.10 *Let* $A_1A_2A_3$ *be a gyrotriangle with a gyrocevian* A_3P_3 *that joins vertex* A_3 *with a point* P_3 *on the interior of side* A_1A_2 *in an Einstein gyrovector space* $(\mathbb{R}_s^n, \oplus, \otimes)$. *Furthermore, let* I_0, I_1 *and* I_2 *be the ingyrocenters of the ingyrocircles of gyrotriangles* $A_1A_2A_3$, $A_1P_3A_3$ *and* $A_2P_3A_3$, *touching the side* A_1A_2 *and the gyrocevian* A_3P_3 *at the tangency points* T_0, T_1 *and* T_2, *respectively, Fig.* 9.7. *Then,*

$$\|\ominus T_0 \oplus P_3\| = \|\ominus T_1 \oplus T_2\|. \tag{9.93}$$

The proof of Theorem 9.10 is left to the reader; see Problem 9.11, p. 299. A synthetic proof of the Euclidean counterpart of Theorem 9.10 was obtained by Vladimir Zajic [75].

9.6 Brocard Gyropoints

Let A_1A_{23}, A_2A_{31} and A_3A_{12} be a set of three concurrent gyrocevians that emanate from the three vertices of a gyrotriangle $A_1A_2A_3$ in an Einstein gyrovector space, and let $\alpha_{k,i}$, $i = 1, 2$, $k = 1, 2, 3$, be the resulting generated gyroangles in the gyrotriangle, as shown in Fig. 9.8. Surprisingly, there is precisely one gyrocevians point of concurrency for which

$$\alpha_{1,1} = \alpha_{2,1} = \alpha_{3,1} := \omega_1. \tag{9.94}$$

Similarly, there is precisely one gyrocevians point of concurrency for which

$$\alpha_{1,2} = \alpha_{2,2} = \alpha_{3,2} := \omega_2 \tag{9.95}$$

satisfying $\omega_1 = \omega_2 := \omega$, ω being the Brocard gyroangle of gyrotriangle $A_1A_2A_3$. These two unique gyrocevian points of concurrency are called the Brocard gyropoints of the gyrotriangle. In this section, we find the gyrocevian feet that determine

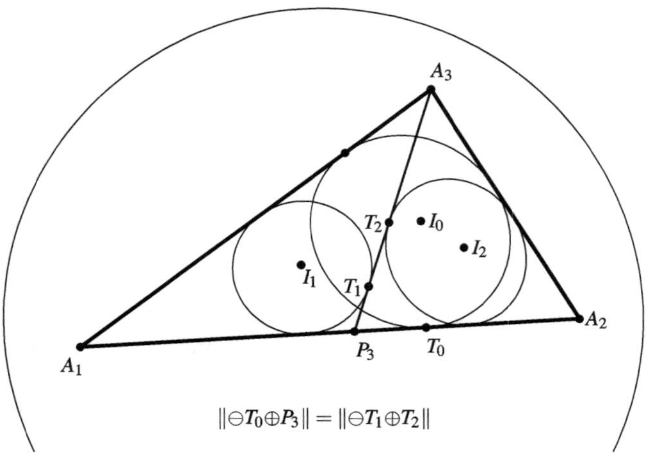

Fig. 9.7 A gyrotriangle $A_1A_2A_3$ in an Einstein gyrovector space $(\mathbb{R}_s^n, \oplus, \otimes)$ split by the gyrocevian A_3P_3 into two gyrotriangles, $A_1P_3A_3$ and $A_2P_3A_3$, resulting in corresponding two ingyrocircles. The two ingyrocircles with gyrocenters I_1 and I_2 touch their generating gyrocevian at the points of tangency T_1 and T_2. The ingyrocircle of gyrotriangle $A_1A_2A_3$, with ingyrocenter I_0, touches the gyrotriangle side A_1A_2 at the tangency point T_0. Surprisingly, the gyrodistance between T_1 and T_2 equals the gyrodistance between T_0 and P_3

the first Brocard gyropoint Ω_1, shown in Fig. 9.8, of any gyrotriangle in an Einstein gyrovector space. The second Brocard gyropoint, Ω_2, can be determined in a similar way. The gyromidpoint of the two Brocard gyropoints of a gyrotriangle will then emerge as an interesting gyrotriangle gyrocenter; see Problem 9.10, p. 299.

In order to determine the gyrotriangle Brocard gyropoints we need the following Lemma, which is extracted and deduced from Theorem 9.5, p. 278:

Lemma 9.11 (Gyrocevian Generated Gyroangles) *Let $A_1A_2A_3$ be a gyrotriangle in an Einstein gyrovector space $(\mathbb{R}_s^n, \oplus, \otimes)$ with the three gyroangles α_k, $k = 1, 2, 3$, and gyrocevians A_1A_{23}, A_2A_{31} and A_3A_{12}, Fig. 9.8, where the gyrocevian feet A_{ij} are given by their gyrobarycentric coordinate representations*

$$
\begin{aligned}
A_{23} &= \frac{m_{12}\gamma_{A_2} A_2 + m_{13}\gamma_{A_3} A_3}{m_{12}\gamma_{A_2} + m_{13}\gamma_{A_3}}, \\
A_{31} &= \frac{m_{21}\gamma_{A_1} A_1 + m_{23}\gamma_{A_3} A_3}{m_{21}\gamma_{A_1} + m_{23}\gamma_{A_3}}, \\
A_{12} &= \frac{m_{31}\gamma_{A_1} A_1 + m_{32}\gamma_{A_2} A_2}{m_{31}\gamma_{A_1} + m_{32}\gamma_{A_2}}
\end{aligned}
\tag{9.96}
$$

with respect to the set $S = \{A_1, A_2, A_3\}$.

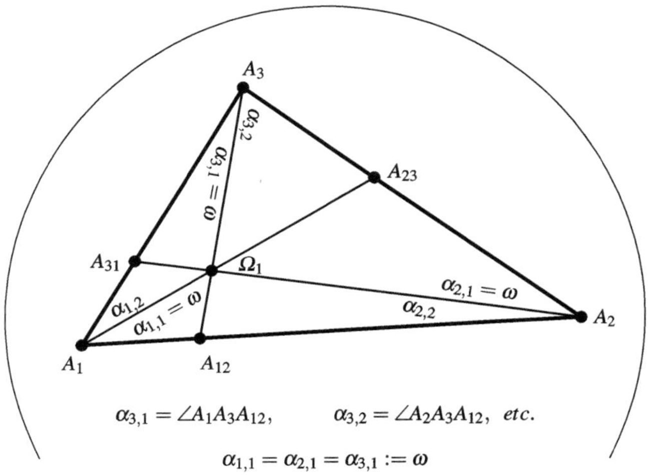

Fig. 9.8 Gyrocevian generated gyroangles, and the Brocard Gyropoints of a gyrotriangle $A_1A_2A_3$ in an Einstein gyrovector space $(\mathbb{R}^n_s, \oplus, \otimes)$. Brocard gyroangle is the gyroangle $\omega := \alpha_{1,1} = \alpha_{2,1} = \alpha_{3,1}$ and its corresponding Brocard gyropoint, Ω_1, is the gyrocevians point of concurrency

Then, the gyrocevian generated gyroangles $\alpha_{k,1}$ and $\alpha_{k,2}$, $k = 1, 2, 3$, Fig. 9.8, are given in terms of the gyrotriangle gyroangles and the gyrocevian feet gyrobarycentric coordinates in (9.96) by the following equations.

1. *Let $\alpha_{1,1} = \angle A_2A_1A_{23}$ and $\alpha_{1,2} = \angle A_3A_1A_{23}$, Fig. 9.8.*
   Then,

$$\begin{aligned}
\sin\alpha_{1,1} &= \frac{1}{D_1} m_{13} \sin\alpha_1 \sin\alpha_2, \\
\cos\alpha_{1,1} &= \frac{1}{D_1} (m_{12} \sin\alpha_3 + m_{13} \sin\alpha_2 \cos\alpha_1), \\
\sin\alpha_{1,2} &= \frac{1}{D_1} m_{12} \sin\alpha_1 \sin\alpha_3, \\
\cos\alpha_{1,2} &= \frac{1}{D_1} (m_{13} \sin\alpha_2 + m_{12} \sin\alpha_3 \cos\alpha_1),
\end{aligned} \tag{9.97a}$$

where $D_1 > 0$ is given by each of the equations

$$\begin{aligned}
D_1^2 &= (m_{12} \sin\alpha_3 - m_{13} \sin\alpha_2)^2 + 2m_{12}m_{13} \sin\alpha_2 \sin\alpha_3 (1 + \cos\alpha_1) \\
&= (m_{13} \sin\alpha_1 \sin\alpha_2)^2 + (m_{12} \sin\alpha_3 + m_{13} \sin\alpha_2 \cos\alpha_1)^2 \\
&= (m_{12} \sin\alpha_1 \sin\alpha_3)^2 + (m_{13} \sin\alpha_2 + m_{12} \sin\alpha_3 \cos\alpha_1)^2.
\end{aligned} \tag{9.97b}$$

2. *Let* $\alpha_{2,1} = \angle A_3 A_2 A_{31}$ *and* $\alpha_{2,2} = \angle A_1 A_2 A_{31}$, *Fig.* 9.8.
Then,

$$\begin{aligned} \sin\alpha_{2,1} &= \frac{1}{D_2} m_{21} \sin\alpha_2 \sin\alpha_3, \\ \cos\alpha_{2,1} &= \frac{1}{D_2} (m_{23} \sin\alpha_1 + m_{21} \sin\alpha_3 \cos\alpha_2), \\ \sin\alpha_{2,2} &= \frac{1}{D_2} m_{23} \sin\alpha_1 \sin\alpha_2, \\ \cos\alpha_{2,2} &= \frac{1}{D_2} (m_{21} \sin\alpha_3 + m_{23} \sin\alpha_1 \cos\alpha_2), \end{aligned} \tag{9.97c}$$

where $D_2 > 0$ *is given by each of the equations*

$$\begin{aligned} D_2^2 &= (m_{23} \sin\alpha_1 - m_{21} \sin\alpha_3)^2 + 2m_{21} m_{23} \sin\alpha_1 \sin\alpha_3 (1 + \cos\alpha_2) \\ &= (m_{12} \sin\alpha_2 \sin\alpha_3)^2 + (m_{23} \sin\alpha_1 + m_{21} \sin\alpha_3 \cos\alpha_2)^2 \\ &= (m_{23} \sin\alpha_1 \sin\alpha_2)^2 + (m_{21} \sin\alpha_3 + m_{23} \sin\alpha_1 \cos\alpha_2)^2. \end{aligned} \tag{9.97d}$$

3. *Let* $\alpha_{3,1} = \angle A_1 A_3 A_{12}$ *and* $\alpha_{3,2} = \angle A_2 A_3 A_{12}$, *Fig.* 9.8.
Then,

$$\begin{aligned} \sin\alpha_{3,1} &= \frac{1}{D_3} m_{32} \sin\alpha_1 \sin\alpha_3, \\ \cos\alpha_{3,1} &= \frac{1}{D_3} (m_{31} \sin\alpha_2 + m_{32} \sin\alpha_1 \cos\alpha_3), \\ \sin\alpha_{3,2} &= \frac{1}{D_3} m_{31} \sin\alpha_2 \sin\alpha_3, \\ \cos\alpha_{3,2} &= \frac{1}{D_3} (m_{32} \sin\alpha_1 + m_{31} \sin\alpha_2 \cos\alpha_3), \end{aligned} \tag{9.97e}$$

where $D_3 > 0$ *is given by each of the equations*

$$\begin{aligned} D_3^2 &= (m_{31} \sin\alpha_2 - m_{32} \sin\alpha_1)^2 + 2m_{31} m_{32} \sin\alpha_1 \sin\alpha_2 (1 + \cos\alpha_3) \\ &= (m_{32} \sin\alpha_1 \sin\alpha_3)^2 + (m_{31} \sin\alpha_2 + m_{32} \sin\alpha_1 \cos\alpha_3)^2 \\ &= (m_{31} \sin\alpha_2 \sin\alpha_3)^2 + (m_{32} \sin\alpha_1 + m_{31} \sin\alpha_2 \cos\alpha_3)^2. \end{aligned} \tag{9.97f}$$

Proof The results in (9.97e) are established in Theorem 9.5, p. 278. The results in (9.97c) and (9.97a) follow from (9.97e) by cyclic permutations of gyrotriangle vertices. □

It follows from Lemma 9.11 that the gyrosines of the gyrocevian generated gyroangles $\alpha_{1,1}$, $\alpha_{2,1}$ and $\alpha_{3,1}$ in a gyrotriangle $A_1 A_2 A_3$ in an Einstein gyrovector space, Fig. 9.8, are given by

$$\sin^2 \alpha_{1,1} = \frac{1}{1 + (\frac{m_{12} \sin\alpha_3 + m_{13} \sin\alpha_2 \cos\alpha_1}{m_{13} \sin\alpha_1 \sin\alpha_2})^2}, \tag{9.98a}$$

$$\sin^2 \alpha_{2,1} = \frac{1}{1 + (\frac{m_{23} \sin\alpha_1 + m_{21} \sin\alpha_3 \cos\alpha_2}{m_{21} \sin\alpha_2 \sin\alpha_3})^2}, \tag{9.98b}$$

$$\sin^2 \alpha_{3,1} = \frac{1}{1 + (\frac{m_{31} \sin\alpha_2 + m_{32} \sin\alpha_1 \cos\alpha_3}{m_{32} \sin\alpha_1 \sin\alpha_3})^2}. \tag{9.98c}$$

Hence, the equations, (9.94),

$$\alpha_{1,1} = \alpha_{2,1} = \alpha_{3,1} \tag{9.99}$$

are equivalent, by means of (9.98a), (9.98b), (9.98c), to the two equations

$$\begin{aligned} \frac{m_{12} \sin\alpha_3 + m_{13} \sin\alpha_2 \cos\alpha_1}{m_{13} \sin\alpha_1} &= \frac{m_{23} \sin\alpha_1 + m_{21} \sin\alpha_3 \cos\alpha_2}{m_{21} \sin\alpha_3}, \\ \frac{m_{12} \sin\alpha_3 + m_{13} \sin\alpha_2 \cos\alpha_1}{m_{13} \sin\alpha_2} &= \frac{m_{31} \sin\alpha_2 + m_{32} \sin\alpha_1 \cos\alpha_3}{m_{32} \sin\alpha_3}. \end{aligned} \tag{9.100}$$

In order to find the gyrocevian feet A_{12}, A_{31} and A_{23} that generate the Brocard gyropoint in Fig. 9.8 we have to determine the gyrobarycentric coordinates m_{12}, m_{13}, m_{21}, m_{23}, m_{31}, and m_{32}, of the gyrocevian feet in (9.96). These are determined as follows:

1. Since gyrobarycentric coordinates are homogeneous, without loss of generality we select the normalization condition

$$m_{13} = m_{21} = m_{32} = 1. \tag{9.101}$$

2. By (9.27), p. 278, of Corollary 9.4, and following (9.101), the condition that the three gyrocevians are concurrent is equivalent to the condition

$$m_{31} = \frac{1}{m_{12} m_{23}}. \tag{9.102}$$

3. Finally, the remaining gyrobarycentric coordinates, m_{12} and m_{23}, are to be determined by the two equations in (9.100).

Substituting (9.101) and (9.102) into (9.100), we obtain two nonlinear equations, resulting in the following determination of the two unknowns m_{12} and m_{23}. The gyrobarycentric coordinate m_{12} is given in terms of the gyrobarycentric coordinate m_{23} and the gyrotriangle gyroangles α_k, $k = 1, 2, 3$, by the equation

$$m_{12} = \frac{\sin^2 \alpha_1}{\sin^2 \alpha_3} m_{23}. \tag{9.103}$$

The gyrobarycentric coordinate m_{23}, in turn, is the unique real-valued solution of the cubic equation

$$c_0 z^3 + c_1 z^2 + c_2 z + c_3 = 0 \tag{9.104}$$

with coefficients

$$\begin{aligned}
c_0 &= \sin^4 \alpha_1, \\
c_1 &= -\sin^2 \alpha_1 \{\sin \alpha_1 \sin(\alpha_2 - \alpha_3) + \sin \alpha_3 \sin(\alpha_2 - \alpha_1)\}, \\
c_2 &= \sin \alpha_1 \sin \alpha_3 \sin(\alpha_1 - \alpha_2) \sin(\alpha_3 - \alpha_2), \\
c_3 &= -\sin^2 \alpha_2 \sin^2 \alpha_3.
\end{aligned} \tag{9.105}$$

9.7 A Gyrocevian Concurrency Condition

Theorem 9.12 (A Gyrocevian Concurrency Condition) *$A_1A_2A_3$ be a gyrotriangle in an Einstein gyrovector space $(\mathbb{R}^n_s, \oplus, \otimes)$ with gyroangles α_k, $k = 1, 2, 3$, and gyrocevians A_1A_{23}, A_2A_{13},A_3A_{12} that generate the gyroangles $\alpha_{k,i}$, $i = 1, 2$, as shown in Fig.* 9.8, *p.* 293. *The gyrocevians are concurrent if and only if*

$$\sin \alpha_{1,1} \sin \alpha_{2,1} \sin \alpha_{3,1} = \sin \alpha_{1,2} \sin \alpha_{2,2} \sin \alpha_{3,2}. \tag{9.106}$$

Proof Let the gyrocevian feet, A_{ij}, be given by their gyrobarycentric coordinate representations, (9.96), p. 292,

$$\begin{aligned}
A_{12} &= \frac{m_{31}\gamma_{A_1} A_1 + m_{32}\gamma_{A_2} A_2}{m_{31}\gamma_{A_1} + m_{32}\gamma_{A_2}}, \\
A_{23} &= \frac{m_{12}\gamma_{A_2} A_2 + m_{13}\gamma_{A_3} A_3}{m_{12}\gamma_{A_2} + m_{13}\gamma_{A_3}}, \\
A_{31} &= \frac{m_{21}\gamma_{A_1} A_1 + m_{23}\gamma_{A_3} A_3}{m_{21}\gamma_{A_1} + m_{23}\gamma_{A_3}}.
\end{aligned} \tag{9.107}$$

Then,

$$\begin{aligned}
\frac{m_{31}}{m_{32}} &= \frac{\sin \alpha_1}{\sin \alpha_2} \frac{\sin \alpha_{3,2}}{\sin \alpha_{3,1}}, \\
\frac{m_{12}}{m_{13}} &= \frac{\sin \alpha_2}{\sin \alpha_3} \frac{\sin \alpha_{1,2}}{\sin \alpha_{1,1}}, \\
\frac{m_{23}}{m_{21}} &= \frac{\sin \alpha_3}{\sin \alpha_1} \frac{\sin \alpha_{2,2}}{\sin \alpha_{2,1}}.
\end{aligned} \tag{9.108}$$

The first equation in (9.108) follows immediately from the first equation in (9.107) by (9.33), p. 280. The second (third) equation in (9.108) follows from the first equation in (9.108) by a cyclic vertex permutation.

Hence, by means of (9.108), the necessary and sufficient condition (9.27), p. 278, for gyrocevian concurrence gives the equivalent condition

$$1 = \frac{m_{31}}{m_{32}} \frac{m_{12}}{m_{13}} \frac{m_{23}}{m_{21}} = \frac{\sin\alpha_{1,1} \sin\alpha_{2,1} \sin\alpha_{3,1}}{\sin\alpha_{1,2} \sin\alpha_{2,2} \sin\alpha_{3,2}} \tag{9.109}$$

for gyrocevian concurrence. The latter, in turn, is clearly equivalent to condition (9.106) for gyrocevian concurrence, as desired. □

If the point of concurrency of the gyrocevians in Theorem 9.12 is the Brocard gyropoint Ω_1 of the gyrotriangle then, by definition, $\omega := \alpha_{1,1} = \alpha_{2,1} = \alpha_{3,1}$, so that the gyrocevian concurrence condition (9.106) implies the identity

$$\sin^3\omega = \sin\alpha_{1,2} \sin\alpha_{2,2} \sin\alpha_{3,2} \tag{9.110}$$

which, in particular, is valid in Euclidean geometry as well.

Brocard gyropoints of a gyrotriangle share remarkable analogies with their Euclidean counterparts, the Brocard points of a triangle, which are studied, for instance, in [26, Chap. 10]. Some Brocard-like points of a triangle, that may be extended to the hyperbolic geometry of Einstein gyrovector spaces, are studied in [1].

9.8 Problems

Problem 9.1 Solving a system of Equations:
Find the equations in (9.8), p. 272, explicitly, and derive the solution in (9.9).

Problem 9.2 A Substitution:
Show that the substitution of γ_{a_1} and γ_{a_3} from (9.42) into (9.52) gives (9.53), p. 283.

Problem 9.3 A Gyrotrigonometric Substitution:
Show that the substitution of gyrotriangle gyrotrigonometric identities from Sect. 7.12, p. 187 into (9.53) result in (9.54), p. 283.

Problem 9.4 Gyrotrigonometric/Trigonometric Addition Laws:

1. Employ the gyrotrigonometric addition law

$$\sin(\varepsilon_1 + \varepsilon_2) = \sin\varepsilon_1 \cos\varepsilon_2 + \cos\varepsilon_1 \sin\varepsilon_2 \tag{9.111}$$

of the gyrosine function, along with substitutions from (9.31) to show that $\sin(\varepsilon_1 + \varepsilon_2) = 0$, as expected from the relation $\varepsilon_1 + \varepsilon_2 = \pi$ shown in Fig. 9.3, p. 279.

2. Employ the gyrotrigonometric addition law

$$\cos(\varepsilon_1 + \varepsilon_2) = \cos\varepsilon_1 \cos\varepsilon_2 - \sin\varepsilon_1 \sin\varepsilon_2 \tag{9.112}$$

of the gyrocosine function, along with substitutions from (9.31) to show that $\cos(\varepsilon_1 + \varepsilon_2) = -1$, as expected from the relation $\varepsilon_1 + \varepsilon_2 = \pi$ shown in Fig. 9.3.

3. Employ the gyrotrigonometric addition law

$$\sin(\alpha_{3,1} + \alpha_{3,2}) = \sin\alpha_{3,1} \cos\alpha_{3,2} + \cos\alpha_{3,1} \sin\alpha_{3,2} \tag{9.113}$$

of the gyrosine function, along with substitutions from (9.32) to show that $\sin(\alpha_{3,1} + \alpha_{3,2}) = \sin\alpha_3$, as expected from the relation $\alpha_{3,1} + \alpha_{3,2} = \alpha_3$ shown in Fig. 9.3.

4. Employ the gyrotrigonometric addition law

$$\cos(\alpha_{3,1} + \alpha_{3,2}) = \cos\alpha_{3,1} \cos\alpha_{3,2} - \sin\alpha_{3,1} \sin\alpha_{3,2} \tag{9.114}$$

of the gyrocosine function, along with substitutions from (9.32) to show that $\cos(\alpha_{3,1} + \alpha_{3,2}) = \cos\alpha_3$, as expected from the relation $\alpha_{3,1} + \alpha_{3,2} = \alpha_3$ shown in Fig. 9.3.

Problem 9.5 Successive Substitutions:
Derive (9.67), p. 285, by successive substitutions into (9.66) from (9.65) and from (9.31)–(9.32), p. 279.

Problem 9.6 Substitutions:
Derive condition (9.89), p. 290, from condition (9.88) by substitutions from (9.31)–(9.32)

Problem 9.7 Substitutions:
Substitute (9.101) and (9.102) into (9.100), p. 295, to obtain the equations (9.103) and (9.104) for the unknowns m_{12} and m_{23}.

Problem 9.8 Establish Equations:
Verify the second and the third equations in (i) (9.97b), p. 293; in (ii) (9.97d), p. 294; and in (iii) (9.97f), p. 294.

Problem 9.9 The Brocard Gyropoints:
Determine the second Brocard gyropoint Ω_2 of a gyrotriangle in an Einstein gyrovector space in a way similar to the determination of the first Brocard gyropoint Ω_1, Fig. 9.8, p. 293, in Sect. 9.6.

Problem 9.10 The Brocard Gyromidpoint:
Determine the Brocard gyromidpoint, which is the gyromidpoint of the two Brocard gyropoints Ω_1, Fig. 9.8, p. 293, and Ω_2 of a gyrotriangle in an Einstein gyrovector space.

Problem 9.11 Generalizing Theorem 9.9:
Prove Theorem 9.10, p. 291, which is a generalization of Theorem 9.9, p. 290.

Chapter 10
Epilogue

Abstract The success of the use of Einstein addition along with Einstein velocity dependent relativistic mass in the determination of various hyperbolic triangle centers is demonstrated in this book. But, looking at the relativistic velocity addition law and its underlying hyperbolic geometry through the lens of cosmological stellar aberration leads to a startling conclusion: relativistic velocities add in the cosmos according to the gyroparallelogram addition law of hyperbolic geometry, which is commutative. This Epilogue of the book may thus serve as the Prologue for the future of Einstein's special relativity theory as a theory regulated by the hyperbolic geometry of Bolyai and Lobachevsky.

10.1 Introduction

The hero of this book is Albert Einstein, whose addition law forms the powerful and elegant tool that enables hyperbolic triangle centers and relations between them to be determined. He introduced his addition law (1.2), p. 4, in his 1905 paper that founded the special theory of relativity [12]. We note that the Euclidean 3-vector algebra was not so widely known in 1905 and, consequently, was not used by Einstein. Indeed, Einstein calculated in [12] the behavior of the velocity components parallel and orthogonal to the relative velocity between inertial systems, which is as close as one can get without vectors to the vectorial version (1.2) that we use in this book. Soon later, in 1908–1910, Vladimir Varičak demonstrated in [66, 67] that Einstein velocity addition law has interpretation in hyperbolic geometry.

Being neither commutative nor associative, Einstein addition law seemed to be structureless until its gyro-structure was discovered by the author in 1988 [55]. Einstein's failure to recognize and advance the extraordinarily rich nonassociative algebraic structure that his addition law encodes contributed to the eclipse of his velocity addition law, creating a void that could be filled only with Minkowskian relativity.

Minkowskian relativity is the reformulation of Einsteinian relativity based on the Lorentz transformation of four-vectors rather than Einstein addition of three-vectors. The term "Minkowskian relativity" was coined by Lewis Pyenson in [43,

A.A. Ungar, *Hyperbolic Triangle Centers*,
Fundamental Theories of Physics 166,
DOI 10.1007/978-90-481-8637-2_10,

p. 146]. The historical struggle between Einsteinian relativity and Minkowskian relativity is skillfully described by the renowned historian of relativity physics Scott Walter in [70] where, for the first time, the term "Minkowskian relativity" appears in a title. According to Walter [72], the reason we have space–time formalism today is that Minkowski's friend, Sommerfeld, took it upon himself to rewrite Minkowski's formalism, and to make it look like ordinary vector analysis. The basic distinction between Einsteinian special relativity and Minkowskian special relativity is that in the former Einstein velocity addition law and its three-vectors form the primitive concept from which the Lorentz transformation group is derived, while in the latter it is the Lorentz transformation group and its four-vectors that form the primitive concept.

The gyrovector space approach to special relativity in this book and in its forerunners [58, 60, 63, 64] tilts the balance away from Minkowskian special relativity toward Einsteinian special relativity. The success of Einstein addition law as a powerful tool in the search for hyperbolic triangle centers is Einstein's triumph. Surprisingly, the success of Einstein addition law in hyperbolic geometry also indicates that this extraordinarily useful law is incomplete, as explained in the sequel.

Ignoring forces, do uniform relativistically admissible velocities in the Universe add according to Einstein velocity addition law? Owing to analogies that (i) the Einstein addition law of Einsteinian, relativistically admissible, velocities and (ii) the vector addition law of Newtonian, classical velocities share, it is commonly accepted that uniform velocities in the Universe add according to Einstein addition law. This belief is partially supported by Fizeau's 1851 experiment [38] that validates experimentally the physical significance of Einstein velocity addition law for parallel velocities.

However, in addition to the remarkable analogies that Einstein velocity addition law of Einsteinian velocities and the common vector addition law of Newtonian velocities share, there is a remarkable disanalogy:

1. The common vector addition law gives rise to the well-known triangle and parallelogram addition law, and the resulting triangle and parallelogram addition law coincide with the common vector addition law.
2. In full analogy, Einstein velocity addition law gives rise to a *gyrotriangle* and a *gyroparallelogram addition law* as well, as we have seen in Sect. 5.3, p. 120, and as one can see, in more detail, in [63]. In contrast, however, the resulting gyrotriangle and gyroparallelogram addition law do not coincide with Einstein addition. Thus, for instance, Einstein gyroparallelogram addition law, (5.11), p. 124, is commutative while, in contrast, Einstein addition law, (1.2), p. 4, is noncommutative.

This remarkable disanalogy raises a natural question: Ignoring forces, do uniform velocities in the Universe add (i) according to Einstein velocity addition law, or (ii) according to the gyrotriangle and the gyroparallelogram addition law to which Einstein addition law gives rise?

The answer to the question lies in the *stellar aberration* effect. Stellar aberration results from the velocity of the Earth in its annual orbit about the Sun, discovered by the English astronomer James Bradley in the 1720s [53].

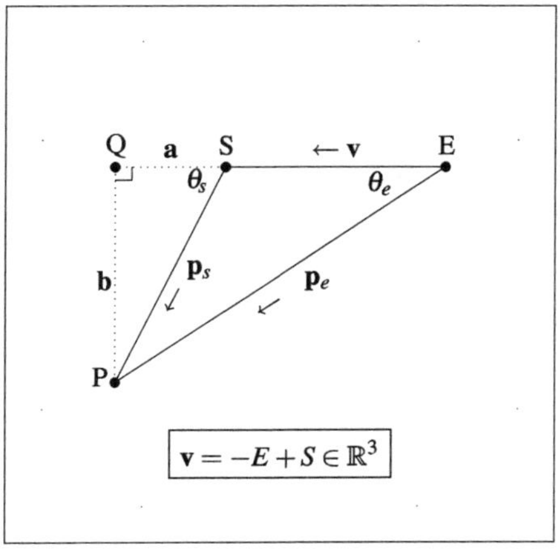

Fig. 10.1 Classical Stellar Aberration and Particle Aberration: Classical interpretation in terms of the triangle law or, equivalently, the parallelogram law, of addition of Newtonian velocities in the standard model of 3-dimensional Euclidean geometry $(\mathbb{R}^3, +)$. Two dimensions are shown for graphical clarity. Here $+$ is the common vector addition in $\mathbb{R}^3$. A particle P moves with Newtonian velocity $\mathbf{p}_e$ ($\mathbf{p}_s$) relative to the Earth E (the Sun S), making an angle θ_e (θ_s) with the Newtonian velocity $\mathbf{v}$ of the Sun S relative to the Earth E. In order to calculate the Newtonian, classical, particle aberration $\theta_s - \theta_e$, the Euclidean triangle ESP is augmented into the Euclidean right-angled triangle EQP, allowing elementary trigonometry to be employed. Points are given by their orthogonal Cartesian coordinates (x, y, z), $x^2 + y^2 + z^2 < \infty$. The coordinates are not shown. The Euclidity of $(\mathbb{R}^3, +)$ is determined by the Euclidean metric in which the distance between two points $A, B \in \mathbb{R}^3$ is $\|-A + B\|$. Classical stellar aberration is a special classical particle aberration when the particle is a photon emitted from a star

10.2 Stellar Aberration

Stellar aberration in the Universe, described in Figs. 10.1–10.2, is our laboratory, and we are the experimenters asking whether relativistic velocities add

(i) According to Einstein velocity addition law (1.2), p. 4, or
(ii) According to Einstein gyroparallelogram addition law (5.11), p. 124

Fortunately, the cosmological phenomenon of *stellar aberration* comes to the rescue. Owing to the validity of well-known relativistic stellar aberration formulas, we will find here that

1. Einsteinian, relativistically admissible velocities that need not be parallel do not add according to Einstein velocity addition law. Rather, they are gyrovectors that add according to the gyroparallelogram addition law (5.11), Fig. 5.4, p. 123, which is commutative, just as
2. Newtonian, classical velocities are vectors that add according to the common parallelogram addition law

To set the stage for presenting the relativistic stellar aberration, we begin with the presentation of the *classical particle aberration*, Fig. 10.1, which will be extended to a presentation of the *relativistic particle aberration*, Fig. 10.2, by gyro-analogies.

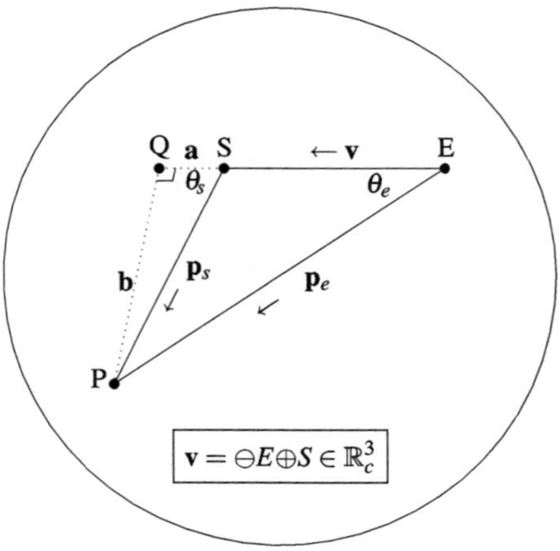

Fig. 10.2 Relativistic Stellar Aberration and Particle Aberration: Relativistic interpretation in terms of the gyrotriangle law or, equivalently, the gyroparallelogram law, of addition of Einsteinian velocities in the Beltrami–Klein ball model of 3-dimensional hyperbolic geometry $(\mathbb{R}_c^3, \oplus)$. Here $\oplus$ is Einstein addition, (1.2), p. 4, in the c-ball $\mathbb{R}_c^3 \subset \mathbb{R}^3$. A particle P moves with Einsteinian velocity $\mathbf{p}_e$ ($\mathbf{p}_s$) relative to the Earth E (the Sun S), making an angle θ_e (θ_s) with the Einsteinian velocity $\mathbf{v}$ of the Sun S relative to the Earth E. In order to calculate the Einsteinian, relativistic, particle aberration $\theta_s - \theta_e$, the gyrotriangle ESP is augmented into the right-gyroangled gyrotriangle EQP, allowing elementary gyrotrigonometry to be employed. Points are given by their orthogonal Cartesian coordinates (x, y, z), $x^2 + y^2 + z^2 < c^2$. The coordinates are not shown. The hyperbolicity of $(\mathbb{R}_c^3, \oplus)$ is determined by the hyperbolic gyrometric in which the distance between two points $A, B \in \mathbb{R}_c^3$ is $\|\ominus A \oplus B\|$. Relativistic stellar aberration is a special relativistic particle aberration when the particle is a photon emitted from a star

Figures 10.1 and 10.2 present, respectively, the Newtonian velocity space $\mathbb{R}^3$ of classical velocities, and the Einsteinian velocity space $\mathbb{R}_c^3$ of relativistically admissible velocities, along with several of their points, where only two dimensions are shown for clarity. The origin, O, of each of these two velocity spaces, not shown in Figs. 10.1–10.2, is arbitrarily selected, representing an arbitrarily selected inertial rest frame $\Sigma_{\mathbf{0}}$.

Points of a velocity space represent uniform velocities relative to the rest frame $\Sigma_{\mathbf{0}}$. In particular, the points E, S and P in Figs. 10.1–10.2 represent, respectively, the velocity of the Earth, the Sun, and a Particle (emitted, for instance, from a star) relative to the rest frame $\Sigma_{\mathbf{0}}$.

Accordingly, the Newtonian velocity vector $\mathbf{v}$ of the Sun relative to the Earth is

$$\mathbf{v} = -E + S \tag{10.1a}$$

and the Einsteinian velocity gyrovector $\mathbf{v}$ of the Sun relative to the Earth is

$$\mathbf{v} = \ominus E \oplus S \tag{10.1b}$$

as shown, respectively, in Figs. 10.1–10.2.

Similarly, the particle P moves uniformly relative to the Earth and relative to the Sun with respective Newtonian velocities, Fig. 10.1,

$$\begin{aligned} \mathbf{P}_e &= -E + P, \\ \mathbf{P}_s &= -S + P. \end{aligned} \tag{10.2a}$$

In full analogy with (10.2a), the Einsteinian velocities of the particle P relative to the Earth and relative to the Sun are, Fig. 10.2,

$$\begin{aligned} \mathbf{P}_e &= \ominus E \oplus P \\ \mathbf{P}_s &= \ominus S \oplus P \end{aligned} \tag{10.2b}$$

The Newtonian velocities (10.2a) of the particle make angles θ_e and θ_s, respectively, with the Newtonian velocity $\mathbf{v}$ in (10.1a), as shown in Fig. 10.1.

Similarly, the Einsteinian velocities (10.2b) of the particle make gyroangles θ_e and θ_s, respectively, with the Einsteinian velocity $\mathbf{v}$ in (10.1b), as shown in Fig. 10.2.

Following Fig. 10.1, *classical particle aberration* is the angular change $\theta_s - \theta_e$ in the apparent direction of a moving particle caused by the motion with Newtonian relative velocity $\mathbf{v}$, (10.1a), between E and S. A relationship between the angles θ_s and θ_e is called a *classical particle aberration formula*.

Similarly, following Fig. 10.2, *relativistic particle aberration* is the gyroangular change $\theta_s - \theta_e$ in the apparent direction of a moving particle caused by the motion with Einsteinian relative velocity $\mathbf{v}$, (10.1b), between E and S. A relationship between the gyroangles θ_s and θ_e is called a *relativistic particle aberration formula*.

In order to uncover classical particle aberration formulas, we draw the altitude PQ from vertex P to side ES (extended if necessary) obtaining the right-angled triangle EQP in Fig. 10.1. The latter, in turn, enables the parallelogram law and the triangle law of Newtonian velocity addition, and the triangle equality and trigonometry, to be applied, obtaining the following two, mutually equivalent, classical particle aberration formulas:

$$\begin{aligned} \cot\theta_e &= \cot\theta_s + \frac{v}{p_s \sin\theta_s}, \\ \cot\theta_s &= \cot\theta_e - \frac{v}{p_e \sin\theta_e}. \end{aligned} \tag{10.3a}$$

The resulting classical particle aberration formulas (10.3a) are in full agreement with formulas available in the literature; see, for instance, [54, (134), p. 147].

The details of obtaining the classical particle aberration formulas (10.3a), illustrated in Fig. 10.1, are presented in [63, Chap. 13] and, hence, will not be presented here.

In full analogy, in order to uncover relativistic particle aberration formulas, we draw the altitude PQ from vertex P to side ES (extended if necessary) obtaining the right gyrotriangle EQP in Fig. 10.2. The latter, in turn, enables the gyrotriangle law and the gyroparallelogram law of Einsteinian velocity addition, and the gyrotriangle equality and gyrotrigonometry, to be applied, obtaining the following two, mutually equivalent, relativistic particle aberration formulas:

$$\begin{aligned}\cot\theta_e &= \gamma_v\left(\cot\theta_s + \frac{v}{p_s\sin\theta_s}\right),\\ \cot\theta_s &= \gamma_v\left(\cot\theta_e - \frac{v}{p_e\sin\theta_e}\right).\end{aligned} \tag{10.3b}$$

The resulting relativistic particle aberration formulas (10.3b) are in full agreement with formulas available in the literature; see, for instance, [45, p. 53], [46, p. 86] and [33, pp. 12–14].

The details of obtaining the relativistic particle aberration formulas (10.3b), illustrated in Fig. 10.2, are presented in [63, Chap. 13] and, hence, will not be presented here.

In Euclidean geometry, the triangle law and the parallelogram law of vector addition are equivalent. In full analogy, their gyro-counterparts are equivalent as well, as explained in [64, Sect. 4.3] in detail.

The equivalence of the two equations in (10.3a) implies $p_s\sin\theta_s = p_e\sin\theta_e$, thus recovering the law of sines

$$\frac{p_s}{\sin\theta_e} = \frac{p_e}{\sin\theta_s} \tag{10.4a}$$

for the Euclidean triangle *ESP* in Fig. 10.1, noting that $\sin\theta_s = \sin(\pi - \theta_s)$.

In full analogy, the equivalence of the two equations in (10.3b) implies the relativistic law of gyrosines (6.44), p. 140,

$$\frac{\gamma_{p_s}p_s}{\sin\theta_e} = \frac{\gamma_{p_e}p_e}{\sin\theta_s} \tag{10.4b}$$

for the gyrotriangle *ESP* in Fig. 10.2, noting that $\sin\theta_s = \sin(\pi - \theta_s)$.

Here, we have described the way to recover the well-known classical particle aberration formulas (10.3a) by employing trigonometry, the triangle equality, the triangle addition law, and the parallelogram addition law of Newtonian velocities.

In full analogy, we have also described here the way to recover the well-known relativistic particle aberration formulas (10.3b) by employing gyrotrigonometry, the gyrotriangle equality, the gyrotriangle addition law, and the gyroparallelogram addition law (5.11), p. 124, of Einsteinian velocities.

In contrast, the well-known relativistic particle aberration formulas (10.3b) are obtained in the literature by employing the Lorentz transformation group of special relativity.

What is remarkable here is that the relativistic particle aberration formulas (10.3b), which are commonly obtained in the literature by Lorentz transformation considerations, are recovered here by gyrotrigonometry and the gyroparallelogram addition law of Einsteinian velocities, in full analogy with the recovery of their classical counterparts. This remarkable way of recovering the particle aberration formulas (10.3a), (10.3b) demonstrates that since special relativity is governed by the Lorentz transformation group, Einsteinian velocities in special relativity add according to the gyroparallelogram addition law, just as Newtonian velocities add according to the parallelogram addition law.

Hence, any experiment that supports the validity of the relativistic particle aberration formulas (10.3b), amounts to an experiment that supports the validity of the gyroparallelogram addition law (5.11), p. 124, of Einsteinian velocities.

In the special case when the particle P in Fig. 10.2 is a photon emitted from a star, the Einsteinian speed of the photon relative to both E and S is $p_e = p_s = c$, and the relativistic particle aberration formulas (10.3b) reduce to the corresponding *stellar aberration formulas*:

$$\begin{aligned} \cot\theta_e &= \gamma_v \frac{\cos\theta_s + v/c}{\sin\theta_s}, \\ \cot\theta_s &= \gamma_v \frac{\cos\theta_e - v/c}{\sin\theta_e}. \end{aligned} \tag{10.5}$$

The discovery of stellar aberration, which results from the velocity of the Earth in its annual orbit about the Sun, by the English astronomer James Bradley in the 1720s, is described, for instance, in [53].

A high precision test of the validity of the stellar aberration formulas (10.5) in special relativity has recently been obtained as a byproduct of the "GP-B" gyroscope experiment. Indeed, the validity of the stellar aberration formulas (10.5) is central for the success of the "GP-B" gyroscope experiment developed by NASA and Stanford University [15] to test two unverified predictions of Einstein's general theory of relativity [14, 20].

The GP-B space gyroscopes encountered two kinds of stellar aberration. Orbital aberration with 97.5-minute period of ± 5.1856 arc-seconds that results from the motion of the gyroscopes around the Earth, and annual aberration with one year period of about ± 20.4958 arc-seconds that results from the motion of the Earth (and the gyroscopes) around the Sun. These aberrations, calculated by methods of special relativity, were used to calibrate the gyroscopes and their accompanying instruments.

If the "GP-B" gyroscope experiment proves successful, it could be considered as an experimental evidence of the validity of the stellar aberration formulas (10.5) and, hence, the validity of the relativistic particle aberration formulas (10.3b) as well. The experimental significance of (10.3b), in turn, could be considered as an experimental evidence of the validity of the gyroparallelogram addition law of Einsteinian velocities. Indeed, the preliminary analysis of data has confirmed the theoretical prediction of the "GP-B" gyroscope experiment [21], so that the experiment seems to prove successful.

The success of "GP-B", thus, establishes experimentally the validity of the gyroparallelogram law (5.11), p. 124, as the commutative addition law of uniform, relativistically admissible velocities. The geometric significance of our special relativistic approach to hyperbolic geometry is, thus, associated with physical-experimental significance as well.

10.3 On the Future of Special Relativity and Hyperbolic Geometry

It is hoped that following the demonstration that Einstein addition forms an extraordinarily powerful, elegant tool for studying hyperbolic geometry, readers of this book will come along and join us in the hunt for more hyperbolic triangle centers and relations between them by means of Einstein addition.

Relativity today is no longer only a matter of pure science aimed at understanding the fundamental laws of Nature and the structure of the Universe. Rather, it has reached the status of applied technology in everyday life, as Claus Lammerzahl points out in detail in [32]. There are many technologies whose good operation requires one to take into account relativistic effects. The best known of these technologies is, perhaps, the Global Positioning System (GPS).

Yet, as Z.K. Silagadze notes in [50],

> "The teaching of special relativity, however, still follows its presentation as it unfolded historically, trying to convince the audience of this teaching that Newtonian physics is natural but incorrect and special relativity is its paradoxical but correct amendment."
>
> Z.K. Silagadze, 2008

Interest in Euclidean triangle centers has long history, indicating that, following this book, hyperbolic triangle centers will prove quite popular and challenging as well [3, 11, 52, 68]. While creating momentum for the exploration of hyperbolic triangle centers, this book contributes to modernizing and popularizing the teaching of Einstein's special relativity theory along with its underlying hyperbolic geometry of Bolyai and Lobachevsky. In the resulting modernized special relativity, Einstein addition is a primitive concept from which the Minkowskian formalism is derived, and the Einstein relativistic mass is a concept that interplays harmoniously with the underlying hyperbolic geometry of the Einsteinian three-vector formalism as well as with the Minkowskian four-vector formalism of special relativity.

Glossary

Brocard gyropoint Ω Sect. 9.6, p. 291
Circumgyrocenter Sect. 7.1, p. 153
Circumgyroradius Sect. 7.4, p. 161
Commutative Group Definition 1.2, p. 11
Exgyrocenter Sect. 8.5, p. 230
Exgyroradius Theorem 8.3, p. 230
Gergonne gyropoint G_e Sect. 7.14, p. 191
Group Definition 1.1, p. 11
Groupoid A non-empty set with a binary operation
Gyration Gyrogroup definition, and Sect. 1.5
Gyroaltitude Theorem 6.14, p. 146
Gyroangle Sect. 2.6, p. 53
Gyroautomorphism Same as gyration
Gyroautomorphism Group Definition 1.18, p. 23
Gyrobarycentric Coordinates Sect. 4.2, p. 88
Gyrocentroid Definition 4.13, p. 104
Gyrocevian Definition 9.1, p. 269
Gyrocircle Sect. 7.3, p. 160
Gyrocommutative Gyrogroup Gyrogroup definition, and Sect. 1.5
Gyrodiagonal Definition 5.2, p. 120
Gyrodistance Sect. 2.2, p. 49
Gyrogroup Definition 1.5, p. 12
Gyromedian Definition 4.13, p. 104
Gyromidpoint Sect. 4.5, p. 97
Gyroparallelogram Definition 5.2, p. 120
Gyrosemidirect Product Sect. 1.18, p. 23
Ingyrocenter Sect. 8.5, p. 230
Ingyroradius Theorem 8.3, p. 230
Nagel gyropoint N_a Sect. 8.9, p. 236
Orthogyrocenter Sect. 7.17, p. 202
Subgroup Definition 1.3, p. 11

A.A. Ungar, *Hyperbolic Triangle Centers,*
Fundamental Theories of Physics 166,
DOI 10.1007/978-90-481-8637-2, © Springer Science+Business Media B.V. 2010

References

1. Abu-Saymeh, S., Hajja, M.: Some Brocard-like points of a triangle. Forum Geom. **5**, 65–74 (2005)
2. Adler, C.G.: Does mass really depend on velocity, dad? Am. J. Phys. **55**, 739–743 (1987)
3. Bottema, O.: On the medians of a triangle in hyperbolic geometry. Can. J. Math. **10**, 502–506 (1958)
4. Brehme, R.W.: The advantage of teaching relativity with four-vector. Am. J. Phys. **36**(10), 896–901 (1968)
5. Collins, G.W.: The Virial Theorem in Stellar Astrophysics, p. 135. Pachart, Tucson (1978)
6. Conselice, C.J.: The universe's invisible hand. Sci. Am. **February**, 35–41 (2007)
7. Copeland, E.J., Sami, M., Tsujikawa, S.: Dynamics of dark energy. Int. J. Mod. Phys. D **15**(11), 1753–1935 (2006)
8. Coxeter, H.S.M.: Introduction to Geometry, 2nd edn., p. 469. Wiley, New York (1969)
9. Coxeter, H.S.M., Greitzer, S.L.: Geometry Revisited. Math. Assoc. Am., New York (1967)
10. Crowe, M.J.: A History of Vector Analysis, p. 270. Dover, New York (1994). The evolution of the idea of a vectorial system. Corrected reprint of the 1985 edition
11. Demirel, O., Soyturk, E.: The hyperbolic Carnot theorem in the Poincaré disc model of hyperbolic geometry. Novi Sad J. Math. **38**(2), 33–39 (2008)
12. Einstein, A.: Zur Elektrodynamik Bewegter Körper. Ann. Phys. (Leipz.) **17**, 891–921 (1905). [On the electrodynamics of moving bodies] (We use the English translation in [13] or in [34], or in http://www.fourmilab.ch/etexts/einstein/specrel/www/)
13. Einstein, A.: Einstein's Miraculous Years: Five Papers that Changed the Face of Physics. Princeton University Press, Princeton (1998). Edited and introduced by John Stachel. Includes bibliographical references. Einstein's dissertation on the determination of molecular dimensions–Einstein on Brownian motion–Einstein on the theory of relativity–Einstein's early work on the quantum hypothesis. A new English translation of Einstein's 1905 paper on pp. 123–160
14. Everitt, C.: In: Fairbank, J.D., Deaver, B.S. Jr., Everitt, C.W.F. (eds.) New Frontiers in Physics, pp. 587–639. Freeman, San Francisco (1988)
15. Everitt, F.C.W., Fairbank, W.M., Schiff, L.I.: Theoretical background and present status of the Stanford relativity—gyroscope experiment. In: The Significance of Space Research for Fundamental Physics. Proceedings of the Colloquium of the European Space Research Organization at Interlaken, Switzerland, Sept. 4, 1969. For current status see the project website at URL: http://einstein.stanford.edu
16. Ferrarese, L., David, M.: A fundamental relation between supermassive black holes and their host galaxies. Astrophys. J. **539**, 9–12 (2000)
17. Feynman, R.P., Leighton, R.B., Sands, M.: The Feynman Lectures on Physics. Vol. 1: Mainly Mechanics, Radiation, and Heat, p. 513. Addison-Wesley, Reading (1963)

A.A. Ungar, *Hyperbolic Triangle Centers,*
Fundamental Theories of Physics 166,
DOI 10.1007/978-90-481-8637-2, © Springer Science+Business Media B.V. 2010

18. Fock, V.: The Theory of Space, Time and Gravitation, 2nd edn., p. 448. Macmillan & Co., New York (1964). Translated from the Russian by N. Kemmer. A Pergamon Press Book
19. Gebhardt, K., Bower, G., Dressler, S., Faber, S., Fillipenko, A., Green, R., Grillmair, C., Ho, L., Kormendy, J., Lauer, T., Magorrian, J.: A relationship between nuclear black hole mass and galaxy velocity dispersion. Astrophys. J. **539**, 13–16 (2000)
20. GP-B: Gravity Probe B. http://www.nasa.gov/mission_pages/gpb/index.html
21. GP-B: Gravity Probe B. http://einstein.stanford.edu
22. Greenberg, M.J.: Euclidean and Non-Euclidean Geometries, 3rd edn., p. 483. Freeman, New York (1993). Development and history
23. Gültekin, K., Richstone, D.O., Gebhardt, K., Lauer, T.R., Tremaine, S., Aller, M.C., Bender, R., Dressler, A., Faber, S.M., Filippenko, A.V., Green, R., Ho, L.C., Kormendy, J., Magorrian, J., Pinkney, J., Siopis, C.: The M-σ and M-L relations in galactic bulges, and determinations of their intrinsic scatter. Astrophys. J. **698**, 198–221 (2009)
24. Hausner, M.: A Vector Space Approach to Geometry, p. 397. Dover, Mineola (1998). Reprint of the 1965 original
25. Hehl, F.W.: Preface: The Minkowski spacetime of special relativity theory—100 years after its discovery. Ann. Phys. (8) **17**(9–10), 617 (2008)
26. Honsberger, R.: Episodes in Nineteenth and Twentieth Century Euclidean Geometry. New Mathematical Library, vol. 37, p. 174. Math. Assoc. Am., Washington (1995)
27. Jackson, J.D.: Classical Electrodynamics, 2nd edn., p. 848. Wiley, New York (1975)
28. Kelly, P.J., Matthews, G.: The Non-Euclidean, Hyperbolic Plane. Universitext, p. 333. Springer, New York (1981). Its structure and consistency
29. Kimberling, C.: Clark Kimberling's Encyclopedia of Triangle Centers—ETC. http://faculty.evansville.edu/ck6/encyclopedia/ETC.html (2010)
30. Kimberling, C.: Triangle Centers and Central Triangles. Congressus Numerantium, vol. 129, p. 295. Utilitas Mathematica, Winnipeg (1998)
31. Kreyszig, E.: Differential Geometry, p. 352. Dover, New York (1991). Reprint of the 1963 edition
32. Lämmerzahl, C.: Relativity and technology. Ann. Phys. (8) **15**(1–2), 5–18 (2006)
33. Landau, L.D., Lifshitz, E.M.: The Classical Theory of Fields. Course of Theoretical Physics, vol. 2, p. 402. Pergamon, Oxford (1975). Fourth revised English edition. Translated from the Russian by Morton Hamermesh
34. Lorentz, H.A., Einstein, A., Minkowski, H., Weyl, H.: The Principle of Relativity, p. 216. Dover, New York. With notes by A. Sommerfeld. Translated by W. Perrett and G.B. Jeffery. A collection of original memoirs on the special and general theory of relativity
35. Maor, E.: Trigonometric Delights, p. 236. Princeton University Press, Princeton (1998)
36. Marx, G.: Is the amount of matter additive? Eur. J. Phys. **12**, 271–274 (1991)
37. McCleary, J.: Geometry from a Differentiable Viewpoint, p. 308. Cambridge University Press, Cambridge (1994)
38. Miller, A.I.: Albert Einstein's Special Theory of Relativity, p. 446. Springer, New York (1998). Emergence (1905) and early interpretation (1905–11), Includes a translation by the author of Einstein's "On the electrodynamics of moving bodies". Reprint of the 1981 edition
39. Moise, E.E.: Elementary Geometry from an Advanced Standpoint, 2nd edn., p. 425. Addison-Wesley, Reading (1974)
40. Møller, C.: The Theory of Relativity, p. 386. Oxford, Clarendon (1952)
41. Mumford, D., Series, C., Wright, D.: Indra's Pearls: The Vision of Felix Klein, p. 396. Cambridge University Press, New York (2002)
42. Okun, L.B.: Energy and Mass in Relativity Theory, p. 311. World Scientific, Hackensack (2009)
43. Pyenson, L.: Relativity in late Wilhelmian Germany: the appeal to a preestablished harmony between mathematics and physics. Arch. Hist. Exact Sci. **27**(2), 137–155 (1982)
44. Rhodes, J.A., Semon, M.D.: Relativistic velocity space, Wigner rotation, and Thomas precession. Am. J. Phys. **72**(7), 943–960 (2004)
45. Rindler, W.: Introduction to Special Relativity, p. 185. Clarendon, New York (1982)

46. Rindler, W.: Relativity: Special, General, and Cosmological, 2nd edn., p. 430. Oxford University Press, New York (2006)
47. Robb, A.A.: A Theory of Time and Space. Cambridge University Press, Cambridge (1914)
48. Ruoff, D.: Why Euclidean area measure fails in the noneuclidean plane. Math. Mag. **78**(2), 137–139 (2005)
49. Sexl, R.U., Urbantke, H.K.: Relativity, Groups, Particles. Springer Physics, p. 388. Springer, Vienna (2001). Special relativity and relativistic symmetry in field and particle physics, Revised and translated from the third German (1992) edition by Urbantke
50. Silagadze, Z.K.: Relativity without tears. Acta Phys. Pol. B **39**(4), 811–885 (2008). MR2399620 (2009b:83009)
51. Sommerfeld, A.: Über die Zusammensetzung der Geschwindigkeiten in der Relativtheorie. Phys. Z. **10**, 826–829 (1909)
52. Sonmez, N.: A trigonometric proof of the Euler theorem in hyperbolic geometry. Algebras Groups Geom. **26**(1), 75–79 (2009)
53. Stewart, A.B.: The discovery of stellar aberration. Sci. Am. **March**, 100–108 (1964)
54. Synge, J.L.: Relativity: The Special Theory, 2nd edn., p. 450. North-Holland, Amsterdam (1965)
55. Ungar, A.A.: Thomas rotation and the parametrization of the Lorentz transformation group. Found. Phys. Lett. **1**(1), 57–89 (1988)
56. Ungar, A.A.: Thomas precession and its associated grouplike structure. Am. J. Phys. **59**(9), 824–834 (1991)
57. Ungar, A.A.: Hyperbolic trigonometry in the Einstein relativistic velocity model of hyperbolic geometry. Comput. Math. Appl. **40**(2–3), 313–332 (2000)
58. Ungar, A.A.: Beyond the Einstein Addition Law and Its Gyroscopic Thomas Precession: The Theory of Gyrogroups and Gyrovector Spaces. Fundamental Theories of Physics, vol. 117, p. 413. Kluwer Academic, Dordrecht (2001)
59. Ungar, A.A.: On the appeal to a pre-established harmony between pure mathematics and relativity physics. Found. Phys. Lett. **16**(1), 1–23 (2003)
60. Ungar, A.A.: Analytic Hyperbolic Geometry: Mathematical Foundations and Applications, p. 463. World Scientific, Hackensack (2005)
61. Ungar, A.A.: Gyrovector spaces and their differential geometry. Nonlinear Funct. Anal. Appl. **10**(5), 791–834 (2005)
62. Ungar, A.A.: Einstein's velocity addition law and its hyperbolic geometry. Comput. Math. Appl. **53**(8), 1228–1250 (2007)
63. Ungar, A.A.: Analytic Hyperbolic Geometry and Albert Einstein's Special Theory of Relativity, p. 628. World Scientific, Hackensack (2008)
64. Ungar, A.A.: A Gyrovector Space Approach to Hyperbolic Geometry. Morgan & Claypool, San Rafael (2009)
65. Ungar, A.A.: Hyperbolic barycentric coordinates. Aust. J. Math. Anal. Appl. **6**(1), 1–35 (2009)
66. Varičak, V.: Beiträge zur nichteuklidischen geometrie. Jahresber. Dtsch. Math.-Ver. **17**, 70–83 (1908). [Contributions to non-Euclidean geometry]
67. Varičak, V.: Anwendung der Lobatschefskjschen Geometrie in der Relativtheorie. Phys. Z. **11**, 93–96 (1910)
68. Vermeer, J.: A geometric interpretation of Ungar's addition and of gyration in the hyperbolic plane. Topol. Appl. **152**(3), 226–242 (2005)
69. Wallace, E.C., West, S.F.: Roads to Geometry, 2nd edn., pp. 362–363. Prentice Hall, New York (1998)
70. Walter, S.: The non-Euclidean style of Minkowskian relativity. In: Gray, J.J. (ed.) The Symbolic Universe: Geometry and Physics 1890–1930, pp. 91–127. Oxford University Press, New York (1999)
71. Walter, S.: Book review: beyond the Einstein addition law and its gyroscopic Thomas precession: the theory of gyrogroups and gyrovector spaces, by Abraham A. Ungar. Found. Phys. **32**(2), 327–330 (2002)

72. Walter, S.: Personal communication (2002)
73. Walter, S.: Hermann Minkowski's approach to physics. Math. Semesterber. **55**(2), 213–235 (2008)
74. Yiu, P.: The uses of homogeneous barycentric coordinates in plane Euclidean geometry. Int. J. Math. Educ. Sci. Tech. **31**(4), 569–578 (2000)
75. Zajic, V.: Property of points where in- and excircles touch a triangle. http://www.cut-the-knot.org/Curriculum/Geometry/TouchPoint.shtml

Author Index

A.A. Ungar, *Hyperbolic Triangle Centers,*
Fundamental Theories of Physics 166,
DOI 10.1007/978-90-481-8637-2,

Subject Index

A.A. Ungar, *Hyperbolic Triangle Centers*,
Fundamental Theories of Physics 166,
DOI 10.1007/978-90-481-8637-2, © Springer Science+Business Media B.V. 2010

MIX
Papier aus verantwortungsvollen Quellen
Paper from responsible sources
FSC® C105338

If you have any concerns about our products,
you can contact us on
ProductSafety@springernature.com

In case Publisher is established outside the EU,
the EU authorized representative is:
Springer Nature Customer Service Center GmbH
Europaplatz 3, 69115 Heidelberg, Germany

Printed by Libri Plureos GmbH
in Hamburg, Germany